WASTEWATER MANAGEMENT FOR IRRIGATION

Principles and Practices

Research Advances in Sustainable Micro Irrigation

VOLUME 8

WASTEWATER MANAGEMENT FOR IRRIGATION

Principles and Practices

Edited by
Megh R. Goyal, PhD, PE and Vinod K. Tripathi, PhD

APPLE
ACADEMIC
PRESS

Apple Academic Press Inc.	Apple Academic Press Inc.
3333 Mistwell Crescent	9 Spinnaker Way
Oakville, ON L6L 0A2	Waretown, NJ 08758
Canada	USA

©2016 by Apple Academic Press, Inc.

First issued in paperback 2021

Exclusive worldwide distribution by CRC Press, a member of Taylor & Francis Group
No claim to original U.S. Government works

ISBN 13: 978-1-77463-542-1 (pbk)
ISBN 13: 978-1-77188-120-3 (hbk)

Typeset by Accent Premedia Services (www.accentpremedia.com)

Library and Archives Canada Cataloguing in Publication

Wastewater management for irrigation : principles and practices / edited by Megh R. Goyal, PhD, PE and Vinod K. Tripathi, PhD.

(Research advances in sustainable micro irrigation; volume 8)
Includes bibliographical references and index.
Issued in print and electronic formats.
ISBN 978-1-77188-120-3 (hardcover).--ISBN 978-1-4987-2739-6 (pdf)

1. Sewage irrigation. 2. Sewage--Management. 3. Microirrigation. I. Goyal, Megh Raj, editor II. Tripathi, Vinod K., author, editor III. Series: Research advances in sustainable micro irrigation; v. 8

| TD760.W42 2015 | 628.3'623 | C2015-905619-5 | C2015-905620-9 |

CIP data on file with US Library of Congress

Apple Academic Press also publishes its books in a variety of electronic formats. Some content that appears in print may not be available in electronic format. For information about Apple Academic Press products, visit our website at **www.appleacademicpress.com** and the CRC Press website at **www.crcpress.com**

CONTENTS

LIST OF CONTRIBUTORS

A. Capra, PhD
Professor, Department of Agro-Forestry and Environmental Sciences, Reggio Calabria Mediterranean University, Reggio Calabria (RC), Italy. E-mail: acapra@unirc.it

Vishal Keshavao Chavan, PhD
Assistant Professor and Senior Research Fellow in SWE, Agriculture University, Akola, Maharashtra, India, Website: www.pdkv.ac.in; E-mail: vchavan2@gmail.com

Hossein Dehghanisanij, PhD
Associate Professor in Irrigation and Drainage Engineering; Head, Irrigation Systems Research Department Agricultural Engineering Research Institute (AERI), P.O. Box 31585–845, Karaj, Iran. Tel: +98-26-32705320, Fax: +98-26-32706277, Mobile: +98-912-167-5238. E-mail: dehghanisanij@yahoo.com

Santosh K. Deshmukh, PhD
Chief Coordinator, Sustainability (Jalgaon Area – India Environmental Services), Jain Irrigation Systems Ltd, Jain Plastic Park, N.H. No. 6, Bambhori, Jalgaon 425001, Maharashtra, India. Tel.: +91–257–2258011, Fax +91–257–2258111, E-mail: jisl@jains.com, Visit at: www.jains.com

Juan C. Durán-Álvarez, PhD
Postdoc Fellow, Center of Applied Science and Technological Development, and Engineering Institute, National Autonomous University of Mexico, 3000 Uni. Ave., Coyoacan, 04510, Mexico, D.F., Mexico. E-mail: carlos.duran@ccadet.unam.mx and azotojcd@hotmail.com

Ebtisam I. Eldardiry, I., PhD
Water Relations and Field Irrigation Department, Agricultural Division and National Research Center, Cairo, Egypt. E-mail: ebtisameldardiry@gmail.com

M. Abd El-Hady, PhD
Water Relations and Field Irrigation Department, Agricultural Division and National Research Center, 12311 El-Dokki, Giza, Cairo, Egypt. hadymnrc60@gmail.com

Ahmed S. El-Henawy, PhD
Associate Professor of Soil Physics, and Chairman of Soils and Water Science Department, Faculty of Agriculture, Kafrelsheikh University Kafr El-Sheikh Egypt. Tel.: +204-73232762 (off.); Mobile: +201-006342335, E-mail: aelhenawy@agr.kfs.edu; aelhenawy@yahoo.com

M. S. Gaballah, PhD
WRFI Department, Agricultural Division, National Research Center, 12311 El-Dokki, Cairo, Egypt.

Megh R. Goyal, PhD, PE
Retired Professor in Agricultural and Biomedical Engineering from General Engineering Department, University of Puerto Rico – Mayaguez Campus; and Senior Technical Editor-in-Chief in Agriculture Sciences and Biomedical Engineering, Apple Academic Press Inc., PO Box 86, Rincon, PR 00677, USA. E-mail: goyalmegh@gmail.com

Cs. Gyuricza, PhD
Institute of Crop Production, Faculty of Agricultural and Environmental Science. SZIU, 2103 Godollo, Hungary. E-mail: gyuricza.csaba@mkk.szie.hu

Marvin E. Jensen, PhD, PE
Retired Research Leader at USDA – ARS. 1207 Spring Wood Drive, Fort Collins, Colorado 80525, USA, E-mail: mjensen419@aol.com

Blanca Jiménez–Cisneros, PhD
International Hydrological Program (IHP), UNESCO, 1 Miollis St., 75732 Paris, France, E-mail: b.jimenez-cisneros@unesco.org

M. R. Khalifa, PhD
Soils and Water Science Department, Fac. of Agric., Kafrelsheikh Univ., Egypt

H. Kosari, M.Sc.
Department of Irrigation and reclamation Engineering, Agricultural and Natural Resources Campus, University of Tehran, Karaj, Iran, E-mail: hk_kosari@yahoo.com

Pradeep Kumar, PhD
Scientist, National Institute of Hydrology, Roorkee, India

Lata, PhD
Scientist, Water Technology Center (WTC), Indian Agricultural Research Institute (IARI), New Delhi, Delhi 110012, India

Hani A. A. Mansour, PhD
Researcher, WRFI Department, Agriculture Division, National Research Center, El-Behouth St., Eldokki, Giza, Postal Code 12311, Cairo, Egypt. Tel.: +201-068989517 and +201-123355393. E-mail: mansourhani2011@gmail.com

V. M., Mayande, PhD
Ex-Vice Chancellor, Dr. Panjabrao Deshmukh Krishi Vidyapeeth, Akola, 444104, Maharashtra, India. Phone: +91-9423174299, E-mail: vmmayande@yahoo.com

Miguel A. Muñoz-Muñoz, PhD
Ex-President of University of Puerto Rico, University of Puerto Rico, Mayaguez Campus, College of Agriculture Sciences, Call Box 9000, Mayagüez, PR. 00681–9000. Tel.: 787–265–3871, E-mail: miguel.munoz3@upr.edu

Neelam Patel, PhD
Senior Scientist, Water Technology Center (WTC), Indian Agricultural Research Institute (IARI), New Delhi, Delhi 110012, India, E-mail: neelam@iari.res.in

Prerna Prasad, PhD
Post graduate scholar, Water Policy and Governance, Tata Institute of Social Sciences, Mumbai, India, E-mail: prerna.ford50@yahoo.com

T. B. S. Rajput, PhD
Principal Scientist, Water Technology Center (WTC), Indian Agricultural Research Institute (IARI), New Delhi, Delhi 110012, India, E-mail: tbsraj@iari.res.in

M. M. Saffan, PhD
Soil and Water Department, Faculty of Agriculture, Kafrelsheikh University, Kafr El-Sheikh, Egypt. http://www.kfs.edu.eg/

B. Scicolone, PhD
Professor, Department of Agro-forestry and Environmental Sciences, Reggio Calabria Mediterranean University, Reggio Calabria (RC), Italy

Gajendra Singh, PhD
Former Vice President, Asian Institute of Technology, Thailand. C-86, Millennium Apartments, Plot E-10A, Sector - 61, NOIDA, U.P., 201301, India, Mobile: (011)-(91) 9971087591, E-mail: prof. gsingh@gmail.com

R. K. Sivanappan, PhD
Former Professor and Dean, College of Agricultural Engineering and Technology, Tamil Nadu Agricultural University (TNAU), Coimbatore. Mailing address: Consultant, 14, Bharathi Park, 4th Cross Road, Coimbatore-641043, India. E-mail: sivanappanrk@hotmail.com

Sushil Kumar Shukla, PhD
Assistant Professor, Center of Environmental Science, Central University of Jharkhand, Ranchi, India, Ph.No.+91–8969271817, E-mail: shuklask2000@gmail.com and sushil.shukl@cuj.ac.in

Ashutosh Tripathi, PhD
Assistant Professor, Amity Institute of Environmental Sciences, Amity University, Noida 125, Uttar Pradesh-201303, India. E-mail: tripathiashutos@gmail.com and atripathi1@amity.edu

Vinod Kumar Tripathi, PhD
Assistant Professor, Center for Water Engineering and Management, Central University of Jharkhand, Ratu-Loharghat road, Brambe, Ranchi, Jharkhand – 835205, India, Mobile: +918987661439; E-mail: tripathiwtcer@gmail.com

LIST OF ABBREVIATIONS

ASABE	American Society of Agricultural and Biological Engineers
CU	coefficient of uniformity
DIS	drip irrigation system
DOY	day of the year
EPAN	pan evaporation
ET	evapotranspiration
ETc	crop evapotranspiration
FAO	Food and Agricultural Organization, Rome
FC	field capacity
FUE	fertilizers use efficiency
gpm	gallons per minute
ISAE	Indian Society of Agricultural Engineers
kc	crop coefficient
kg	kilograms
Kp	pan coefficient
lps	liters per second
lph	liter per hour
msl	mean sea level
PE	polyethylene
PET	potential evapotranspiration
pH	acidity/alkalinity measurement scale
PM	Penman-Monteith
ppm	one part per million
psi	pounds per square inch
PVC	poly vinyl chloride
PWP	permanent wilting point
RA	extraterrestrial radiation
RH	relative humidity
RMAX	maximum relative humidity
RMIN	minimum relative humidity

RMSE	root mean squared error
RS	solar radiation
SAR	sodium absorption rate
SDI	subsurface drip irrigation
SW	saline water
SWB	soil water balance
TE	transpiration efficiency
TEW	total evaporable water
TMAX	maximum temperature
TMIN	minimum temperature
TR	temperature range
TSS	total soluble solids
TUE	transpiration use efficiency
USDA	US Department of Agriculture
USDA-SCS	US Department of Agriculture-Soil Conservation Service
WSEE	weighed standard error of estimate
WUE	water use efficiency

LIST OF SYMBOLS

A	cross sectional flow area (L^2)
AW	available water
Cp	specific heat capacity of air, in J/(g·°C)
CV	coefficient of variation
D	accumulative intake rate (mm/min)
d	depth of effective root zone
D	depth of irrigation water in mm
Δ	slope of the vapor pressure curve (kPa°C^{-1})
e	vapor pressure, in kPa
e_a	actual vapor pressure (kPa)
E	evapotranspiration rate, in g/(m^2·s)
Ecp	cumulative class A pan evaporation for two consecutive days (mm)
eff	irrigation system efficiency
E_i	irrigation efficiency of drip system
E_p	pan evaporation as measured by Class-A pan evaporimeter (mm/day)
E_s	saturation vapor pressure, in kPa
E_{pan}	class A pan evaporation
ER	cumulative effective rainfall for corresponding two days (mm)
e_s	saturation vapor pressure (kPa)
e_s-e_a	vapor pressure deficit (kPa)
ET	evapotranspiration rate, in mm/year
ETa	reference ET, in the same water evaporation units as Ra
ETc	crop-evapotranspiration (mm/day)
ET_o	the reference evapotranspiration obtained using the Penman-Monteith method, (mm/day)
ET_{pan}	the pan evaporation-derived evapotranspiration
EU	emission uniformity

F	flow rate of the system (GPM)
F.C.	field capacity (v/v, %)
G	soil heat flux at land surface, in W/m^2
H	plant canopy height in meter
h	soil water pressure head (L)
I	infiltration rate at time t (mm/min)
IR	injection rate, GPH
IRR	irrigation
K	unsaturated hydraulic conductivity (LT^{-1})
K_c	crop-coefficient for bearing 'Kinnow' plant
Kp	pan factor
K_p	pan coefficient
n	number of emitters
P	percentage of chlorine in the solution*
Pa	atmospheric pressure, in Pa
P.W.P.	permanent wilting point
Q	flow rate in gallons per minute
q	mean emitter discharges of each lateral (lh^{-1})
R	rainfall
r_a	aerodynamic resistance ($s\ m^{-1}$)
R_a	extraterrestrial radiation, in the same water evaporation units as ETa
R_e	effective rainfall depth (mm)
R_i	individual rain gauge reading in mm
R_n	net radiation at the crop surface ($MJ\ m^{-2}day^{-1}$)
R_s	incoming solar radiation on land surface, in the same water evaporation units as ETa
RO	surface runoff
r_s	bulk surface resistance ($s\ m^{-1}$)
S	sink term accounting for root water uptake (T^{-1})
Se	effective saturation
S_p	plant-to-plant spacing (m)
S_r	row-to-row spacing (m)
SU	statistical uniformity (%)
S_ψ	water stress integral (MPa day)
t	the time that water is on the surface of the soil (min)

T   time in hours
V   volume of water required
V_{id} irrigation volume applied in each irrigation (liter tree^{-1})
V_{pc} plant canopy volume (m^3)
W   canopy width
W_p fractional wetted area
z vertical coordinate positive downwards (L)

Greek Symbols

α inverse of a characteristic pore radius (L^{-1})
γ psychrometric constant (kPa°C^{-1})
θ volumetric soil water content (L^{3}L^{-3})
$\theta(h)$ soil water retention (L^{3}L^{-3}),
θ_r residual water content (L^{3}L^{-3})
θ_s saturated water content (L^{3}L^{-3})
θ_{vol} volumetric moisture content (cm^3/cm^3)
λ latent heat of vaporization (MJ kg^{-1})
λE latent heat flux, in W/mo
ρa mean air density at constant pressure (kg m^{-3})

PREFACE

Due to increased agricultural production, irrigated land has increased in the arid and sub-humid zones around the world. Agriculture has started to compete for water use with industries, municipalities and other sectors. This increasing demand along with increments in water and energy costs have made it necessary to develop new technologies for the adequate management of water. The intelligent use of water for crops requires understanding of evapotranspiration processes and use of efficient irrigation methods.

Every day, news on water scarcity appears throughout the world, indicating that government agencies at central/state/local levels, research and educational institutions, industry, sellers and others are aware of the urgent need to adopt micro irrigation technology that can have an irrigation efficiency up to 90%, compared to 30–40% for the conventional gravity irrigation systems. I stress the urgent need to implement micro irrigation systems in water scarcity regions.

Irrigation has been a central feature of agriculture since the start of civilization and the basis of the economy and society of numerous societies throughout the world. Among all irrigation systems, micro irrigation has the highest irrigation efficiency and is most efficient. Micro irrigation is sustainable and is one of the best management practices. The water crisis is getting worse throughout the world, including in the Middle East and Puerto Rico where I live. We can, therefore, conclude that the problem of water scarcity is rampant globally, creating the urgent need for water conservation. The use of micro irrigation systems is expected to result in water savings, and increased crop yields in terms of volume and quality. The other important benefits of using micro irrigation systems include expansion in the area under irrigation, water conservation, optimum use of fertilizers and chemicals through water, and decreased labor costs, among others. The worldwide population is increasing at a rapid rate, and it is imperative that food supply keeps pace with this increasing population.

Micro irrigation, also known as trickle irrigation or drip irrigation or localized irrigation or high frequency or pressurized irrigation, is an irrigation method that saves water and fertilizer by allowing water to drip slowly to the roots of plants, either onto the soil surface or directly onto the root zone, through a network of valves, pipes, tubing, and emitters. It is done through narrow tubes that deliver water directly to the base of the plant. It supplies controlled delivery of water directly to individual plants and can be installed on the soil surface or subsurface. Micro irrigation systems are often used for farms and large gardens but are equally effective in the home garden or even for houseplants or lawns.

Water and fertigation management are important practices for the success of drip/micro/trickle irrigation. Water management is the activity of planning, developing, distributing and optimum use of water resources under defined water policies and regulations. It includes: management of water treatment of drinking water/industrial water; sewage or wastewater; management of water resources in agriculture; management of flood protection; management of irrigation; management of the water table; and management of drainage, etc.

Reuse of wastewater in irrigation is being practiced only recently to solve water scarcity problems in agriculture. Management of water, soil, crop and operational procedures, including precautions to protect farm workers, play an important role in the successful use of sewage effluent for irrigation. Appropriate water management practices must be followed to prevent salinization, irrespective of whether the salt content in the wastewater is high or low. If salt is not flushed out of the root zone by leaching and removed from the soil by effective drainage, salinity problems can build up rapidly. Leaching and drainage are thus two important water management practices to avoid salinization of soils. One of the options that may be available to farmers is the blending of treated sewage with conventional sources of water, canal water or ground water, if multiple sources are available. It is possible that a farmer may have saline ground water and, if he has non-saline treated wastewater, can blend the two sources to obtain a blended water of acceptable salinity level. Further, by blending, the microbial quality of the resulting mixture can be superior to that of the unblended wastewater. Another strategy is to use the treated wastewater alternately with the canal water or groundwater, instead of blending.

From the point of view of salinity control, alternate applications of the two sources will be superior to blending.

Our goal is to improve permanently existing land and soil conditions in order to make irrigation with wastewater easier. Typical activities include leveling of land to a given grade, establishing adequate drainage (both open and sub-surface systems), deep ploughing and leaching to reduce soil salinity.

The procedures involved in preparing plans for effluent irrigation schemes are similar to those used in most forms of resource planning and summarized the main physical, social and economic dimensions (Appendix K in this book).

Adopting a mix of effluent use strategies is normally advantageous in respect of allowing greater flexibility, increased financial security and more efficient use of the wastewater throughout the year, whereas a single-use strategy will give rise to seasonal surpluses of effluent for unproductive disposal. Therefore, in site and crop selection, the desirability of providing areas for different crops and forestry so as to utilize the effluent at maximum efficiency over the whole yearly cycle of seasons must be kept in mind.

Among the soil properties important from the point of view of wastewater application in irrigation are: physical parameters (such as texture, grading, liquid and plastic limits, etc.), permeability, water-holding capacity, pH, salinity and chemical composition. After elimination of marginal sites, each site under serious consideration must be investigated by on-site borings to ascertain the soil profile, soil characteristics and location of the water table. When a site is developed, a long-term groundwater-monitoring program should be an essential feature of its management.

The degree to which the use of treated effluent influences crop selection will depend on government policy on effluent irrigation, the goals of the user and the effluent quality. Government policy will have the objectives of minimizing the health risk and influencing the type of productivity associated with effluent irrigation. Regulations must be realistic and achievable in the context of national and local environmental conditions and traditions. At the same time, planners of effluent irrigation schemes must attempt to achieve maximum productivity and water conservation through the choice of crops and effluent application systems.

A multiple-use strategy approach will require the evaluation of viable combinations of the cropping options possible on the land available. This will entail a considerable amount of survey and resource budgeting work, in addition to the necessary soil and water quality assessments. The annual, monthly and daily water demands of the crops, using the most appropriate irrigation techniques, have to be determined. Domestic consumption, local production and imports of the various crops must be assessed so that the economic potential of effluent irrigation of the various crop combinations can be estimated. Finally, the crop irrigation demands must be matched with the available effluent so as to achieve optimum physical and financial utilization throughout the year.

The mission of this compendium is to serve as a reference manual for graduate and undergraduate students of agricultural, biological and civil engineering; horticulture, soil science, crop science and agronomy. I hope that it will be a valuable reference for professionals who work with micro irrigation/wastewater and water management and for professional training institutes, technical agricultural centers, irrigation centers, agricultural extension services, and other agencies that work with micro irrigation programs.

After my first textbook, *Drip/Trickle or Micro Irrigation Management* published by Apple Academic Press Inc., and response from international readers, I was motivated to bring out for the world community this 10-volume series on Research Advances in Sustainable Micro Irrigation. This book series will complement other books on micro irrigation that are currently available on the market, and my intention is not to replace any one of these. This book series is unique because of its worldwide applicability to irrigation management in agriculture. This series is a must for those interested in irrigation planning and management, namely, researchers, scientists, educators and students.

The contribution by the authors to this book series has been most valuable in the compilation of these volumes. Their names are mentioned in each chapter and in the list of contributors of each volume. This book would not have been written without the valuable cooperation of these investigators, many of whom are renowned scientists who have worked in the field of micro irrigation throughout their professional careers. I am glad to introduce Dr. Vinod Kumar Tripathi, Assistant Professor and

Distinguished Research Scientist in Wastewater Use in Micro Irrigation at Center for Water Engineering and Management, Central University of Jharkhand Ranchi-Jharkhand, India. He joins as co-editor for this volume. Without his support and extraordinary job, readers will not have this quality publication.

I will like to thank editorial staff, Sandy Jones Sickels, Vice President, and Ashish Kumar, Publisher and President at Apple Academic Press, Inc. (http://appleacademicpress.com/contact.html) for making every effort to publish the book when the diminishing water resources is a major issue worldwide. Special thanks are due to the AAP production staff for the quality production of this book.

We request the reader to offer us constructive suggestions that may help to improve the future editions.

I express my deep admiration to my family for understanding and collaboration during the preparation of this ten- volume book series. With my whole heart and best affection, I dedicate this volume to the late Dr. Gordon L. Nelson who was an Emeritus Professor in the Food, Agricultural and Biological Engineering Department at the Ohio State University (OSU). During my MSc/PhD studies at OSU, he was instrumental in shaping my professional career. Under his leadership, I am the first person to apply "Similitude and Dimensional Analysis in Seedling Emergence through Crusted Soils." Throughout my postgraduate career, I have been able to apply my experience at OSU in Micro irrigation technology to come up with new ideas/developments, etc. in order alleviate problems of water scarcity and salinity. My salute to Dr. Nelson for his devotion and vocation.

As an educator, there is a piece of advice to one and all in the world: *"Permit that our Almighty God, our Creator and excellent Teacher, irrigate the life with His Grace of rain trickle by trickle, because our life must continue trickling on..."*

—Megh R. Goyal, PhD, PE, Senior Editor-in-Chief

March 17, 2015

WARNING/DISCLAIMER

The goal of this compendium, **Wastewater Management for Irrigation: Principles and Practices**, is to guide the world community on how to manage efficiently for economical crop production. The reader must be aware that dedication, commitment, honesty, and sincerity are the most important factors in a dynamic manner for complete success. This reference is not intended for a one-time reading; we advise you to consult it frequently. To err is human. However, we must do our best. Always, there is a place for learning from new experiences.

The editor, the contributing authors, the publisher, and the printer have made every effort to make this book as complete and as accurate as possible. However, there still may be grammatical errors or mistakes in the content or typography. Therefore, the contents in this book should be considered as a general guide and not a complete solution to address any specific situation in irrigation. For example, one size of irrigation pump does not fit all sizes of agricultural land and will not work for all crops.

The editor, the contributing authors, the publisher, and the printer shall have neither liability nor responsibility to any person, organization, or entity with respect to any loss or damage caused, or alleged to have caused, directly or indirectly, by information or advice contained in this book. Therefore, the purchaser/reader must assume full responsibility for the use of the book or the information therein.

The mention of commercial brands and trade names are only for technical purposes and does not imply endorsement. The editor, contributing authors, educational institutions, and the publisher do not have any preference for a particular product.

All weblinks that are mentioned in this book were active on December 31, 2014. The editors, the contributing authors, the publisher, and the printing company shall have neither liability nor responsibility if any of the weblinks are inactive at the time of reading of this book.

ABOUT THE SENIOR EDITOR-IN-CHIEF

Megh R. Goyal, PhD, PE

Megh R. Goyal, PhD, PE, is at present a retired professor in Agricultural and Biomedical Engineering from the General Engineering Department in the College of Engineering at the University of Puerto Rico, Mayaguez Campus, and Senior Acquisitions Editor and Senior Technical Editor-in-Chief in Agricultural and Biomedical Engineering for Apple Academic Press Inc. (AAP).

He received his BSc degree in engineering in 1971 from Punjab Agricultural University, Ludhiana, India; his MSc degree in 1977; his PhD degree in 1979 from the Ohio State University, Columbus; and his Master of Divinity degree in 2001 from Puerto Rico Evangelical Seminary, Hato Rey, Puerto Rico, USA.

Since 1971, he has worked as Soil Conservation Inspector (1971); Research Assistant at Haryana Agricultural University (1972–1975) and the Ohio State University (1975–1979); Research Agricultural Engineer/Professor at the Department of Agricultural Engineering of UPRM (1979–1997); and Professor in Agricultural and Biomedical Engineering at General Engineering Department of UPRM (1997–2012). He spent 1-year sabbatical leave in 2002–2003 at Biomedical Engineering Department, Florida International University, Miami, USA.

He was the first agricultural engineer to receive the professional license in Agricultural Engineering in 1986 from the College of Engineers and Surveyors of Puerto Rico. On September 16, 2005, he was proclaimed as "Father of Irrigation Engineering in Puerto Rico for the twentieth century" by the ASABE, Puerto Rico Section, for his pioneer work on micro irrigation, evapotranspiration, agroclimatology, and soil and water engineering. During his professional career of 45 years, he has received awards such as Scientist of the Year, Blue Ribbon Extension Award, Research Paper Award, Nolan Mitchell Young Extension Worker Award, Agricultural Engineer of the Year, Citations by Mayors of Juana Diaz and Ponce, Membership Grand Prize for ASAE Campaign, Felix Castro Rodriguez Academic Excellence, Rashtrya Ratan Award and Bharat Excellence Award and Gold Medal, Domingo Marrero Navarro Prize, Adopted Son of Moca, Irrigation Protagonist of UPRM, Man of Drip Irrigation by Mayor of Municipalities of Mayaguez/Caguas/Ponce, and Senate/Secretary of Agriculture of ELA, Puerto Rico.

He has authored more than 200 journal articles and textbooks, including Elements of Agroclimatology (Spanish) by UNISARC, Colombia, and two Bibliographies on Drip Irrigation. Apple Academic Press Inc. (AAP) has published his books, namely *Management of Drip/Trickle or Micro Irrigation, Evapotranspiration: Principles and Applications for Water Management,* and *Sustainable Micro Irrigation Design Systems for Agricultural Crops: Practices and Theory,* among others. He is the editor-in-chief of the book series Innovations and Challenges in Micro Irrigation and the 10-volume Research Advances in Sustainable Micro Irrigation. Readers may contact him at goyalmegh@gmail.com.

ABOUT THE CO-EDITOR

Vinod Kumar Tripathi, PhD

Vinod Kumar Tripathi, PhD is working as Assistant Professor in the Center for Water Engineering & Management, School of Natural Resources Management, Central University of Jharkhand, Brambe, Ranchi, Jharkhand, India

He has developed a methodology to improve the quality of produce by utilizing municipal wastewater under drip irrigation during Ph.D. research. His area of interest is geo-informatics, hydraulics in micro-irrigation and development of suitable water management technologies for higher crop and water productivity. He has taught hydraulics, design of hydraulic structures, water & wastewater engineering, geo-informatics to the post-graduate students at Central University of Jharkhand, Ranchi, India.

He is a National Level Expert to evaluate the government of India funded schemes for rural water supply and sanitation. He obtained his B.Tech. Degree in Agricultural Engineering in 1998 from Allahabad University, India; his M.Tech. Degree from G. B. Pant University of Agriculture & Technology, Pantnagar, India in Irrigation & Drainage Engineering; and his PhD. Degree in 2011, from Indian Agricultural Research Institute, New Delhi, India.

He is a critical reader, thinker, planner and fluent writer to publish more than thirty two peer-reviewed research publications, bulletins, and attended several national/International Conferences. Without his support and extraordinary job, readers will not have this quality book. Most of the research studies in this volume were conducted by Dr. Tripathi and his colleagues. Readers may contact him at: tripathiwtcer@gmail.com

BOOK ENDORSEMENTS

This book series is user-friendly and is a must for all irrigation planners to minimize the problem of water scarcity worldwide.

— Miguel A Muñoz, PhD,
Ex-President of University of Puerto Rico;
and Professor/Soil Scientist

I recall my association with Dr. Megh Raj Goyal while at Punjab Agricultural University, India. I congratulate him on his professional contributions and his distinction in micro irrigation. I believe that this innovative book series will aid the irrigation fraternity throughout the world.

— A. M. Michael, PhD,
Former Professor/Director, Water Technology Center – IARI;
Ex-Vice-Chancellor,
Kerala Agricultural University, Trichur, Kerala

Recycled wastewater use in drip irrigation poses challenges of potential clogging of drippers and non-uniformity of water application in the field. In providing this book volume, the editors have addressed these issues.

— Gajendra Singh, PhD,
Former Vice Chancellor, Doon University, Dehradun, India;
Former Deputy Director General (Engineering),
Indian Council of Agricultural Research (ICAR), New Delhi;
and Former Vice-President/Dean/Professor and Chairman, Asian
Institute of Technology, Thailand

Recently recycled and/or treated wastewater (WW) is being used as an alternative to freshwater to deal with problems of freshwater scarcity in agriculture. However, use of wastewater in micro irrigation cannot only help but also requires careful attention with regard to emitter health hazards and social acceptance. The information provided in this book series will go a long way in bringing large areas under micro irrigation in the world, especially in water scarcity countries.

— Professor (Dr.) R. K. Sivanappan,
Former Dean-cum-Professor of College of Agricultural Engineering
and Founding Director of Water Technology Center at Tamil Nadu
Agricultural University (TAMU), Coimbatore, India.
Recipient of Honorary PhD degree by Linkoping University –
Sweden and the honorary DSc degree by the TAMU, India

There is a need to address fundamental issues of wastewater management for irrigation use in India. Recycling of industrial and sewage water will make enough water available for agriculture in India. Micro-irrigation can mitigate abiotic stress situation by saving over 50% irrigation water.

—— V. M. Mayande, PhD,
Former Vice Chancellor,
Dr. Panjabrao Deshmukh Krishi Vidyapeeth,
Akola–444104, Maharashtra, India

OTHER BOOKS ON MICRO IRRIGATION TECHNOLOGY FROM AAP

Management of Drip/Trickle or Micro Irrigation
Megh R. Goyal, PhD, PE, Senior Editor-in-Chief

Evapotranspiration: Principles and Applications for Water Management
Megh R. Goyal, PhD, PE, and Eric W. Harmsen, Editors

BOOK SERIES: RESEARCH ADVANCES IN SUSTAINABLE MICRO IRRIGATION
Senior Editor-in-Chief: Megh R. Goyal, PhD, PE

Volume 1: Sustainable Micro Irrigation: Principles and Practices
Senior Editor-in-Chief: Megh R. Goyal, PhD, PE

Volume 2: Sustainable Practices in Surface and Subsurface Micro Irrigation
Senior Editor-in-Chief: Megh R. Goyal, PhD, PE

Volume 3: Sustainable Micro Irrigation Management for Trees and Vines
Senior Editor-in-Chief: Megh R. Goyal, PhD, PE

Volume 4: Management, Performance, and Applications of Micro Irrigation
Senior Editor-in-Chief: Megh R. Goyal, PhD, PE

Volume 5: Applications of Furrow and Micro Irrigation in Arid and Semi-Arid Regions
Senior Editor-in-Chief: Megh R. Goyal, PhD, PE

Volume 6: Best Management Practices for Drip Irrigated Crops
Editors: Kamal Gurmit Singh, PhD, Megh R. Goyal, PhD, PE, and
Ramesh P. Rudra, PhD, PE

Volume 7: Closed Circuit Micro Irrigation Design: Theory and Applications
Senior Editor-in-Chief: Megh R. Goyal, PhD; Editor: Hani A. A. Mansour, PhD

Volume 8: Wastewater Management for Irrigation: Principles and Practices
Editor-in-Chief: Megh R. Goyal, PhD, PE; Coeditor: Vinod K. Tripathi, PhD

Volume 9: Water and Fertigation Management in Micro Irrigation
Senior Editor-in-Chief: Megh R. Goyal, PhD, PE

Volume 10: Innovations in Micro Irrigation Technology
Senior Editor-in-Chief: Megh R. Goyal, PhD, PE;
Coeditors: Vishal K. Chavan, MTech, and Vinod K. Tripathi, PhD

BOOK SERIES: INNOVATIONS AND CHALLENGES IN MICRO IRRIGATION

Senior Editor-in-Chief: Megh R. Goyal, PhD, PE

Volume 1: Principles and Management of Clogging in Micro Irrigation
Editors: Megh R. Goyal, PhD, PE, Vishal K. Chavan, and Vinod K. Tripathi

Volume 2: Sustainable Micro Irrigation Design Systems for Agricultural Crops: Methods and Practices
Editors: Megh R. Goyal, PhD, PE, and P. Panigrahi, PhD

Volume 3: Performance Evaluation of Micro Irrigation Management: Principles and Practices
Editors: Megh R. Goyal, PhD, PE

PART I

IRRIGATION MANAGEMENT

CHAPTER 1

QUALITY OF IRRIGATION WATER RESOURCES: EGYPT

A. S. EL-HENAWY, M. R. KHALIFA and M. M. SAFFAN

CONTENTS

1.1 INTRODUCTION

Throughout the world, irrigated agriculture faces challenge of using less water, in many cases of poor quality, to provide food and fiber for growing population. Water resources in Egypt are limited. Consequently improving irrigation system, increasing water use efficiency and reuse of drainage water for irrigation are must. Water supply from irrigation canals is not sufficient enough, especially in the North of Nile Delta, and therefore, farmers use drainage water in irrigated the fields [2].

The mandatory use of supplemental irrigation water from sources other than the Nile needs knowledge for the factors that govern the water consumption. The suitability of any water for irrigation is determined by the amount and kind of salts present, and content of some heavy metals, and soil properties [6].

In recent years the use of wastewater in irrigation is considered a major source for heavy metals for the soil and plant, especially in arid and semi-arid zones, where crop production depends mainly on irrigation. The recycled water provides the soils with heavy metals, which may exceed the permissible limits for safe consumption by animals and humans [11, 13, 17].

This chapter focuses on the research study: (i) to evaluate the canal irrigation and drainage water in three districts of Egypt (Beiala, El-Hamoul and El-Borullus) in Kafr El-Sheikh Governorate; and (ii) To evaluate the effects of some heavy metal and micronutrient contents in soil and selected field crops.

1.2   MATERIALS AND METHODS

During October 1999 to June 2000, a field survey field was conducted in three districts of Egypt (Beiala, El-Hamoul and El-Borullus), each with 10 irrigation canals and two farms with clover and wheat crops, which were located adjacent to each irrigation canal.

Water samples were collected from each irrigation canal six times per month during the irrigation duration (between off and on). These water samples were analyzed to determine EC, pH, and soluble Ca^{++}, Na^+ and Mg^{++} using the methods described by Klute [15], and then SAR (Sodium adsorption ratio) was calculated. Soluble heavy metals and micronutrients in water samples (Zn, Mn, Cd, Ni, Co and Cu) were determined using procedures using atomic absorption spectrophotometer [16].

At harvesting, representative samples of grains and straws for wheat and shoots of clover plants were oven dried (70°C) and wet digested in $HClO_4$ + H_2SO_4 mixture according to Chapman and Pratt [7]. Concentration of Zn, Mn, Cd, Ni, Co and Cu were determined with an atomic absorption spectrophotometer [16]. Composite soil samples were taken at the depth of 0–30 cm to determine extractable Zn, Mn, Cd, Ni, Co and Cu using atomic absorption spectrophotometer [16]. The data were subjected to statistical analysis using Irristat program. Mean values were compared using Duncan's multiple range tests.

1.3 RESULTS AND DISCUSSION

1.3.1 EVALUATION OF WATER RESOURCES

1.3.1.1 Total Soluble Salts (EC, dS/m) and Sodium Adsorption Ratio (SAR)

Tables 1.1–1.3 indicate that irrigation water sources and duration of irrigation rotations had a highly significant effect on the values of EC and SAR of irrigation waters in Beiala and El-Borullus districts, whereas in El-Hamoul district only canals of irrigation had affected the these parameters. The observations also indicate that values of EC and SAR of the irrigation water in most of irrigation canals in districts of Beiala and El-Hamoul ranged from 0.35 to 0.63 dS/m for EC and from 1.40 to 2.29 for SAR, respectively. These values indicate: good water quality based on classification by Ayers and Westcot [6]; water classification class (C_2-S_1). This implies that this irrigation water is suitable for irrigation without causing any detrimental effects in these soil types and crops (clover and wheat).

On the other hand, EC values of irrigation water ranged from 0.94 to 2.08 dS/m in drain No. 5 (Biealla district), Kitchener drain, Bahr El-Mansoura, Fom El-Khalleg, El-Hallab canals (El-Hamoul district), and all irrigation canals of El-Borullus district. This range of EC indicates high salinity according to Ayers and Westcot [6] classification and it lies in class C_3-S_1.

This implies that high salinity water with low sodicity hazard can cause increasing salinity problems. This may be due to because the irrigation water in these canals is considered a mixture of drainage, waste and freshwaters. Therefore, when this water is used for irrigation, it must be adequately controlled and managed with good tillage, addition of amendments, and good cropping. Finally, EC and SAR values of irrigation canals in the three districts can be arranged in the ascending order: El-Borullus > El-Hamoul > Beiala districts. Regarding, pH values of irrigation water in all irrigation canals for the three districts was about 8 implying that it was slightly alkaline. Our results agree with those reported by other investigators [11, 14].

TABLE 1.1 Mean Values of Chemical Properties and Soluble Heavy Metals of Different Irrigation Water Canals in Beiala District (Kafr El-Sheikh Governorate), During 1999/2000 Season

Name of canal, C	Irrigation rotation, R	Chemical properties			Soluble heavy metals, ppm					
		pH	EC, dS/m	SAR	Mn	Zn	Cu	Cd	Ni	Co
Drain no. 5	On	8.17	1.18[a]	5.67[a]	0.060	0.081	0.001	0.006	N.D	0.028
	Off	8.20	1.36[a]	4.60[a]	0.013	0.075	0.000	0.001	N.D	0.001
Ebshan canal	On	8.20	0.38[b]	1.51[b]	0.029	0.051	0.001	0.001	N.D	0.001
	Off	8.16	0.47[bc]	1.53b	0.036	0.078	0.001	0.001	N.D	0.000
El-Sharkawia canal	On	8.18	0.35[b]	1.53[b]	0.00	0.048	0.001	0.010	N.D	0.042
	Off	8.28	0.42[C]	1.62[bc]	0.035	0.073	0.000	0.002	N.D	0.001
Bahr terra	On	8.50	0.36[b]	1.54[b]	0.007	0.055	0.000	0.003	N.D	0.000
	Off	8.31	0.37[C]	1.41[C]	0.030	0.056	0.000	0.002	N.D	0.001
Fouda canal	On	8.23	0.37[b]	1.67[b]	0.028	0.065	0.001	0.012	N.D	0.042
	Off	8.32	0.38[C]	1.42[C]	0.029	0.067	0.000	0.001	N.D	0.001
Garrd El-Agamy canal	On	8.24	0.35[b]	1.52[b]	0.020	0.053	0.000	0.009	N.D	0.042
	Off	8.32	0.41[C]	1.80[bc]	0.046	0.081	0.001	0.002	N.D	0.001
Bahr El-Nour	On	8.29	0.35[b]	1.42[b]	0.020	0.051	0.000	0.001	N.D	0.001
	Off	8.30	0.37[C]	1.53[bc]	0.034	0.067	0.001	0.000	N.D	0.000

TABLE 1.1 Continued.

Name of canal, C	Irrigation rotation, R	Chemical properties			Soluble heavy metals, ppm					
		pH	EC, dS/m	SAR	Mn	Zn	Cu	Cd	Ni	Co
Bahr Beiala	On	8.27	0.35[b]	1.55[b]	0.029	0.050	0.001	0.001	N.D	0.000
	Off	8.29	0.39[C]	1.56[bc]	0.024	0.066	0.001	0.001	N.D	0.000
El-Shorafa canal	On	8.12	0.44[b]	2.17[b]	0.033	0.057	0.001	0.000	N.D	0.000
	Off	8.18	0.63[b]	2.61[b]	0.029	0.075	0.000	0.009	N.D	0.000
Marrzoka canal	On	8.22	0.38[b]	1.68[b]	0.034	0.078	0.001	0.002	N.D	0.000
	Off	8.27	0.54[bc]	2.29[bc]	0.020	0.075	0.001	0.000	0.00	0.001
F. Test	C	NS	*	*	NS	NS	NS	NS	-	NS
	R	NS	*	NS	NS	*	NS	NS	-	NS
	CxR	NS	NS	NS	NS	NS	NS	NS	-	NS

Note: NS = not significant; N.D. = no data; C = canal; R = irrigation rotation; C x R = interaction.

TABLE 1.2 Mean Values of Chemical Properties and Soluble Heavy Metals in Irrigation Water From Canals in El-Hamoul District (Kafr El-Sheikh Governorate), During 1999/2000 Season

Name of Canal, C	Irrigation rotation, R	Chemical properties					Soluble heavy metals, ppm			
		pH	EC, dS/m	SAR	Mn	Zn	Cu	Cd	Ni	Co
Ketshenar drain	On	7.89[b]	1.30[a]	4.17[a]	0.039	0.082[ab]	0.000	0.041[a]	N.D	0.022[ab]
	Off	8.29	1.27[b]	4.77[ab]	0.004	0.094[ab]	0.001	0.00[a]	N.D	0.000[s]
El-Kafr El-Sharki canal	On	8.24	0.44[b]	1.42[b]	0.039	0.052[ab]	0.001	0.001[ab]	N.D	0.069[a]
	Off	8.21	0.49[C]	2.08[c]	0.00	0.092[ab]	0.015	0.00[a]	N.D	0.000[a]
Bahr El-Banawan canal	On	8.25	0.37[b]	1.49[b]	0.043	0.045[b]	0.002	0.009[ab]	N.D	0.028[ab]
	Off	8.17	0.41[c]	1.70[c]	0.001	0.090[ab]	0.000	0.00[a]	N.D	0.000[a]
Ragheeb canal	On	8.26	0.36[b]	1.40[b]	0.044	0.054[ab]	0.001	0.065[b]	N.D	0.000[b]
	Off	8.25	0.38[C]	1.64[c]	0.004	0.131[a]	0.015	0.001[a]	N.D	0.001[a]
El-Ganabia El-Sabaa	On	8.24	0.39[b]	1.44[b]	0.030	0.052[ab]	0.001	0.012[ab]	N.D	0.001[b]
	Off	8.25	0.40[c]	1.61[c]	0.027	0.082[b]	0.001	0.001[a]	N.D	0.000[a]
Zouba canal	On	8.29	0.37[b]	1.43[a]	0.034	0.042[b]	0.001	0.00[ab]	N.D	0.029[ab]
	Off	8.28	0.39[c]	1.71[C]	0.020	0.063[b]	0.001	0.002[a]	N.D	0.001[a]
El-Wallda canal	On	8.24	0.37[b]	1.39[b]	0.055	0.045[b]	0.000	0.014[ab]	N.D	0.012[ab]
	Off	8.17	0.41[C]	1.62[c]	0.020	0.087[b]	0.001	0.003[a]	N.D	0.000[a]
Bahr El-Mansour	On	7.97	0.94[a]	3.83[a]	0.004	0.081[ab]	0.001	0.010[ab]	N.D	0.000[b]
	Off	8.08	1.22[a]	4.34[b]	0.044	0.074[b]	0.001	0.001[a]	N.D	0.000[a]

TABLE 1.2 Continued.

Name of Canal, C	Irrigation rotation, R	Chemical properties			Soluble heavy metals, ppm					
		pH	EC, dS/m	SAR	Mn	Zn	Cu	Cd	Ni	Co
Fom El-Khaleg	On	8.12	1.08[a]	4.35[a]	0.023	0.079[ab]	0.000	0.015[ab]	N.D	0.028[ab]
	Off	8.10	1.74[a]	6.54[a]	0.050	0.067[b]	0.000	0.001[a]	N.D	0.001[a]
El-Hallab canal	On	8.09	1.09[a]	4.58[a]	0.034	0.091[a]	0.001	0.015[ab]	N.D	0.028[ab]
	Off	8.20	1.17[b]	4.58[b]	0.053	0.068[b]	0.000	0.001[a]	N.D	0.000[a]
F. Test	C	N.S	**	**	N.S	N.S	N.S	N.S	-	N.S
	R	N.S	N.S	N.S	N.S	**	N.S	*	-	*
	C x R	N.S	N.S	N.S	N.S	**	N.S	N.S	-	N.S

Note: NS = not significant; N.D. = no data; C = canal; R = irrigation rotation; C x R = interaction.

TABLE 1.3 Mean Values of Chemical Analysis and Soluble Heavy Metals of Different Irrigation Water Canals in El-Borullus District (Kafr El-Sheikh Governorate), 1999/2000 Season

Name of canal, C	Irrigation rotation, R	Chemical properties			Soluble heavy metals, ppm					
		pH	EC, dS/m	SAR	Mn	Zn	Cu	Cd	Ni	Co
Branch of Terra 1 canal	On	8.11	1.21[cd]	4.16[a]	0.024	0.091	0.002	0.003	N.D	0.001
	Off	8.09	1.47[C]	4.60[c]	0.025	0.080	0.001	0.001	N.D	0.001
Terra 2 canal	On	8.12	1.14[d]	4.34[a]	0.021	0.081	0.001	0.004	N.D	0.001
	Off	8.01	1.81[ab]	6.79[abc]	0.304	0.080	0.001	0.001	N.D	0.001
El-Magazz canal	On	8.10	1.16[d]	4.16[a]	0.019	0.081	0.000	0.008	N.D	0.023
	Off	7.97	1.93[ab]	5.88[bc]	0.033	0.084	0.000	0.005	N.D	0.000
El-Hellmyia canal	On	8.10	129[bcd]	4.63[a]	0.020	0.086	0.000	0.014	N.D	0.001
	Off	8.09	1.88[ab]	6.89[abc]	0.110	0.069	0.000	0.001	N.D	0.000
Terra 4 canal	On	8.08	1.19[cd]	3.98[a]	0.009	0.090	0.001	0.005	N.D	0.001
	Off	8.20	1.69[bc]	6.00[bc]	0.030	0.085	0.001	0.001	N.D	0.000
El-Khashaa canal	On	8.15	1.61[ab]	6.48[a]	0.027	0.119	0.001	0.011	N.D	0.001
	Off	8.16	2.06[a]	6.80[abc]	0.339	0.091	0.000	0.014	N.D	0.000
El-ganabia El-gharbia canal	On	8.23	1.51[abc]	5.80[a]	0.012	0.097	0.002	0.010	N.D	0.001
	Off	8.41	1.87[ab]	7.30[abc]	0.043	0.0102	0.001	0.001	N.D	0.001
El-Nahda canal	On	8.20	1.65[a]	6.35[a]	0.026	0.095	0.002	0.015	N.D	0.001
	Off	8.11	1.96[ab]	8.55[ab]	0.043	0.090	0.001	0.001	N.D	0.000

TABLE 1.3 Continued.

Name of canal, C	Irrigation rotation, R	Chemical properties			Soluble heavy metals, ppm					
		pH	EC, dS/m	SAR	Mn	Zn	Cu	Cd	Ni	Co
Balteem El-Gedida canal	On	8.09	1.59[ab]	6.05[a]	0.024	0.090	0.003	0.012	N.D	0.006
	Off	8.15	1.93[ab]	7.73[abc]	0.426	0.086	0.001	0.001	N.D	0.001
Nyhaite Bahr Terra (Balteem) canal	On	8.22	1.62[ab]	6.40[a]	0.017	0.102	0.001	0.008	N.D	0.006
	Off	7.96	2.08[a]	9.64[a]	0.357	0.091	0.001	0.000	N.D	0.00
F. Test	C	N.S	**	N.S	N.S	N.S	N.S	N.S	-	N.S
	R	N.S	**	**	N.S	N.S	N.S	**	-	N.S
	C x R	N.S	N.S	N.S	N.S	N.S	N.S	N.S	-	N.S

Note: NS = not significant; N.D. = no data; C = canal; R = irrigation rotation; C x R = interaction.

1.3.1.2   Soluble Heavy Metals and Micronutrients Contents in Water Sources

The data are shown in Tables 1.1–1.3 for different sources of irrigation water according to content of heavy metals. These observations indicate that there was no significant effect of both irrigation canals and irrigation rotations on the concentrations of soluble heavy metals and micronutrients in all districts. However, Zn-concentration in the irrigation canals (Beiala and El-Hamoul districts) and Cd-concentration (El-Hamoul and El-Borullus districts) were significantly affected only with irrigation rotations.

On the other hand, concentration of Mn, Zn and Cd elements were high in irrigation canals of the three districts, especially in case of period of irrigation rotation (on) for El-Wallda, Bahr El-Mansour, Fom El-Khallieg, Kitchener drain and El-Hallab canals (El-Hamoul district). This is because the water in these irrigation canals is agriculture drainage water mixed with fresh water or drainage water mixed with wastewater from human activity. Meanwhile, concentration of these elements was high in period of irrigation rotation (off) in irrigation canals of Fara Terra, El-Khashaa, Balteem El-Gedida and Neyhaite Bahr terra (El-Borullus district), Kitchener drain and El-Kafr El-Shareki (El-Hamoul district).

Finally, the concentration of most tested heavy metals and micronutrients in irrigation water of most canals were less than the safe limits, recommended by Cottenie et al. [9] and Alloway [4]. The concentration was 2.0 ppm of Zn, 0.2 ppm of Mn, 0.2 ppm of Cu, 0.01 ppm of Cd, 0.2 ppm of Ni, and 0.05 ppm of Co. However, Co-concentration in El-Kafr El-Sharki canal (El-Hamoul district) and Mn-concentration in El-Khashaa and Balteem El-Gadida canals (El-Borullus district) exceeded the safe limits. These results are in agreement with those reported by and El-Henawy [11] and El-Wakeel [12].

1.3.2   AVAILABLE CONTENT OF HEAVY METALS AND MICRONUTRIENTS IN SOIL

The Table 1.4 indicates that the range of available content of heavy metals in soil after harvesting clover crop was: 2.26–4.17 of Mn, 0.72–1.41 of Zn,

0.19–2.03 of Cu, 0–0.25 of Ni, 0–0.126 of Cd, and 0.054 to 0.216 ppm of Co in Beiala district. The highest content of available Mn, Cu, Cd and Co was found in the soil, which was irrigated from Marrzoka canal, which had accepted wastewater during the winter season. Meanwhile, the highest content of available Zn and Ni was found in soil, which was irrigated from Ebshan canal, which received wastewater from the human activity.

In the soil samples after harvesting of wheat crop (Table 1.4), the concentration of available elements ranged between 2.37–3.079 of Mn, 0.8–2.45 of Zn, 0.38–1.78 of Cu, 0.17–0.25 of Ni, 0–0.063 of Cd and 0.54–0.216 ppm of Co, respectively. The highest content of available Mn, Zn and Ni, Cu and Co was found in the soils, which were irrigated with water from El-Shorafaa, Bahr El-Nour, Ebshan and El-Sharkawia canals, respectively.

In El-Hamoul district (Table 1.4), observations reveal that available content of all heavy metals in soils were found in all irrigated canals, after harvesting of clover and wheat crops. The highest contents of available Zn, Cu, Ni and Cd were found in soils which was irrigated from Kitchener drain after wheat, while the highest content of Mn and Co elements in the soils was found in case of El-Kafr El-Sharki and Bahr El-Mansour canals, respectively. Furthermore after harvesting of clover crop, the highest content of available Mn, Zn, Cu, Ni, Cd and Co in soils was found with irrigation water from Rajheeb, Bahr El-Banawaan, Zoubaa, El-Kafr El-Sharki and El-Hallab canals, respectively.

Regarding El-Borullus district, data in Table 1.4 revealed that after harvesting clover soil samples had highest concentration of available Mn, Zn, Cu and Cd, with irrigation from Farah Terra 1 canal. However, the highest concentration of Ni and Co were found in the soils adjacent to Farah Terra 4 and El-Helmia canals.

After harvesting of wheat crop, the highest content of available Mn and Cu were found with irrigation water from El-Helmia canal, while the highest values of Zn, Ni and Co were found in the soils with irrigation from El-Khashaa El-gedida canal. Also, the highest value of Cd element was found in the soil adjacent to Farah Terra 1.

We can conclude that available heavy metal concentrations in soils can be arranged in the ascending order: Borullus > El-Hamoul > Biela districts. This may due to because the water from all irrigated canals in El-Borullus

TABLE 1.4 DTPA Extractable of Heavy Metals (mg.kg^{-1}) in the Soil Samples After Harvesting of Clover and Wheat Crops, Which Were Adjacent to Irrigation Canals

Irrigation water source	After clover						After wheat					
	Mn	Zn	Cu	Ni	Cd	Co	Mn	Zn	Cu	Ni	Cd	Co
Beiala district (Kafr El-Sheikh Governorate)												
Bahr Beialla canal	2.26	0.96	0.19	0.00	0.063	0.054	2.78	0.96	0.38	0.17	0.063	0.162
Bahr El-Nour canal	3.06	0.93	0.32	0.17	0.063	0.00	2.37	2.45	0.57	0.25	0.00	0.127
Card El-Agamy canal	3.27	0.80	0.76	0.08	0.053	0.108	2.61	0.85	0.44	0.25	0.021	0.108
Drain No. 5	3.27	0.72	0.83	0.25	0.053	0.054	3.03	1.87	0.95	0.25	0.021	0.216
Ebshan canal	3.13	1.41	0.76	0.25	0.00	0.00	2.54	1.07	1.78	0.17	0.01	0.108
El-Sharkawia canal	3.65	1.25	1.14	0.08	0.074	0.0544	2.43	0.80	1.27	0.17	0.053	0.216
El-Shorafa canal	3.30	0.69	0.89	0.08	0.011	0.216	3.76	0.91	1.21	0.17	0.021	0.054
Fouda canal	3.48	1.07	1.59	0.08	0.042	0.162	3.48	0.80	0.95	0.17	0.00	0.216
Marzouka canal	4.17	1.33	2.03	0.17	0.126	0.324	2.61	0.83	1.27	0.25	0.042	0.108
El-Hamoul district (Kafr El-Sheikh Governorate)												
Bahr El-Banawan canal	4.52	1.52	1.78	0.00	0.011	0.108	2.33	0.8	0.63	0.25	0.00	0.054
Bahr El-Mansour canal	2.30	0.67	0.76	0.17	0.011	0.00	2.54	0.59	0.44	0.17	0.053	0.216
El-Ganabia El-Sabaa canal	3.20	0.64	0.76	0.08	0.032	0.00	2.61	0.96	0.06	0.08	0.053	0.00
El-Hallab canal	4.87	0.67	0.83	0.00	0.084	0.108	2.10	0.80	0.89	0.17	0.032	0.054
El-Kafr El-Sharki canal	4.45	1.12	1.59	0.25	0.042	0.054	3.48	1.01	0.32	0.17	0.053	0.108
El-Walda canal	2.61	0.69	0.32	0.17	0.053	0.054	2.78	0.75	1.14	0.08	0.032	0.054
Fom El-Khalieg canal	2.78	0.80	0.13	0.08	0.00	0.00	2.19	0.48	0.44	0.08	0.012	0.108

TABLE 1.4 Continued.

Irrigation water source	After clover						After wheat					
	Mn	Zn	Cu	Ni	Cd	Co	Mn	Zn	Cu	Ni	Cd	Co
Kitchener Drain	4.43	0.85	0.44	0.08	0.021	0.108	2.16	2.80	1.40	0.25	0.084	0.054
Ragheb canal	5.04	1.07	1.46	0.17	0.042	0.054	2.37	0.85	0.63	0.17	0.042	0.162
Zouba canal	3.65	0.96	2.03	0.00	0.011	0.054	2.23	0.53	0.44	0.00	0.032	0.00
El-Borullus district (Kafr El-Sheikh Governorate)												
El-Ganabia El-Gharbia canal	3.90	1.47	1.59	0.25	0.00	0.170	2.61	1.65	0.87	0.25	0.021	0.162
El-Hellmya canal	4.52	0.61	0.76	0.25	0.00	0.187	6.43	1.044	2.86	0.08	0.011	0.00
El-Khashaa El-Gedida canal	2.68	0.75	0.95	0.08	0.00	0.162	3.09	2.05	0.94	0.25	0.00	0.160
El-Magazz canal	4.07	0.64	1.14	0.17	0.042	0.162	3.62	0.67	1.14	0.17	0.00	0.00
El-Nahda canal	4.00	0.56	0.63	0.17	0.032	0.054	3.23	0.69	0.63	0.080	0.032	0.054
Farha Terra 1 canal	5.11	1.89	2.16	0.08	0.042	0.054	3.20	1.07	0.70	0.17	0.021	0.108
Farha Terra 2 canal	4.34	0.67	0.89	0.17	0.021	0.162	3.8	1.41	0.32	0.25	0.011	0.108
Farha Terra 4 canal	5.57	0.85	1.90	0.25	0.032	0.162	3.76	0.59	0.32	0.25	0.011	0.054

district is a mixture of water source from the drainage, sewage and fresh waters, which contained more concentrations of heavy metals. Also, it was noticed that the concentration of the most available heavy metals was high in soils after harvesting of clover than in soils after harvesting of wheat. This may be due to variation of biological activity which is accompanied by the cultivated plants that led to the increasing the of organic acidity thus causing the increment of the availability of these elements.

These results are in agreement with those obtained by Abou Hussin et al. [3], Amer et al. [5], El-Henway [11] and Salt et al. [17]. They reported that use of the poor water quality and wastewater in irrigation increased the content of total and available heavy metals in soil.

1.3.3 CONCENTRATION OF HEAVY METALS AND MICRONUTRIENTS IN THE SHOOTS OF CLOVER

Table 1.5 indicates that the concentration (mg.kg^{-1} of dry matter plant) of heavy metals (Ni, Cd and Co) and micronutrients (Zn, Cu and Mn) varied from site to site, depending on the available concentration of these elements in soil samples (see Table 1.4), soil pH and the concentrations of these elements in irrigation water for the crop (see Tables 1.1–1.3). Data in Table 1.5 shows that Ni-concentration in shoot was higher than the safety limit of 8 ppm in all locations of the three districts [8], except in locations adjacent to Fouda and El-Shorafa canals (Biealla district) and Bahr El-Banawan canal (El-Hamoul district). Cd-concentration in shoot was higher than the safety limits (0.01–1.23 ppm) in most of the locations in the three districts [4]. Whereas, the highest concentration of Cd was 11.58 in locations adjacent to Bahr El-Banwan (El-Hamoul district) and 20.53 in locations adjacent to Farah terra 4 (El-Borullus district), respectively. Co-concentration in all locations was less than the safety limits (5 – 20 ppm) according to Cottenie et al. [9]. Zn, Mn and Cu concentrations in shoot for all locations were less than the safety limits (50 ppm of Zn, 100 ppm of Mn and 20 ppm of Cu) [4], except Cu-concentration was higher than the safety limit for locations adjacent to Garrd El-Agamy, Bahr Biealla canals (Biealla district); Farah terra 1, El-Khashaa El-Gedida

TABLE 1.5 Concentrations of Heavy Metals in the Shoots of Clover Plants That Have Been Irrigated From Locations Adjacent to Irrigation Canals in Three Districts of Egypt

Irrigation water source	Mn	Zn	Cu	Ni	Cd	C
Beiala district (Kafr El-Sheikh Governorate)						
Bahr Beialla canal	37.8	29.3	23.5	8.33	0.00	7.31
Bahr El-Nour canal	28.1	34.7	11.8	25.0	0.00	5.00
Drain No. 5	31.3	40.0	16.7	8.33	0.00	1.07
Ebshan canal	40.6	18.7	17.6	8.33	44.21	2.23
El-Sharkawia canal	28.1	32.0	17.6	8.33	1.05	0.00
El-Shorafa canal	34.4	37.3	0.00	0.00	8.42	2.11
Fouda canal	34.4	32.0	5.88	0.00	2.11	7.01
Gard El-Agamy canal	37.5	32.0	29.4	8.33	9.47	3.17
Marzouka canal	31.3	24.0	17.6	33.3	9.47	2.15
El-Hamoul district (Kafr El-Sheikh Governorate)						
Bahr El-Banawan canal	25.0	32.0	11.8	0.00	11.58	4.82
Bahr El-Mansour canal	31.3	16.0	5.88	8.33	8.42	0.00
El-Ganabia El-Sabaa canal	31.3	26.7	17.6	16.7	2.11	1.15
El-Hallab canal	28.1	29.3	0.00	8.33	0.00	5.40
El-Kafr El-Sharki canal	37.5	56.0	17.6	8.33	1.05	7.23
El-Walda canal	25.0	26.7	17.6	8.33	2.26	0.00
Fom El-Khalieg canal	28.1	44.0	5.88	8.33	1.05	0.00
Kitchener Drain	37.5	26.7	0.00	16.7	0.00	2.12
Ragheb canal	25.0	37.3	11.8	25.0	2.11	0.00
Zouba canal	25.0	10.7	0.00	8.33	0.00	5.40
El-Borullus district (Kafr El-Sheikh Governorate)						
El-Ganabia El-Gharbia canal	62.5	53.3	11.8	16.7	5.26	6.82
El-Hellmya canal	40.6	42.7	17.6	8.33	5.26	3.22
El-Khashaa El-Gedida canal	37.8	37.3	29.4	16.7	0.00	5.40
El-Magazz canal	31.3	40.0	11.8	8.33	3.16	7.72
El-Nahda canal	62.5	26.7	29.4	25.0	6.32	1.97
Farha Terra 1 canal	40.6	32.0	29.4	33.3	4.21	0.00
Farha Terra 2 canal	34.0	32.0	17.6	25.0	9.47	3.17
Farha Terra 4 canal	18.8	2.0	17.6	16.7	20.53	0.00

and El-Nahdaa canals (El-Burullus district). These results are in harmony with those reported by Abd El-Naiem et al. [1] and El-Henawy [11].

We can conclude that the frequent utilization of agriculture drainage water or mixture of drainage water with different waste water for irrigating clover may cause an accumulation of some heavy metals (Ni and Cd), thus leading to detrimental effects on animals and humans.

1.3.4 CONCENTRATIONS OF HEAVY METALS AND MICRONUTRIENTS IN THE STRAW AND GRAINS OF WHEAT PLANT

The Table 1.6 reveals that the concentration of heavy metals (Ni and Cd) and Cu-element in straw of wheat plant was higher than the values in the grains for most of the locations. The highest values of Cu (23.5 mg.kg^{-1}), Ni (50 mg.kg^{-1}) and Cd (6.32 mg.kg^{-1}) were found in locations adjacent to Marzouka and Bahr El-Nour canals (Beialla district) and Farah terra 2 (El-Borullus district) for Cu; Fouda and Marrzoka canals (Beiala district) and El-Banawan and El-Halabe canals (El-Hamoul district) for Ni and Kitchener drain for Cd.

On the other hand, concentrations of Mn, Zn and Co in grains of wheat were higher than the corresponding values of straw for all locations. However, the highest value of Mn (38.4 mg kg^{-1}), Zn (45.3 mg kg^{-1}) and Co (6.17 mg kg^{-1}) were found in Zobaa, Bahr El- Mansour and El-Magazz canals, respectively.

Furthermore, data indicates that concentrations of Mn, Zn and Cu in both straw and grains of wheat were less than the safety limits reported by Cottenie et al. [9]. However, Cu-concentration in some locations was above safety levels and was below the phytotoxic levels. The Cd and Ni concentrations in both straw and grains were higher than permitted levels, according to Cottenie et al. [8]. This implies that frequent use of agriculture drainage water or mixture of drainage water and wastewater for irrigation of wheat plant caused an accumulation of some heavy metals (Ni and Cd), which in turn was harmful to animals and human. Similar results have been reported by Davis and Smith [10] and El-Henawy [11].

TABLE 1.6 Concentrations of Heavy Metals and Micronutrients in the Grains and Straw of Wheat Plant Which Had Been Irrigated From Adjacent Irrigation Canals, During 1999/2000

Irrigation water source	grains						straw					
	Mn	Zn	Cu	Ni	Cd	Co	Mn	Zn	Cu	Ni	Cd	Co
Beiala district (Kafr El-Sheikh Governorate)												
Bahr Beialla canal	31.3	34.7	11.7	0.00	1.05	2.67	28.1	10.7	11.8	41.7	4.21	0.22
Bahr El-Nour canal	31.3	24.0	11.7	25.0	3.16	2.18	15.6	2.67	23.5	25.0	2.11	0.41
Drain No. 5	28.1	37.3	5.88	0.00	2.11	5.12	28.1	10.7	5.88	16.7	4.02	0.77
Ebshan canal	28.1	24.3	11.7	25.0	0.00	1.17	18.8	10.7	23.5	41.7	2.11	0.12
El-Sharkawia canal	34.4	21.3	5.88	25.0	0.00	4.21	18.8	2.67	23.5	33.3	1.05	1.41
El-Shorafa canal	37.5	40.0	11.7	16.70	0.00	4.21	21.9	8.00	17.6	25.0	3.16	0.32
Fouda canal	28.1	24.0	5.88	16.70	1.05	0.00	18.8	10.7	17.6	50.0	1.05	0.00
Gard El- Agamy canal	25.0	26.7	11.7	25.0	0.00	0.165	12.5	2.67	0.00	33.3	2.00	1.23
Marzouka canal	37.5	26.7	0.00	16.7	2.11	4.21	21.7	2.67	23.3	50.0	1.05	0.00
El-Hamoul district (Kafr El-Sheikh Governorate)												
Bahr El-Banawan canal	25.0	26.7	5.88	0.00	44.21	2.16	21.9	13.3	11.8	50.0	3.16	0.15
Balir El-Mansour canal	28.1	45.3	17.6	8.33	0.00	3.21	21.9	10.7	0.00	8.33	5.26	1.22
El-Ganabia El-Sabaa canal	21.9	16.0	4.70	16.7	3.16	5.12	25.0	13.3	5.88	25.0	4.21	0.17
El-Hallab canal	21.9	16.0	0.00	0.00	0.00	3.98	18.8	10.7	17.6	50.0	0.00	0.00
El-Kafr El-Sharki canal	15.6	29.3	5.88	8.33	0.00	0.00	18.8	13.3	11.8	8.33	2.11	0.23
El-Walda canal	31.3	24.0	0.00	8.33	0.00	4.16	16.6	5.33	0.00	0.00	0.00	1.11
Fom El-1 Chalieg canal	25.0	21.3	4.70	0.00	0.00	1.15	21.9	13.3	0.00	0.00	3.16	0.52

TABLE 1.6 Continued.

Irrigation water source	grains						straw					
	Mn	Zn	Cu	Ni	Cd	Co	Mn	Zn	Cu	Ni	Cd	Co
Kitchener Drain	25.0	42.7	0.00	8.33	0.00	3.00	18.8	10.7	11.8	8.33	6.32	0.12
Ragheb canal	34.4	32.0	5.88	0.00	0.00	0.00	25.0	2.67	17.6	8.33	0.00	0.00
Zouba canal	38.4	24.0	5.88	8.33	0.00	1.15	21.9	13.3	0.00	16.7	1.05	1.00
El-Borullus district (Kafr El-Sheikh Governorate)												
El-Ganabia El-Gharbia canal	25.0	32.0	23.5	25.0	2.11	3.00	18.8	10.7	11.8	16.7	2.11	0.17
El-Hellmya canal	31.3	26.7	0.00	8.33	0.00	4.85	28.1	16.0	11.8	41.7	3.16	1.17
El-Khashaa El-Gedida canal	31.3	34.7	0.00	0.00	2.11	4.27	25.0	6.67	11.8	0.00	1.05	0.52
El-Magazz canal	34.4	32.0	11.7	8.33	1.05	6.17	18.8	13.3	17.6	0.00	3.16	1.35
El-Nahda canal	28.1	40.0	17.6	33.3	1.16	0.00	15.6	10.67	5.88	33.3	3.25	1.37
Farha Terra 1 canal	31.4	21.3	0.00	8.33	2.11	3.98	15.6	13.3	17.6	41.7	4.21	0.00
Farha Terra 2 canal	34.4	40.0	17.6	16.70	3.16	6.17	31.3	13.3	23.5	0.00	4.21	1.17
Farha Terra 4 canal	34.4	21.3	11.7	3.33	3.16	0.00	12.5	8.0	17.6	33.3	4.21	1.07

1.4 SUMMARY

An experimental survey was carried out to evaluate the irrigation water resources in three districts (Beiala, El-Hamoul and El-Borullus) of Kafr El-Shiekh Governorate of Egypt, during season 1999/2000. Ten irrigation canals were selected in each district. Chemical analysis for heavy metal concentrations and micronutrients in water, soil and plant (wheat and clover crops) was performed.

The values of salinity index (EC) and Sodium adsorption ratio (SAR) of irrigation water in all three districts had highly significant differences among irrigation canals and irrigation rotations, except EC values in El-Hamoul district and SAR values in El-Borullus district had only significant differences between irrigation canals and irrigation rotation, respectively.

Data shows that the irrigation water had a medium quality class (C_2-S_1) for all irrigation canals in Beiala district, except drain No. 5, which had a lower quality class (C_3-S_2). El-Borullus district had a low quality class (C_3-S_1) and (C_3-S_2) for all canals. Irrigation water in El-Hamoul district had a medium quality class (C_2-S_1) in most of canals, while canals of Kitchener, Bahr El-Mansour, Fom El-Khalieg and El-Hallab had a low quality class (C_3-S_2).

There was no significant difference among heavy metal concentrations for either canals or rotations. However, Zn concentration in both Beiala and El-Hamoul districts as well as Cd concentration in El-Borullus and El-Hamoul districts had a significant difference. It is concluded that all concentrations of heavy metals were less than the permitted limits.

Heavy metals concentration in soil after cultivation of clover and wheat were in the order of Borullus > El-Hamoul > Beiala districts. Zn, Ni, Cd and Cu had the highest concentration in Beiala (Marzouka canal), El-Hamoul (Kitchener) and in El-Borullus (Tiera No. 4 and El-Gannabia El-Gharbia canals). Also the concentration of all heavy metals in soil was less than the hazard levels. Ni and Cd showed a high level concentration above the permitted level in both wheat and clover crops, in some areas adjacent to irrigation canals in three districts, which was harmful to animals and humans.

KEYWORDS

- clover
- crop production
- Egypt
- electrical conductivity, EC
- heavy metals
- micro nutrients
- Nile Delta
- pH
- salts
- sodium adsorption ratio, SAR
- Tanta University
- water pollution
- water quality
- wheat

REFERENCES

1. Abd El-Naim, M., Abu El-Enien, R., Fahmiry, S., Said, A. (1998). Risk assessment of soil and water pollution on agriculture crops in Egypt. World Congress of Soil Science, Montpellier – France, 20–26 August.
2. Abo-Soliman, M. S. M., Khalifa, M. R., El-Sabry, W. S., Sayed, K. H. (1992). Use of drainage water in irrigation at North of Nile Delta, its effects on soil salinity and wheat production. *J. Agric. Res. Tanta Univ.*, 18(2), 425–435.
3. Abou-Hussien, E. A., El-Koumey, B. Y. (1997). Influence of waste water on some soils and plant in Menofiya Governorate. *Menofiya J. Agric. Res.*, 22(6), 1733–1748.
4. Alloway, B. J. (1995). *Heavy metals in soils*. Blackie Academic professional: Imprint of Chapman & Hull, Eastern Cleddens Rood, UK.
5. Amer, A. A., Abd El-Wahab, S. A., Abou El-Soud, M. A. (1997). Effect of water quality on soil salinity and some crops production. *J. Agric. Sci. Mansoura Univ.*, 22(4), 1287–1265.
6. Ayers, R. S., Wescot, D. W. (1985). *Water quality for agriculture*. Irrigation and Drainage Paper 29, FAO, Rome, 174 pages.

7. Chapman, H. D., Pratt, P. F. (1980). *Methods of analysis of soil, plants and waters.* Univ. of Calif. Divison of Agric. Sci., pages 60 – 69.

8. Cottenie, A., Dhaese, A., Camerlynek, R. (1976). Plant quality response to uptake of polluting elements. *Qualities Planetarium, Plant Food for Human's Nutrition*, 26: 293.

9. Cottenie, A., Verloo, M., Velghe, G., Kiekens, L. (1982). Biological and analytical aspects of soil pollution. Laboratory of Analytical Agro. State Univ., Ghent -Belgium.

10. Davis, R. D., Smith, C. (1980). Crops as indicators of the significance of contamination of soil by heavy metals. Technical Report TR 140 by Water Research Center, Tevenage, U.K. 43 pages.

11. El-Henawy, A. S. (2000). Impact of available water sources at North Delta on soil and some field crops. M. Sc. Thesis, Fac. of Agric. Kafr El-Sheikh, Tanta Univ. Egypt.

12. El-Wakeel, A. F., El-Mowelhi, N. M. (1993). Reuse of drainage water on farm with reference to soil properties", crop production and heavy metals contents in plant Tissues and diary products. *Egypt. J. Agric. Res.,* 71(4), 845–861.

13. Khalifa, M. R. (1990). Evaluation of suitability of drainage water for irrigation purpose, its effect on some soil properties of clay soils in North Delta. *J. Agric. Res. Tanta Univ.,* 16(3), 573–586.

14. Khalifa, M. R., Rabie, A., Youssef, S. M., El-Henawy, A. S. (2003). Evaluation of available sources of irrigation water at North Delta and its effect on soil salt storage under some field crops. J. Agric. Sci. Mansoura Univ., Special issue for Scientific Symposium on "Problems of soils and waters in Dakahlia and Damietta Governorates". March 18.

15. Klute, A. (1986). Methods of Soil Analysis (Part 1). American Society of Agronomy, Madison, Wisconsin, USA, 3rd edition.

16. Lindsay, W. L., Norvell, W. A. (1978). Development of DTPA soil Test for zinc, iron, Manganese and copper. *Soil Science. Soc. Amer. Proc.,* 42:421–428.

17. Salt, D. E., Blaylock, M., Kumar, N. P., Dushenkov, V., Ensley, B. D., Chet, I., Raskin, I. (1995). Phytoremediation; A novel strategy for removed of toxic metals from the environment using plants. *Bio-Technology*, 13(5), 468–474.

CHAPTER 2

EMITTER CLOGGING: A MENACE TO DRIP IRRIGATION

VISHAL K. CHAVAN

CONTENTS

2.1 INTRODUCTION

Drip irrigation is being widely accepted in the world and becoming increasingly popular in areas with water scarcity and salt problems. In regions where good quality water is either insufficient or not available, saline water irrigation is inevitable; thus drip irrigation has gained immense value in some specific situations with not only good quality water but saline water, too. Drip irrigation is being used in many areas with surface water too, that is, water from streams/canals and rivers, which contain various sizes of sand particles and sand concentrations. Drip irrigation is based on a very slow and frequent application of water from relatively small nozzles or orifices, with discharge rate of 1 to 10 lph. In order to achieve such a small discharge, the passage and orifices must be very small. In most cases, the diameter of an emitter orifice is less than 1 mm.

A formidable obstacle to the successful operation of the system over its intended life of service is clogging of emitters. High water application uniformity is one of the significant advantages that a properly designed and maintained drip system can offer over other methods of irrigation. In many cases, the yield of crops may be directly related to the uniformity of water application. Partial or complete clogging drastically affects water application uniformity and, hence, may put a complete system out of operation, causing heavy loss to the crop and damage to the system itself. Thus, emitter clogging can nullify all the advantages of drip irrigation.

Since drip irrigation is expensive, its longevity must be maximized to assure a favorable benefit-cost ratio. If emitters get clogged in a short time after their installation, reclamation procedures to correct clogging increase maintenance cost and unfortunately may not be permanent. Clogging problems often discourage operators and consequently cause them to abandon the system and return to less efficient methods of irrigation.

Emitter clogging is directly related to the quality of irrigation water. Large quantities of irrigation water are obtained from underground sources. Since calcareous formations are considered to be good aquifers, water pumped from wells in these areas is generally rich in calcium carbonate and bicarbonate. Similar enrichment of water occurs as it comes in contact with other minerals, but dissolution and precipitation of calcium carbonates is most common. Precipitation of calcium carbonates may occur through the drip system, but the problem is most acute when it occurs within the narrow passages of the emitters or at the outlets. Fertilizers added to irrigation water greatly change the precipitation properties of the water if it is calcareous. Mineral precipitates (often seen as scale deposits), algae and bacteria clog drip emitters. Clogged emitters result in variable distribution during irrigation and uneven fertilizer application during fertigation, and thus hinder uniform crop development, reduce yield and jeopardize quality. For growers to effectively use drip technology, they must prevent clogging of drip emitters.

This chapter discusses potential problems of emitter clogging.

2.2   CAUSES OF CLOGGING

Clogging is normally caused by one of the following factors:

1.   Solid particles in suspension (physical clogging).
2.   Chemical precipitation (chemical clogging).
3.   Microorganisms and organic matter (biological clogging).

The clogging agents are summarized in Table 2.1. The clogging problem can be classified as minor, moderate and severe (Table 2.2). The physical contributors include mineral particles of sand, silt and clay, and debris that are too large to pass through the small openings of filters and emission devices. Silt and clay particles that are usually much smaller than the smallest passages are often deposited in the low velocity areas of the laterals where they coagulate to form masses large enough to clog emission devices. Coating of clay particles in filters and emission devices can also reduce water flow.

TABLE 2.1   Water Quality Factors Affecting the Clogging of Drip Irrigation Systems

Physical (Suspended Solids)	Chemical (precipitation)	Chemical (Bacteria and Algae)
Inorganic particles: Sand, Silt, Clay	Calcium or magnesium carbonates	Filaments
Plastic	Calcium sulphate	Slime
Organic particles	**Heavy metals**	**Microbial decomposition**
Aquatic plants (phytoplankton/algae)	Oxides, hydroxides	Iron
Aquatic animals (Zooplankton)	Carbonates	Sulfur
Bacteria	Silicates and Sulphides	Manganese
	Oils or other lubricants; Fertilizers	
	Phosphate, Aqueous ammonia, Iron, Copper, Zinc, Manganese	

TABLE 2.2   Severity of Clogging Hazard

Types of agents	Clogging hazard		
	Minor	Moderate	Severe
Physical			
Suspended solids	<50	50–100	>100
Chemical			
pH	<7.0	7.0–8.0	>8.0
Dissolved solids[a]	<500	500–2000	>2000
Manganese[a]	<0.1	0.1–1.5	>1–5
Total iron[a]	<0.2	0.2–1.5	>1.5
Hydrogen sulphide[a]	<0.2	0.2–2.0	>2.0
Biological			
Bacterial population[a]	<10,000	10,000–50,000	>50,000

[a.] Maximum measured concentration from a representative number of water samples using standard procedures for analysis (mg/L).

[b.] Maximum number of bacteria per milliliter can be obtained from portable field samplers and laboratory analysis. Bacteria populations do reflect increased algae and microbial nutrients.

Salinity is an important water quality factor in irrigation and does not contribute to emitter clogging unless the dissolved ions interact with each other to form precipitates or promote slime growth. When irrigation water contains soluble salts, crusts of salt often form on emission devices as the water evaporates between irrigations. If the salt does not dissolve during the subsequent irrigation, crust accumulation will continue and clogging of the emission device will usually result. The factors conducive to chemical precipitation are high concentrations of calcium and magnesium and bicarbonate ions and relatively high pH of water. Temperature is also a factor because the solubility of calcium carbonate precipitates decrease with an increase in temperature.

Favorable environmental conditions within drip systems can cause rapid growth of several species of algae and bacteria. These micro organisms often become large enough to cause complete clogging.

Biological oxidation processes involving certain species of bacteria and water with even very low concentration of ferrous and manganese ions can produce deposits of iron and manganese oxides that promote clogging. Clogging would not be a problem if the water sources were free of organic carbon, which is an energy source for bacteria. Microorganisms such as ants, spiders, fleas and fresh water crustaceans can also contribute to clogging problems.

Two or more of these factors may occur at the same time. Apart from the customary practices of overcoming these problems, it is standard practice to tint all the drip system components black during manufacture in order to prevent algal growth and reduce bacterial development.

The corrective treatment for controlling clogging depends on the types of clogging agents. Many agent scan be removed using settling basins, water filtration and/or periodic flushing of filters, mainlines, laterals and emission devices. Injection of acids, algaecides and bactericides are common treatments used to control chemically and biologically caused clogging

By making certain water analyses, possible problems can be estimated. This is especially advisable before a new drip system is installed. The factors are rated in terms of an arbitrary clogging hazard, ranging from minor to severe and are presented in Table 2.2. Following precautionary measures are mostly related to the physical and chemical properties of water:

Clogging factors:	Remedial measures:
Poor physical quality water	Filtration
Poor chemical quality water	Flushing

2.3 FILTRATION

Adequate filtration requires the processing of water entering the system. The particle size, which can be tolerated in the system, depends on the emitter construction. Typically, the recommendation is for the removal of particles larger than one-tenth the diameter of the orifice or the flow passage of the emitter. One reason for this is that several particles may group together and obstruct the passageway. This is typical with organic particles having about the same density as water. Another reason is that heavier inorganic particles (fine and very fine sands) tend to settle and deposit in slow-flow zones, particularly inside walls of laminar flow emitters where the flow rate is slow. The result of clogging may not be rapid, but it is inevitable. It may be necessary to use a 200-mesh screen, which has a 0.074 mm (0.0029 inch) hole size even with a passageway of one mm (0.04 inch) in cross-section. Most manufacturers recommend removing particles larger than 0.075 mm or 0.15 mm (0.003 or 0.006 inch), but some allow particles as large as 0.6 mm (0.024 inch). Table 3 summarizes the minimum-sized particles that can be removed by several of these devices. Removal of suspended particles is usually required to get optimum performance of the drip system, since irrigation water is rarely free of suspended material. Settling basins, sand/media filters, screens.

Cartridge filters, disc filters and centrifugal separators are the primary devices used to remove suspended material. The best way to reduce or prevent clogging is by adequate filtration.

1.   Suspended organic matter and clay particles may be separated with gravel filters or screen filters.
2.   Filter cleaning becomes necessary if pressure drops significantly between the entry and the exit sides of the filter. As a rule it is customary to clean the filters when the allowable pressure drops (about 4 m or 0.4 kg/cm^2 or 6 psi).

Table 3. Filter effectiveness.

Filter type	Size range (micro-m)
Sand media	5–100
Sand media	>20
Screen	>75
Screen (100–200 mesh)	75–150
Screen (200 mesh)	>100
Sediment basins	>40
Separator[a]	>74
Separator[a] (two stage)	>44
Slotted cartridge	>152

[a] Separators remove 98% of particles larger than size indicated.

2.4   SETTLING BASIN

A settling basin, settling pond or decant pond is an earthen or concrete structure using sedimentation to remove settleable matter and turbidity from wastewater. The basins are used to control water pollution in diverse industries such as agriculture, aquaculture, and mining. Turbidity is an optical property of water caused by scattering of light by material suspended in that water. Although turbidity often varies directly with weight or volumetric measurements of settleable matter, correlation is complicated by variations in size, shape, refractive index, and specific gravity of suspended matter. Settling ponds may be ineffective at reducing turbidity caused by small particles with specific gravity low enough to be suspended by Brownian motion.

Settling Basins are designed to retain water long enough so that suspended solids can settle to obtain a high purity water in the outlet and also provide the opportunity for pH adjustment. Other processes that could be used: thickeners, clarifiers, hydro-cyclones and membrane filtration are highly used techniques in the field. Compared to those processes, settling basins have a simpler and cheaper design, with fewer moving parts; demanding less maintenance, despite requiring cleaning and vacuuming of the quiescent zones at least once every two weeks. Also there can be more than one settling basins in series.

Settling basins or reservoirs (Fig. 2.1) can remove large volumes of sand and silt. The minimum size of a particle that can be removed depends on the time that sediment-laden water is detained in the basin. Longer detention times are needed to remove smaller particles. The basin should be constructed so that water entering the basin takes at least a quarter of an hour to travel to the system intake. In this length of time, most inorganic particles larger than 80 microns (about 200 mesh) will settle. A basin 1.2 m deep x 3.3 m wide x 13.7 m long (4 x 10 x 45 ft.) is required to provide a quarter hour retention time for a 57 lps (900 gpm) stream. Removal of clay-sized particles requires several days and is not practical unless flocculating agents such as alum and/or poly electrolytes are used. Settling basins may need to be cleaned several times a year when large quantities of water with high concentrations of sediment are being passed through the basin.

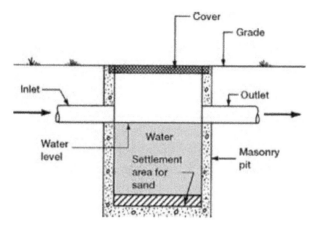

FIGURE 2.1 Setting basin to intercept sand particles.

Algae growth and wind-blown contaminants can be severe problems in settling basins. The sides and bottom of the basin should be lined to discourage vegetative growth and chemical treatment with chlorine or copper sulphate may be required to control algae. Because of these problems, settling basins are recommended for use with drip systems only in extreme circumstances.

2.5   SAND MEDIA FILTERS

Sand or gravel or media filters (Fig. 2.2) are used for filtering out heavy loads of very fine sands and organic materials. These consist of layered beds of graduated sand and gravel placed inside one or more pressurized tanks. They effectively remove suspended sands, organic minerals and most other suspended substances from the surface and groundwater. Also, long, narrow particles, such as some algae or diatoms, can be caught in the multi-layered sand bed than on the surface of a screen. They do not remove very fine particles (that is, silt and clay) or bacteria. Media filters are relatively inexpensive and easy to operate. It is generally recommended that the filter material be as coarse textured as possible but fine enough to retain all particles larger than one-sixth the size of the smallest passageway in the drip irrigation system. Filter materials should be large enough not to be removed during filter cleaning processes. A recommended practice is to

FIGURE 2.2   A typical sand media filters and backwash process for cleaning (bottom figure).

use a screen filter downstream from the media filter to pick particles that escape during back-washing.

Factors that affect filter characteristics and performance are water quality, type and size of sand media, flow rate and allowable pressure drop. A sand media filter can handle larger loads of contaminants than a screen of comparable fineness. It can do it with less frequent back flushing and a smaller pressure drop.

However, sand filters are considerably more expensive. They are generally used only when a screen filter would require very frequent cleaning and attention or to remove particles smaller than 0.075 mm (0.003 inch).

The sand media used in most drip irrigation filters is designated by numbers. Numbers 8 and 11 are crushed granite, and numbers 16, 20 and 30 are silica sands. The mean granule size in microns for each media number is approximately 1900, 1000, 825, 550 and 349 for numbers 8, 11, 16, 20 and 30, respectively.

At a flow velocity of 17 liters/sec/m² (25 gpm/ft²) of bed, the numbers 8 and 11 crushed granite removes most particles larger than one-twelfth of the mean granule size or approximately 160 and 80 microns, respectively. The silica sand numbers 16, 20 and 30 remove particles approximately one-fifteenth the mean granule size or approximately 60, 40 and 20 microns, respectively.

Typically, the initial pressure drop across the number 8, 10 and 16 media is between 14 and 21 kPa (2 and 3 psi). For the number 20 and 30 media, it is approximately 34 kPa (5 psi). The rate of pressure drop increase is usually linear with time for a given quality of water and flow rate. Assuming 1.0 unit of pressure drop per unit of time for a number 11 media, the units of pressure drop per unit in time across the other media would be: 0.2 unit for number 8 media, 2 units for number 16 media, 8 units for number 20 media and 15 units for number 30 media. For example, if it takes 24 hours for the pressure drop to increase by 34 kPa (5 psi) across a number 11 media, it would take only about three hours for the same increase across a number 20 media. The maximum recommended pressure drop across a sand filter is generally about 70 kPa (10 psi or 0.6 kg/cm²).

2.6   SUMMARY

The orifices in the drip lines or the emitters emit water to the soil. The emitters allow only the discharge of few liters or gallons per hour. The emitters have small orifices and these can be easily obstructed. For a trouble free operation, one should follow these considerations: Pay strict attention to filtration and flushing operation. Maintain an adequate operating pressure in the main, sub main and lateral lines. Flushing and periodic inspection of the drip irrigation system are must.

For effective filtration efficiency, we must maintain the system in good condition and it is not obstructed by the clogging agents. For this, pressure gages are installed at the entrance and the exit of a filter. The frequency of flushing depends on the water quality. Some recommendations for an adequate maintenance are cleaning with pressurized air, acids and chlorine. This chapter discusses clogging agents, prevention of clogging, types of filters, and principle of operation of filters, selection of filters, and procedure to solve problems of clogging. We must select an adequate filtration system to provide water free of clogging agents.

KEYWORDS

- backwashing
- benefit–cost ratio
- biological clogging
- calcium carbonates
- chemical clogging
- clogging
- clogging agents
- drip irrigation
- emission devices
- emitter
- emitter clogging
- fertigation
- filtration
- hydrocyclone filters
- irrigation
- physical clogging
- precipitation
- saline water
- sand media filters
- screen filters
- settling basin
- water application uniformity
- water scarcity

CHAPTER 3

MECHANISM OF EMITTER CLOGGING: A REVIEW

VISHAL K. CHAVAN, S. K. DESHMUKH, and M. B. NAGDEVE

CONTENTS

3.1   INTRODUCTION

Indian agriculture mainly depends upon vagaries of monsoon rains, which are unevenly distributed in space and time and not adequate to meet the moisture requirement of the crops for successful farming. India with only 2.4% of the world's total area and 4% of the total available fresh water supports about 17% of the world's population. The agricultural sector consumes over 80% of the available water in India for irrigation of crops and would continue to be the major water-consuming sector due to the intensification of agriculture [68, 77, 106]. Further, considering the fact that the population of the country is estimated to touch 1.4 billion by 2020 with the food requirement of 280 million-tons, the agricultural sector must grow by 4% and augment about 3–4 million-tons per year. Though the ultimate irrigation potential of the country has been assessed at 140 million-ha planned to be achieved by 2050, even after achieving the same, approximately half the cultivated land would still remain rain-fed, and therefore, water would continue to be the most critical resource limiting agricultural growth.

The water resources of the country are varied and limited, but still most of the area is irrigated using the conventional methods of irrigation with the efficiency of 35–40%. Considering the daunting task of achieving the food production targets, it is imperative that efficient irrigation methods like drip and sprinkler irrigation systems are adopted in large scale for judicious use and management of water to cope up with increasing demand for water in agriculture in order to enhance and accelerate the agricultural production in the country.

Drip irrigation, also called trickle or daily irrigation, is a localized irrigation method that slowly and frequently provides water directly to the plant root zone [45] and is the most efficient irrigation method with an application efficiency of more than 90%. However, not until the innovation of polyethylene plastics in the 1960s did drip irrigation begin to gain momentum. Traditionally, irrigation had relied upon a broad coverage of water to an area that may or may not contain plants. Promoted for water conservation, drip irrigation does just the opposite. It applies small amounts of water (usually every two or three days) to the immediate root zone of plants. In drip irrigation, water is delivered to individual plants at a low pressure and delivery rate to specific areas or zones in the landscape or garden. The slow application promotes a thorough penetration of the water to individual plant root zones and reduces potential runoff and deep percolation. The depth of water penetration depends on the length of time the system is allowed to operate and the texture of the soil.

The suitability of any irrigation system mainly depends upon its design, layout and performance. Due to its merits and positive effects, drip irrigation has become rapidly popular in India and also the state governments are promoting drip irrigation on a large scale by providing subsidy. The advantage of using a drip-irrigation system is that it can significantly reduce soil evaporation and increase water use efficiency by creating a low, wet area in the root zone. World over, the studies indicate that drip irrigation results in 30–70% water saving and yield increase by about 40–100% or even more compared to surface irrigation methods. Due to water shortages in many parts of the world today, drip irrigation is becoming quite popular [96, 107]. In 2000, more than 73% of all agricultural fields in Israel were irrigated using drip irrigation systems and 3.8 million-ha worldwide were irrigated using drip irrigation systems [65]. By 2008, total world agriculture area was 1,628 million-ha and 277 million-ha was under irrigation and 6 million-ha were drip irrigated [67]. In India, there has been a tremendous growth in the area under drip irrigation during the last 15 years. In India, the area under drip increased from a mere 1,500 ha in 1985 to 70,859 ha in 1991–1992 and at present, around 3.51×10^5 ha area is under drip irrigation with the efforts of the Governments of India and the States. The National Committee on Plasticulture Applications in Horticulture (NCPAH), Ministry of Agriculture, Government of India (GoI) has estimated a total of 27 million-ha area in the country that has the potential of drip irrigation application, thus, there is a vast scope for increasing the area under drip irrigation [66].

Due to limited water resources and environmental consequences of common irrigation systems, drip irrigation technology is getting more attention and playing an important role in agricultural production, particularly with high value cash crops such as greenhouse plants, ornamentals and fruits. Therefore, use of drip irrigation systems is rapidly increasing around the world. Despite its advantages, in drip irrigation system, emitter clogging is one of the major problems, which can cause large economic losses to the farmers. Emitter clogging is directly related to the quality of the irrigation water, which includes factors such as suspended solid particles, chemical composition and microbes, and also insects and root activities within and around the tubing can also cause problems. The major operational difficulties in drip irrigation method arise from the clogging of dripper, which reduces the efficiency and crop yield [21].

Emitter clogging continues to be a major problem in drip irrigation systems. For high-valued annual crops and perennial crops, where the longevity of the system is especially important, emitter clogging can cause large economic losses. Even though information is available on the factors causing clogging, control measures are not always successful. These Problems can be minimized by appropriate design, installation, and operational practices. Reclamation procedures to correct clogging increase maintenance costs, and unfortunately, may not be permanent. Clogging problems often discourage the operators, and consequently cause abandonment of the system and return to a less efficient irrigation application method.

Emitter clogging is directly related to the quality of the irrigation water, which includes factors such as suspended particle load, chemical composition, and microbial type and population. Insect and root activities within and around the tubing can cause similar problems. Consequently, these factors dictate the type of water treatment or cultural practices necessary for clogging prevention. Clogging problems are often site-specific and solutions are not always available or economically feasible [82]. No single foolproof quantitative method is available for estimating the clogging potential. However, by analyzing the water for some specific constituents, possible problems can be anticipated and control measures formulated.

Most tests can be made in the laboratory. However, some analyses must be made at the sampling sites because rapid chemical and biological changes can occur after the source of water is introduced into the drip irrigation system. Water quality can also change throughout the year so that samples should be taken at various times over the irrigation period. These are further rated in terms of an arbitrary clogging hazard ranging from minor to severe. Clogging problems are diminished with lower concentrations of solids, salts and bacteria in the water. Additionally, clogging is aggravated by water temperature changes.

The causes of clogging differed based on emitter dimension [5, 31] and positions in lateral. The tube emitter system with laminar flow suffered more severe clogging than the labyrinth system with turbulent flow, because laminar flow is predisposed to clogging [41]. Emitter clogging was recognized as inconvenient and one of the most important concerns for drip irrigation systems, resulting in lowered system performance and water stress to the non-irrigated plants [95]. Partial and total plugging of emitters is closely related to the quality of the irrigation water, and occurs as a result of multiple factors, including physical, biological and chemical agents [55, 94]. Favorable environmental conditions in drip irrigation systems can cause rapid growth of several species of algae and bacteria resulting in slime and filament build up, which often becomes large enough to cause biological clogging [54]. On the other hand, some of the bacterial species may cause emitter clogging due to the precipitation of iron, manganese and sulfur minerals dissolved in irrigation water [69, 94]. Filtration, chemical treatment of water and flushing of laterals are means generally applied to control emitter clogging [57, 84]. Physical clogging can be eliminated with the use of fine filters and screens. Emitter clogging is directly related to irrigation water quality, which is a function of the amount of suspended solids, chemical constituents of water and microorganism activities in water. Therefore the above mentioned factors have a strong influence on the precautions that will be taken for preventing the plugging of the emitters. During irrigation some clogging due to microorganism activities take place in cases when wastewater is used [24, 88].

In micro-irrigation systems that are characterized by a number of emitters with narrow nozzles, irrigation uniformity can be spoilt by the clogging of the nozzles with particles of chemical character [42, 78]. Chemical problems are due to dissolved solids interacting with each other to form precipitates, such as the precipitation of calcium carbonate in waters rich in calcium and bicarbonates [117]. In locations where the amount of the ingredients as dissolved calcium, bicarbonate, iron, manganese and magnesium are excessive in irrigation water, the emitters are clogged by the precipitation of these solutes [59]. Chemical precipitation can be controlled with acid injection. However, biological clogging is quite difficult to control. Chlorination is the most common practice used in the prevention and treatment of emitter clogging caused by algae and bacteria [61, 121]. Calcium hypochlorite, sodium hypochlorite and particularly chlorine are the most common and inexpensive treatments for bacterial slimes and for inhibition of bacterial growth in drip irrigation systems [9, 45, 63]. However, continuous chlorination would increase total dissolved solids in the irrigation water and would contribute to increased soil salinity [62].

3.1.1 NATURE AND SCOPE OF PRESENT INVESTIGATION

The various components of drip irrigation are made up of plastic and polymer materials and due to their flexibility and other advantages over metals. There are more than 200 various components and materials used in drip irrigation installed at farm level. But, the major component coming in contact of water includes poly tube (material: linear low density poly ethylene), dripper (material: poly propylene), pipes (material: high density poly ethylene and polyvinyl chloride) and silicon diaphragm in emitters (material: silicon). It is found that the initiation of clogging starts at molecular level. Considering this, it is very necessary to study the clogging mechanism and initial adsorption mechanism in drip irrigation system. The phenomenon of adsorption is the collection or accumulation of one substance on the surface of another substance. In adsorption mainly the surface of solids is involved and accumulated substances remain on the surface. Adsorption therefore is said to be a surface phenomenon as it occurs because of attractive forces exerted by atoms or molecules present at the surface of the adsorbent. These attractive forces may be of two types: (i) physical forces (cohesive forces or Vander Waals forces), (ii) chemical forces (chemical bond forces). Thus, an attempt on study of adsorption parameters of these materials of silicon diaphragm, poly vinyl chloride (PVC), poly propylene (PP), high density poly ethylene (HDPE) and linear low density poly ethylene (LLDPE) used in drip irrigation would give insight into which material is more susceptible for adsorption and possible solutions to reduce clogging mechanism in the initial stage itself.

In this Chapter the past research studies are reviewed, presented and discussed on: causes of clogging, water quality impacts, preventive measures and that related to adsorption mechanism in physical, chemical and biological clogging of drip irrigation.

3.2   CAUSES OF CLOGGING

The major disadvantage and the only real concern with drip irrigation systems is the potential for emitter clogging. The causes of clogging fall into three main categories: (i) physical (suspended solids), (ii) chemical (precipitation) and (iii) biological (bacterial and algal growth). Emitter clogging however is usually the result of two or more of these elements working together [80]. Emitter clogging is a major concern in these systems because of the high suspended solids and nutrient values associated with treated wastewater effluent. Early research on drip irrigation systems investigated two different but related issues regarding emitter clogging. One focused on the clogging mechanism of emitter and the other focused on adsorption mechanism in emitters.

In order to avoid emitter clogging one could use larger emitter orifices, and to maintain the application rate at optimum, the system had to be operated by impulse methods [6]. To prevent clogging problem due to bacteria or algae, adding of chlorine on daily basis with concentration 10–20 mg/L is recommended. This will also remove precipitates from bacterial activity or algae growth if chlorine is injected with high concentration (500 mg.L^{-1}) inside the system for 24 h before washing [113]. The clogging problems are often site-specific and the solutions are not always available or economically feasible.

3.2.1 PHYSICAL CLOGGING

The biological plugging of emitter in drip irrigation system installed in various citrus groves in central and south Florida. They found that the clogging problems in drip irrigation system were due to isolation of algae in the irrigation lines and emitters [50].

The micro-tube emitters clogged more fastly than any other type of emitters and the chemical treatments were effective on emitters [93]. They further conducted a study to test the effect of water temperature variation upon the discharge rates of micro tube, spiral passage, orifice and vortex type emitters.

The dominant causes of emitter clogging and observed flow reduction due to physical factors *viz.*, sand grain, plastic particles, sediment, body parts of insects and animals, deformed septa which contributed to 55% of the total clogging that occurred during their study followed by the combined development of biological and chemical factors like microbial slime, plant roots and algae mats, carbonate precipitate, iron -magnesium precipitates which contributed to 16% of the total clogging [53]. They tabulated the dominant causes of clogging with their relative percentage occurrence in trickle irrigation emitter as shown in Table 3.1.

A field study evaluated the effects of emitter design and influent quality on emitter clogging using eight different emitters in combination with six water treatments of Colorado River water USA [52]. The treatments were varied by the degree of filtration for removal of suspended solids and the amount of chemical additives (sulfuric acid and calcium hypochlorite) for controlling pH and preventing biological growth. They found

TABLE 3.1   Dominant Causes of Clogging and Their Relative Percentage Occurrence in Drip Irrigation Emitters [57]

Causes of clogging [57]	Percent of occurrence	
	Individual	Total
Physical factors		
Sand grains	17	
Plastic particles	26	
Sediment	02	
Body parts of insects and animals	03	
Deformed septa	07	55
Biological factors		
Microbial slime	11	
Plants root and algae mats	03	14
Chemical factors		
Carbonate precipitate	02	
Iron magnesium precipitate	00	02
Combined factors		
Physical/biological	08	
Physical/chemical	02	
Chemical/biological	06	
Physical/biological/chemical	02	18
Non-detectable (probably physical)	–	11

that the emitter performance was reduced during average flow rates for each experimental subplot. They recommended a combination of sand and screen filtration to remove suspended solids, acid treatment to reduce chemical precipitation and flushing of laterals to eliminate sediments.

Emitter clogging of particulate removal by granular filtration and screen filtration of wastewater stabilization pond effluent has been discussed by Adin [3]. He showed that deep-bed granular media controlled suspended particles, with larger particles (10 μm) removed at a higher rate than smaller particles and very little removal occurred in the 1–2 μm size range; further 10–60 μm size particles were removed by 40–50 % "in-depth filtration" and by more than 80% when bio-mats developed and surface filtration occurred. He found that the particulate removal efficiency increased as the filter media size approached 1.20 mm and the filter bed depth increased to 0.45 m and decreased (especially for small particles) with increased filtration velocity from 2.2 $L.m^{-2}.s^{-1}$. He also observed that the release of large particles from the screen into the effluent during the formation of "filter cake" actually increased the chance of clogging in the irrigation drip lines and emitters. He suggested the manufacturers of wastewater drip irrigation systems on numerous modifications to emitter design and other system components to incorporate the suggestions of early researchers to reduce clogging and maintain distribution uniformity.

An experiment was conducted at Farm Research Institute in China on subsurface drip irrigation system (SDI) related to emitter clogging for 8 years [48]. The clogging rates of labyrinth emitter, mini-pipe and orifice reached 16.67, 25, and 63.89%, respectively. Clogging was found to be mainly caused by attached granules and some suggestions they put forward to solve this problem were enhancing filtration, flushing timely and improving the route of water in the emitter. Causes of clogging are listed in Table 3.1.

The efficiency of using nitric acid and sodium hypochlorite was evaluated to unclog drippers due to the use of water with high algae content in a rose field of Holambra, Sao Paulo State, Brazil [11]. Researchers conducted the research study in six, 4,216 m^2 greenhouses, each with two sectors comprising of 10 spaces or lines, totaling 12 sectors of a dripper irrigation system. Their results on irrigation water quality in the greenhouses indicated that the pH and iron presented moderate risk for clogging. Chemical and physical analyses and the bacteriological count in water were carried out in three water sources that supplied the irrigation water to check the factors causing clogging. Evaluations were carried out on water distribution uniformity in all sectors before and after the chemical treatment in order to evaluate efficiency. The treatment improved water distribution uniformity and a lead to a reduction of the coefficient of variation for dripper flow in all sectors. There was a good correlation between the coefficient of variation (Cv) and the water distribution uniformity index.

According to them, this was an excellent method to be used to unclog drippers due to biological problems.

A laboratory experiment was conducted to evaluate emitter clogging in drip irrigation by solid liquid two-phase turbulent flow model simulations describing the flow within drip emitters [97]. The moving trace and depositing feature of suspending solids in emitter channels by computational fluid dynamics (CFD) were based on the turbulent model established which provided some visual and direct evidences for predicting the clogging performance of drip emitters. Three types of emitters were used with novel channel form, including eddy drip-arrows, pre-depositing drippers and round-flow drip-tapes. The simulation results showed that the solids moved along a helical path in the eddy drip-arrow, but no obvious deposition existed in its interior channel. In the pre-depositing dripper, some solids concentrated in the parts of "depositing pones." In the round-flow drip-tape, a small number of solids adhered to outer-edges of every channel corner, which was a potential factor for the occurrence of emitter clogging. To verify the predictions from the CFD simulations, series of "short-cycle" clogging tests for the three emitters were conducted in laboratory. The statistical data of discharge variation caused by emitter clogging were in good agreement with the two-phase flow CFD simulations.

3.2.2 CHEMICAL CLOGGING

At the Tamil Nadu Agricultural University (TNAU), the average clogging was found to be 20 to 25% in the nozzles, 34% in the micro tube and 40% in hole and socket type of emitters [92]. The study concluded that the extent of reclamation of partially clogged emitters treated with hydrochloric acid resulted in an increase of 0.2 L h^{-1} when acid concentration increased from 0.5 to 2% by volume.

A field experiment on drip irrigation was installed in gold coast farms of Santa Maria near coastal and southern part of California [79]. Researchers divided a 10 acre field of strawberry into six plots. They used the maleic anhydride polymer injected at 2 mg L^{-1} and continuous chlorine of 1 mg L^{-1} in four plots, and used only continuous chlorine injection in two plots. They found that emission evaluation of polymer treated water showed only slight decrease over the 6 month growing period, but untreated well waters indicated a decrease of 50%. They finally concluded that the system injected with 2 mg L^{-1} maleic anhydride polymer supplied the actual amount of water required for the plant needs, while the untreated tubings output decreased.

The effects of chemical oxidants on effluent constituents in drip irrigation were evaluated with synthetic effluents rather than authentic effluents in order to realize the role of oxidants in these processes and to obtain meaningful and reproducible results [99]. Researchers studied the effects of Cl_2 and ClO_2 and their constituents. The demand of these effluents for Cl_2 was 5–8 mg.L^{-1} and for ClO_2 3–4 mg.L^{-1}. They found that 2 mg.L^{-1} of either oxidant caused a very fast bacteria inactivation that reached four orders of magnitude after 1 min. However, with respect to algae, concentrations up to 20 mg.L^{-1} of either oxidant did not affect the number of algae cells, although they caused a remarkable decrease in algal viability as expressed by its chlorophyll content and replication ability. Both oxidants demonstrated a notable aggregation effect on the effluents. The conclusions of the results described above were examined in a pilot system. They suggested that continued chlorination by 5–10 mg.L^{-1} Cl_2 applied directly to the drippers was not very effective. The reason for this was the presence of clogging agents, "immune" to low Cl_2 concentrations, produced as early as in the reservoir, and carried down to the drippers by the effluent stream. Finally, they suggested that batch treatment combined with settling was much more efficient as it reduced the clogging significantly, because in this case the Cl_2 reacted not only as a disinfectant, but also as a coagulant due to the oxidation of humic constituents.

Enciso and Porter [46] worked on maintenance program that included cleaning the filters, flushing the lines, adding chlorine and injecting acids. They reported that if these preventive measures were done, the need for major repairs such as replacing damaged parts, often can be avoided and the life of the system extended.

Precipitation of salts such as calcium carbonate, magnesium carbonate or ferric oxide can cause either partial or complete blockage of the drip system [12]. Acid treatment was applied to prevent precipitation of such salts in drip irrigation system; and it was also found effective in cleaning and unclogging the irrigation system, which was already blocked with precipitates of salts.

The past research as suggested that the irrigation pumping plant and the chemical injection pump should be interlocked; because if the irrigation pumping plant were to stop, the chemical injection pump would also stop. This would prevent chemicals from the supply tank from entering the irrigation lines, when the irrigation pump stops [57].

The drip system should be flushed as regularly as determined by water quality, monitoring and recording system. Flushing process should start from the pump onwards to make sure: the filters are clean and pressure is set correctly; and systematically the mainline, sub-mains, laterals and flushing manifold should be cleaned [58].

Five types of emitters with different nominal discharges were evaluated with or without self-flushing system and with or without pressure compensating system under three management schemes: untreated well water, acidic treated water and magnetic treated water in order to reduce chemical clogging [1].

In order to save money, concentrated and inexpensive technical acids should be used such as: concentrated commercial grade hydrochloric, nitric or sulfuric acid [87]. Phosphoric acid, applied as fertilizer through the drip system, can also act as a preventive measure against the formation of precipitates under certain conditions.

A comparison for clogging in emitters was made during application of sewage effluent and groundwater for investigating the temporal variations of the emitter discharge rate and the distribution of clogged emitters in the system and quantifying the impact of emitter clogging on system performance [73]. In the experiment, six types of emitters (with or without a pressure-compensation device) and two types of water sources (stored secondary sewage effluent and groundwater) were assessed by measuring the emitter discharge rate of the system at approximately ten-day intervals. The total duration of irrigation was 83 days, with a daily application for 12 h. The water source had a very significant influence on the level of drip emitter clogging. Of all the emitters tested over the entire period of the experiments, the emitters applying sewage effluent were clogged much more severely, producing a lower average mean discharge rate 26% than those applying groundwater. They found that different types of emitters had different susceptibilities to clogging, but the clogging sensitivity was inversely proportional to the pathway area in the emitter and the emitter's manufacturing coefficient of variability

(Mfg Cv). For groundwater application, the clogged emitters tended to be generally located at the terminals of the laterals in case of emitters without a pressure-compensation device, while randomly distributed clogged emitters were found for pressure-compensating emitters. A more random distribution of clogged emitters was found for the sewage application. Clogging of emitters could seriously degrade system performance. They found the values of the uniformity coefficient (CU) and the statistical uniformity coefficient (Us) decreased linearly with the mean clogging degree of emitters. Finally they suggested that to maintain a high system performance, more frequent chemical treatments should be applied to drip irrigation system applying the sewage effluent than to systems applying the ground water.

3.2.3 BIOLOGICAL CLOGGING

3.2.3.1 Iron Bacteria

Emitter clogging problem occurring in certain areas of Florida were caused by iron bacteria [49]. Iron bacteria frequently thrive in iron-bearing water. It is unclear whether the iron bacteria exist in groundwater before well construction and multiply as the amount of iron increases due to pumping, or whether the bacteria are introduced into the aquifer from the subsoil during well construction. Well drillers should use great care to avoid the introduction of iron bacteria into the aquifer during the well drilling process. All drilling fluid should be mixed with chlorinated water at 10 mg.L^{-1} free chlorine residual.

3.2.3.2 Sulfur Bacteria

Florida micro irrigation systems ceased to function properly because of filter and emitter clogging caused by sulfur bacteria [51]. *Thiothrix nivea* is usually found in high concentrations in warm mineral springs and contributes to this problem. This bacterium oxidizes hydrogen sulphide to sulfur and can clog small openings within a short period. *Beggiatoa,* another sulfur bacterium, is also often found in micro irrigation systems. Continuous chlorine or copper treatment can control sulfur bacterial problems.

3.2.3.3   Microorganisms

Clogging of micro sprinklers in a Florida citrus grove was noted to be inversely related to the orifice dimension of the emitters [19]. Clogging occurred when irrigation water from a pond was used where the water was chemically conditioned, and filtered through a sand media filter. Emitter clogging was caused by algae 46%, ants and spiders, 34%, snails 16%, and solid particles 4% such as sand and bits of PVC. About 20% of the 0.76 mm orifice emitters required cleaning or replacement during each quarter compared with about 14% for the 1.02 mm, 7% for 1.27 mm, and 5% for 1.52 mm orifice emitters.

Pests can damage or clog emitters or system components and increase maintenance costs. Some pests can also cause leakage problems and others can cause clogging. Coyotes and burrowing animals were observed to damage micro irrigation tubing. Similarly, rats and mice chewed holes in micro irrigation tubing [112]. *Tortricidae* (Lepidoptera) larvae and pupae of *Chrysoperla externa* (Hagan) caused emitter clogging [30]. Emitter clogging by spiders and ants is a problem in many surface micro irrigation and micro sprinkler systems. Additional problems include damage to spaghetti tubing and destruction of diaphragms in pressure compensating emitters.

3.2.3.4   Larvae

The larvae of *Selenisa sueroides* (Guenee) damaged spaghetti tubing of micro irrigation systems in south Florida citrus groves [22]. Larvae of *S. sueroides* used the spaghetti tubing on micro sprinkler assemblies as an alternate host to native plant species with hollow stems. The *S. sueroides* caterpillars chewed holes in the spaghetti tubing in order to enter and pupate. The deterioration appeared to be selective as *S. sueroides* damaged stake assemblies constructed with black tubing to a larger extent than with colored spaghetti tubing [17]. In addition, tubing composition could also affect the severity of damage, with a higher incidence of damage on polyethylene tubing compared with vinyl spaghetti tubing, and even less damage on assemblies made with colored tubing.

Problems caused by the *S. sueroides* caterpillars can be controlled by methods other than pesticide treatment of the emitters. Elimination of known host plants in early autumn through herbicide application and mowing is probably more cost effective than pesticide treatment. When infestations of *S. sueroides* caterpillars persist after mid-September, spray applications of liquid Teflon® to the emitter assemblies provide some protection from *S. sueroides* larvae.

3.2.3.5   Ants

Ants frequently cause clogging of emitters leading to non-uniform water application. In addition to causing clogging, ants can also physically damage certain types of emitters causing increased flow rates. For example, red fire ants (*Solenopsis invicta*, Buren) chewed and damaged the silicone diaphragms that control flow of micro-pulsators. Ant activities decreased the diaphragm mass (partial to complete removal) and increased the orifice diameters. Extensive damage occurred within a 16-month period and more than 9,000 m of micro irrigation tubing with internal compensating emitters had to be replaced [18].

3.2.3.6   Plant Roots

Problems with plant roots affecting water flow in subsurface micro irrigation systems were observed. This was a case of "the root biting the hand that feeds it." Roots were found penetrating into cracks and grow within the tubing or the emitter and restrict water flow. A control of root activity was done by chemical injection or impregnation of the tubing, emitter, or filter with trifluralin [104]. Massive root growths were observed to physically pinch flexible tubing, and thus restricting water flow.

3.2.3.7   Bio-Film Structure

The bacteria in a bio-film are found to aggregate at different horizontal and vertical sites with the highest concentration of cells occurring at the base or at the top, resulting in a mushroom-like shape (Fig. 3.1). The bio-films are highly hydrated with cells making up about 1/3 of the volume; the rest is mostly water. The horizontal and vertical voids provide

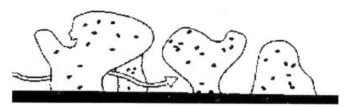

FIGURE 3.1   Bio-film structure.

for the flow of water though the network of cells. This simple type of circulatory system that carries nutrients to and waste product away from the cells is yet another example of how bio-films may function more like eukaryotic multi-cellular organisms. A wide variety of microenvironments exist within the cell matrix. This encourages phenotypic variations among genetically identical cells and provides for a greater diversity of species within the bio-film community [34].

3.2.3.8   Role of Bio-Films in Emitter Clogging

A field study of water quality and preventative maintenance procedures were conducted to reduce clogging using a variety of treatment methods and emitter systems [23]. The examination of clogged emitters concluded that three factors contributed to emitter clogging: (i) organic and inorganic suspended solids, (ii) chemical precipitation of calcium, magnesium, heavy metals and fertilizer constituents, and (iii) bacterial filaments, slimes, and depositions. Authors concluded however, that these contributors to clogging were closely interrelated. For example, by reducing biological slime, suspended solids were observed less likely to stick to the slime and agglomerate inside the tubing and emitters. Microorganisms were found also responsible for the chemical precipitation of iron, sulfur, and magnesium. For successful operation of drip irrigation, it is recommended to use water filtration, chemical treatment, and pipeline flushing.

A research study was conducted in the field and laboratory using oxidation pond effluent aimed at defining the clogging factors and mechanisms of blockage within the emitter for developing technical measures to overcome the problem [2]. Two types of influents were tested using three different emitter types. The clogged emitters were carefully removed from the lines and were carefully opened to examine the contents. Researchers reported that slimy gelatinous deposits of amorphous shape (bio-films) served as triggers for serious emitter blockage. Particles of definable shape (inorganic solids) were found in the matrix of the gelatinous substance and formed sediment in the emitters. It was concluded that suspended solids in the influent primarily caused emitter clogging but the clogging process was initiated by bacterial bio-films.

Effect of treflan injection on winter wheat growth and root clogging of subsurface drippers were investigated [120]. To attempt to solve the problem of root clogging of drippers, a series of field experiments were performed in the growing seasons of 2006–2008, to investigate the effects of treflan injection on dripper clogging by roots, and on root distribution, yield, and the quality of winter wheat (*Triticum aestivum* L.) under subsurface irrigation system (SDI). For each growing season, two treflan injection dates (March 6 and April 15 for the 2006–2007 growing season, and March 6 and April 15 for the 2007–2008 growing season) and three injection concentrations (0, 3, and 7 mg.L^{-1}) were arranged in a randomized block experimental design. The experimental results showed that treflan injection effectively reduced root density in areas adjacent to drippers, thereby significantly decreasing the potential of root clogging. In 2007, 4 out of 35 drippers were found with root intrusion problems in the control (without treflan injection) while no root clogging existed in any dripper in treflan application treatments. In 2008, 6 drippers from the control but only 1 dripper from those treated with treflan application showed root clogging. In addition, within the range of concentrations used by the current experiment, treflan concentrations had no significant effects on winter wheat root distribution, yield, and quality. Injection date, however, influenced the vertical root distribution significantly. By injection of treflan late in the growing season influenced the root distribution only in the areas close to the drippers and the influenced areas increased if treflan was injected early in the growing season.

3.2.3.9 Bio-Film Development

Bio-film formation is thought to begin when environmental signals trigger the transition from a planktonic lifestyle to a sessile lifestyle. It appears that flagella22 mediated motility are important in establishing the initial cell surface contact that leads to colonization of solid surfaces [89]. Once a cell makes contact with the surface, adhesion rather than motility is the key to successful colonization. The production of specific outer membrane proteins that provide for stable attachments is necessary for bio-film formation [37]. Once the surface is colonized with a monolayer of cells, the bacteria move across the surface organizing themselves into micro colonies. This coordinated movement called "twitching motility" was accomplished through the contraction of type IV pili. Cells within the micro-colony secrete the exopolysaccharide (EPS) matrix that functions like glue holding the bacterial cells together [38].

An experiment study was conducted to evaluate the performance of various irrigation emitters widely used in Israel, using waste water from a storage reservoir [100]. Fine particulate matter agglomerated by microbial by products and inclined developed biomass were the principal clogging agents. They reported that the clogging process started with emitters located at far end of the lateral and partial emitter clogging was more common than complete clogging.

Both field survey and experimental studies were carried out to identify the causes of clogging [105] by collecting the samples of clogging material and water for microscopic and chemical analysis in order to. Researchers found that the colonial protozoa occupied by the irrigation equipment only at a location, where the water flow velocity was slower than 2 m.s^{-1} and the colonies of protozoa and sulfur bacteria were found attached to the walls of the irrigation equipment and growing in a direction perpendicular to the surface to which they were attached. Colonial protozoa developed in wide range of water qualities: temperature (17–32°C), pH (6.8–9.5), dissolved oxygen (0.5–10.4 mg.L^{-1}), BOD (0.33–55 mg.L^{-1}). *Epystilys balanarum*, from the order of *Peritricha* species of colonial protozoa, was found. It was further concluded that the size of colonies varied from single or a few cells up to several hundred cells in a colony.

The performance was evaluated for two different types of emitters under various filtration methods as well as the effect of chemical treatments designed to control clogging using reservoir water containing secondary sewage effluents and storm water runoff [101]. The influent was filtered using a media filter of a uniform bed of gravel with a mean size of 1 mm, a 140-mesh disc filter or a 155–200 mesh screen filter. Emitter performance was determined by automatic measurements of pressure and flow rate in each drip lateral along with manual measurements of the discharge rate from individual emitters. It was observed that the emitter clogging was associated mainly with mucous products of microbial communities including colonial protozoa, bryozoa, and bacteria.

Treated effluent was evaluated for drip irrigated eggplant at As-Samra experimental site [85]. The study included determination of soil characteristics prior to irrigation and the accumulation of salts and heavy metals in the soil as well as concentration of the nutrients and heavy metal accumulation in the plant tissues. Clogging of the irrigation system was evaluated and treated and the yield was determined. The results of the study showed that the effluent had low heavy metal content and moderate restriction for surface trickle irrigation. The soil analysis indicated that after eggplant harvest there was slight increase in heavy metals and salt accumulation at the periphery of the wet zone. Finally it was suggested that the clogging can be successfully controlled with acid and chlorine.

A laboratory experiment in Irrigation Department of Rural Engineering at ESALQ, USA was performed to evaluate the use of different chlorine doses to recover the original flow rate for total and partial emitters clogging [33]. Scientists evaluated five different types of Netafim's drippers: Streamline 100, Ram 17L, Drip line 2000, Tiran 17 and Typhoon 20. Four chlorine levels of 150, 300, 450 and 600 mg.L^{-1} were tested maintaining the irrigation water pH between 5.0 and 6.0 in the recovering process of blocked drippers. A regular water source for irrigation equipment was used presenting a bacterial population about 50,000 bacteriam/liter. Sodium hypochlorite (12%) was used as a chlorine supply. The chlorination process took place during 60 min and after this period the solution remained inside the hose for 12 h without flow. Then the flow of each emitter was measured. Average flow, coefficient of variation of average flow and percentage of number of emitters at different classes of flow reduction were analyzed. It was concluded that for all kinds of drippers, except the Streamline, the chlorine improved the flow rate.

The performance was assessed for five different drip line types receiving filtered but untreated wastewater from a beef feedlot runoff lagoon [116]. Study included five drip line types, each with a different emitter flow rate. The emitter flow rates tested were 0.15, 0.24, 0.40, 0.60, and 0.92 gal.h^{-1} per emitter. Each drip line was replicated three times in plots of 20 x 450 ft. Acid and chlorine were injected in drip lines and flushed as needed to prevent algae and bacteria from growing and accumulating in the drip lines. After three years of study, it was concluded that the three largest emitter sizes (0.4, 0.6, and 0.92 gal.h^{-1} per emitter) showed little sign of clogging and had less than (5%) reduction in flow.

Bio-fouling were identified as a major contributor to emitter clogging in drip irrigation systems that distributed reclaimed wastewater [118]. Two types of drip emitters were evaluated for use with reclaimed wastewater in the study. Microbial bio-film accumulations, including proteins, poly-saccharides, and phospholipids fatty acids (PLFAs), were tested to determine the bio-film development and diversity in the emitter flow path. The microbial bio-film structure was analyzed using scanning electron microscopy. The results showed that rapid growth of the bio-film and accumulated sediments led to eventual reduction of emitter discharge, and the bio-film played an inducing role in the clogging process and biomass growth in the emitter flow path fluctuated as biomass was scoured off the surface areas. This study provided some suggestions for control of clogging and a framework for future investigations into the role of bio-films in the clogging of drip emitters that distribute reclaimed wastewater.

3.3   IMPACTS OF WATER QUALITY ON CLOGGING

Emitter clogging and crop response were evaluated in a replicated field study of Netafim labyrinth emitters distributing primary and trickling filter wastewater plant effluent compared to a tap water control [90]. Emitter clogging was determined by measuring the flow rate at the end of one irrigation season. Data from this study shows that more emitters clogged at the beginning of laterals than at the ends. It was speculated that may be from the differences in pressure associated velocity heads throughout the system. It was concluded that distributing wastewater could clog emitters, which can lead to poor water distribution and severe crop damage. It was recommended: (i) the effluent be screened with automatic flushing filters; (ii) chemical reagents, such as hypochlorite, be added to dissolve obstructions in the 6 emitters; and (iii) uniform emitter clogging should be maintained along the laterals by maintaining consistent pressure heads throughout the laterals.

An experiment on trickle irrigated citrus orchard was conducted with Colorado River water in south-western Arizona to develop water treatment methods for preventing emitter clogging and maintaining long-term operation of the system under actual field conditions [81]. The study included eight emitter systems in combination with six water treatments during a comprehensive 4-year study. Authors found that emitters with flexible membranes either failed after a few months of use with chemically conditioned water or showed serious deterioration and decomposition after 4 years. Finally it was observed that the dominant causes of emitter clogging and flow reduction were physical particles, next and minor in comparison, was the combined development of biological and chemical deposits.

An experiment was carried out at the IFAS Lake Alfred Citrus Research and Education Center for irrigation and fertigation study of citrus [111]. In the study, authors measured monthly clogging percentages for 5-yr study of trickle irrigation of citrus. They found that clogging percentages were much greater for drip emitters as compared to spray-jets. Further, they observed that most drip emitters were clogged by iron deposits whereas most spray-jets were clogged by insects and drip emitters clogged much more frequently during low-use winter months as compared to high-use summer months. Finally, they concluded that clogging percentages increased annually from 2.5% in 1979 to 21.3% in 1983. Clogging percentages of spray-jets were low (2.5%) and were unaffected by time or season throughout their study.

It has been observed that lime precipitate clogged the buried drip irrigation systems and which was difficult to detect and could cause problems where water quality was poor [108]. Based on the field trials, it was suggested that injection of phosphonate (organophosphorus compounds) was as effective as acid against clogging and might cost less. The experimental procedure consisted of preparing a tank with chemical constituents appropriate to the treatment and pumping at constant pressure into four buried drip irrigation lines corresponding to the treatment. As the treatments were being irrigated, the flow rate to each of the four drip lines was monitored.

The procedure was continued for each of the 16 treatments. All the treatments were irrigated for 2 to 3 days and each was irrigated 10 times during the experiment. Finally it was concluded that the treatments with low levels of calcium (1 meq.L^{-1}) showed no significant clogging problem. This study indicated that acid injection reduced clogging.

An experiment was carried out at Ft. Pierce Agricultural Research and Education Center to evaluate clogging rates for 10 models of micro irrigation emitters for a period of 3–5 years [16]. The 5 spray emitter models and 5 spinner models were used in a randomized complete block design with 5 replications. It was found that the clogging was caused by ants, spiders, or bacteria and algae. The average clogging rate per inspection period ranged from 2 to 38% averaging 19%. It was observed that the emitter, which had a relatively large orifice and a mechanism to plug the orifice when not in use, had the lowest clogging rate.

A field study was conducted to evaluate emitter clogging under different methods of filtration for removal of suspended solids, and chlorination to prevent bio-film formation using secondary wastewater treatment plant effluent [114]. It was concluded that sand filtration and intermittent chlorination (2 mg.L^{-1} for 1 h every 22 h of operation) prevented clogging in pressure compensating emitters. Finally it was suggested that an intermittent chlorination of 2 mg.L^{-1} with screen filtration and a continuous chlorination of 1 mg.L^{-1} with sand filtration prevented clogging in turbulent flow labyrinth emitters.

All irrigation systems require proper maintenance and subsurface drip irrigation systems were no exception [8]. The major cause of failures in subsurface drip irrigation and other micro-irrigation systems worldwide was clogging.

Many producers use drippers for trickle irrigation systems for flower production in the field and in protected environments [102]. A frequent problem in this type of irrigation system is the clogging of drippers, which is directly related to water quality and filtering system efficiency.

3.4 PREVENTIVE MEASURES FOR CLOGGING IN DRIP IRRIGATION SYSTEM

An experiment in south-west Portugal was carried out to investigate the causes of emitter clogging in waste stabilization pond effluents used for drip irrigation of crops [115]. The emitter that operated most successfully utilized a long water path labyrinth to reduce flow to required level. The clogging was shown to result from the deposition and entrapment of sand particles within the emitter. It was concluded that pond micro algae alone did not constitute a major hazard to the operation of drip irrigation equipment and that waste stabilization pond effluents might be used for drip irrigation.

The effluents of different qualities for drip irrigation were examined at the Ohio State University Extension, Ohio Agricultural Research and Development Center Site [103]. The experimental emitters were designed for use with treated wastewater and contained antimicrobial agents to prevent emitter clogging. It was observed that many clogged emitters recovered to near the original flow rates after the end of experiment.

The microbial organisms were evaluated to prevent the clogging in drip irrigation system caused by biological factors [107]. In the study, three antagonistic bacterial strains were used in the *bacillus spp* (ERZ, OSU-142) and *burkholdria* spp (OSU-7) for treatment of biological clogging of emitters. It was concluded that antagonistic bacterial strains have the potential to be used as anti-clogging agents for treatments of emitters in drip irrigation system. Finally it was suggested that the use of antagonistic bacterial strains in drip irrigation may reduce or completely eliminate the need for repetitive chemical applications to treat emitter clogging and these strains have the potential to be used not only for cleaning of biologically clogged emitters, but also for biological control of pathogenic microorganisms that causes diseases in plants watered with drip irrigation systems.

A study in Japan under Tohaku Irrigation Project was undertaken to reduce emitter clogging induced by biological agents such as algae and protozoa (AP) to enhance the performance of drip irrigation using chlorination [39]. The main objective of the study was to quantify the impact of AP induced changes on discharge rate and uniformity from different types of emitter under two management schemes of without and with chlorine injection into irrigation water. The assessment also included different orifice area. It was observed that there was reduction in emitter discharge induced by AP, due to chlorine injection.

The clogging was evaluated by measuring through head loss across filters, and the filtration quality of different filters using different effluents [14].

It was observed that with the meat industry effluent, the poorest quality efflu-ent, disc filters clogged more than the other filter types. It was also found that the parameter that explained the clogging, expressed as Boucher's fil-terability index, was different depending on the type of effluent and filter. They suggested that the best quality of filtration was achieved with a sand filter when the meat industry effluent was used. No significant differences were observed between the quality of filtration of disc and screen filters when operating with the secondary and tertiary effluents.

Hydraulic performance of three drip irrigation subunits were tested [15] using effluents: Suspended solids and microorganism from WWTP (Waste Water Treatment Plant) of Castell-Platja d' Aro (Girona, Spain). All the subunits were operated intermittently for a total of 10 h per day, 5 days per week. The influence of different strategies of effluent treatment was evaluated on irrigation uniformity at the subunit level. It was con-cluded that with the secondary effluents, uniformity in subunit diminished considerably due to clogging of emitters. It was also observed that clog-ging occurred due to biological aspects.

The clogging mechanism of labyrinth channel in the emitter was exam-ined [122], using a three-dimensional numerical model of clogging analy-sis. Reynolds stress model with wall function was used to simulate the fluid flow in the Eulerian frame, and stochastic trajectory model was adopted to track the motion of the particles in a Lagrangian co-ordinate system without taking into account the agglomerating behavior of particles. The analytical results showed that in the labyrinth channel, low velocity region developed ahead of each saw tooth and large vortex is shaped just behind it. Small particles were apt to deposit in those regions due to their better following behaviors than those of large ones. It was found that the poten-tial clogging regions predicted by simulation were reasonably consistent with the experimental results. Further, it was also found that the particles ranging from 30 to 50 μm behaved best when passing through the laby-rinth channel, and particle densities have a remarkable effect on the pen-etration only when their diameters were larger than 50 μm.

Root intrusion into emitters poses a threat to the long-term success of subsurface drip irrigation systems, particularly in fibrous-rooted crops [44]. In this study, a Bermuda grass was grown in a greenhouse to examine the effectiveness of chemicals in preventing root intrusion into subsurface drip emitters in two-year, two-part experiments. During the first year of study, two acids (sulfuric and phosphori) and two pre-emergence herbicides (trifluralin and thiazopyr) were tested on Bermuda grass grown in small pots. As an initial step for the emitter clogging experiment, the first-year experiment focused on the effectiveness of the chemicals in preventing overall root growth in pots saturated with either trifluralin or thiazopyr. It was found that only thiazopyr significantly inhibited root growth, and visual quality of shoot growth in the thiazopyr-treated pots was lower than the observed quality in the rest of the treatments and in non-treated Bermuda grass. During the second year, nine treatments were prepared based on the first-year study, and were examined for control of root intrusion into actual subsurface drip emitters. It was observed that emitters were completely free of roots with thiazopyr treatment at the highest concentration, and with the trifluralin-impregnated emitter treatment under water stress. Authors concluded that root and rhizome growth was generally unaffected by treatment.

The effect of increasing sediment concentration in irrigation water and aperture size of screen filter were used on the sensitivity of some kinds of emitters to clogging [43]. The study included four concentrations of sediments in irrigation water (0, 70, 230, and 315 ppm) with aperture sizes of screen filter of 428.6, 179.3, 152, and 125 micron. The results indicated that the ratio of clogging differed from emitter to another under the same treatment due to the variations of emitter types and specifications. It was observed that the emitter LL (laminar, long path, type on-line) was the most sensitive to increasing of suspended solids in irrigation water.

Experimental trials were conducted on the behavior of several kinds of filter and drip emitters using poor quality municipal wastewater [24]. The performance of the emitters and filters depended on the quality of the wastewater. It was suggested that the total suspended solids (TSS) influenced the percentage of totally clogged emitters, the mean discharge emitted, the emission uniformity, and the operating time of the filter between cleaning operations. Vortex emitters were more sensitive to clogging than labyrinth emitters and no significant difference was observed between the same kind of emitter placed on soil or sub-soil. Gravel media and disk filters assured better performance than screen filters. Finally, it was found that the use of wastewater with a TSS greater than 50 mg.L^{-1} did not permit optimal emission uniformity to be achieved.

The water from surface and underground sources was examined if it picked up particulate matter during conveyance of sands, silts, plant fragments, algae, diatoms, larvae, snails, fishes, etc. [26]. It was found that as the flow slowed down and/or the chemical background of the water changes, chemical precipitates and/or microbial flocs and slimes began to form and grow, thus micro irrigation emitter clogging occurred. It was suggested that section delineating the occurrences of chemical precipitated and the chemistry of acidification was employed to mitigate clogging caused by chemical precipitates. Finally it was concluded that the clogging which resulting from formation of microbial flocs and slimes was controllable by acidification as well as chlorination.

In a field experiment conducted at Hasanabad, Iran by James [69], five types of emitters with different nominal discharges, with or without self-flushing system and with or without pressure compensating system were evaluated under three management schemes: untreated well water (S_1), acidic treated water (S_2) and magnetic treated water (S_3) in order to reduce chemical clogging. Flow reduction rate, statistical uniformity coefficient (Uc), emission uniformity coefficient (Eu) and variation coefficient of emitters' performance in the field (Vf) were monitored. The emitter performance indices (Uc and Eu) decreased during the experiment due to emitter clogging. The Uc and Eu values in different management schemes confirmed that the acidification had better performance than the magnetic water in order to control emitter clogging and keep high distribution uniformity. Regarding Vf values, the priority of untreated and treated water was as $S_2 > S_3 > S_1$ for each emitter.

An experiment was conducted to evaluate the effects of emitter clogging of four filtrations and six emitter types placed in laterals 87 m long using two different effluents with low suspended solid levels from a wastewater treatment plant for 1,000 h [40]. It was found that only with the effluent that had a higher number of particles did the filter and the interaction of filter and emitter location had a significant effect.

A laboratory experiment was conducted to study the performance of three common emitter types with application of freshwater and treated sewage effluent (TSE). The three types of emitters were the inline-labyrinth types of emitters with turbulent flow (E_1) and laminar flow (E_2) and online pressure compensating type of emitter (E). It was found that for both freshwater and TSE treatment the emitter clogging was more severe for emitter type E_2 due to its smaller flow path dimension and higher manufacturing coefficient of variation. Authors reported that main reason for emitter clogging was due to high pH and ions concentration in TSE treatments [75].

The temporal variations of emitter discharge rate and the distribution of clogged emitters were studied in the drip irrigation system to quantify the impact of emitter clogging on system performance [74]. In the experiment, six types of emitters with or without pressure compensating device and two types of water sources were considered. It was observed that different types of emitters had different susceptibility to clogging. A more random distribution for clogged emitter was found to be suitable for sewage application. It was also reported that clogging of emitters deteriorated the system performance seriously.

The effect of three drip line flushing frequency treatments (no flushing, one flushing at the end of each irrigation period, and a monthly flushing during the irrigation period) were evaluated in surface and subsurface drip irrigation system operated using a waste water treatment plant effluent for three irrigation periods of 540 h each [13]. It was found that drip line flow of the pressure compensating emitter increased 8% over time and 25% in case of non-pressure compensating emitter and 3% decrease in subsurface drip lines by emitter clogging. It was concluded that emitter clogging was affected by the interaction between emitter location, emitter type, and flushing frequency treatment and number of completely clogged emitter were affected by the interaction between irrigation system and emitter type. The results of this study showed that the application of well saline water in drip irrigation system had the potential to induce emitter clogging. The concentration of Fe and Mg in well water was lower than the hazardous levels that could clog emitters. It was found that the flow rate reduction in emitters was affected by emitter characteristics and water treatment methods. Further, the acid injection treatment provided better performance than the magnetic field. On the other hand, less flow rate reduction occurred in emitters using acidic water.

A survey on clogging level of emitters was conducted at some agricultural farms situated in Cannakale Turkey Onsekiz Mart University [119]. In the study, authors tested the emitters under pressures of 50, 100, 150,

200, 250, 300 kPa in laboratory. It was found that some of the emitters were plugged on laterals used for 2 to 3 years in consequence of the tests. The laboratory test showed that 15.6% of 3-year used emitters collected from drip irrigated land did not have any flow under operating pressure of 100 kpa. Finally it was suggested that drip irrigation system must be executed under prescribed pressure (100 kpa).

Laboratory and field testing were conducted in the hydraulic laboratory of the Agricultural Research Council at Institute for Agricultural Engineering in South Africa [47]. Authors evaluated the drippers from two dripper companies with three dripper types and 10 dripper models under controlled conditions. The field evaluation involved 42 surface drip systems throughout South Africa where different water quality conditions and management practices are present. Performance of these practices was evaluated in the field twice a year for two consecutive years and after each evaluation, one dripper line was sampled and also tested in the laboratory. It was reported that the performance was affected by clogging due to water quality and lack of proper maintenance schedules.

3.5 ADSORPTION MECHANISM

A study was conducted on adsorption of radio labeled infectious poliovirus type-2 by 34 well-defined soils and mineral substrates [80]. Also these samples were analyzed in a synthetic freshwater medium containing 1 mM of $CaCl_2$ and 1.25 mM of $NaHCO_3$ at pH 7. It was found that in a model system, adsorption of poliovirus by Ottawa sand was rapid and reached equilibrium within 1 h at 4°C. Near saturation, the adsorption was described by the Langmuir equation. The apparent surface saturation was 2.5×10^6 plaque-forming units of poliovirus per mg of Ottawa sand. At low surface coverage, adsorption was described by the Freundlich equation. It was observed that most of the substrates adsorbed more than 95% of the virus. Among the soils, muck and Genesee silt loam were the poorest adsorbents. Among the minerals, montmorillonite, glauconite, and bituminous shale were the least effective and the most effective adsorbents were magnetite sand and hematite, which are predominantly oxides of iron. Correlation coefficients for substrate properties and virus adsorption revealed that the elemental composition of the adsorbents had little effect on poliovirus uptake. Substrate surface area and pH were not significantly correlated with poliovirus uptake. A strong negative correlation was found between poliovirus adsorption and both the contents of organic matter and the available negative surface charge on the substrates as determined by their capacities for adsorbing the cationic polyelectrolyte, polydiallyldimethyl ammonium chloride.

The adsorption processes were studied for the fabrication of layer by layer films using poly-o-methooxyaniline (POMA) [98]. It was concluded that the amount of material adsorbed in any given layer depended on experimental parameters. It was observed that the H-bonding played a fundamental role in the adsorption of polyanilines on a glass substrate when the polymers were charged and electrostatic attraction was expected to predominate. The probability of adsorption increased in sites where some polymers were already adsorbed, which caused the roughness to increase with the number of layers. Also the electrostatic attraction was the predominant factor in the films.

Adsorption isotherm of Q-cresol from aqueous solution by granular activated carbon [76] were studied to investigate the equilibrium adsorption isotherms of Q-cresol from aqueous solution by a series of laboratory batch studies. A commercial norit granular activated carbon was used to evaluate the adsorption characteristic of Q-cresol at different temperatures of 30, 38 and 48°C. The effect of various initial concentrations (25–200 mg.L^{-1}) and time of adsorption on Q-cresol adsorption process were studied. The evaluated the isotherm data using Langmuir and Freundlich isotherms in order to estimate the monolayer capacity values of activated carbon was used in the sorbate-sorbent system. The results revealed that the empirical Langmuir isotherm matched the observed data very well as compared to Freundlich isotherm. It was also found that the adsorption capacity of Q-cresol decreased with the increasing the adsorption temperature. The maximum adsorption capacity of 270 mg.g^{-1} was obtained by Q-cresol at temperature of 30°C, 120 rpm and 24 h of adsorption time.

In Sao Paulo State of Brasil [7], the relationships were evaluated between sulphate adsorption and physical, electrochemical and mineralogical properties of representative soils. The experimental results were subjected to variance, correlation and regression analyses. When the adsorption was evaluated in the clay fraction, the kaolinite content was associated with low capacities of sulphate adsorption. However, no relationship was observed between the kaolinite soil content and the sulphate adsorption by the whole soil. No significant effects on sulphate adsorption were observed for individual hematite and goethite soil contents. On the other hand, the sum of hematite and goethite contents were related to sulphate adsorption.

Hussain et al. [64] studied phosphorous adsorption by five saline sodic soil samples collected from Faisalabad district. They prepared 0.01M of $CaCl_2$ solution with different concentrations of P and placed 3 g of soil in 30 mL solution of all P concentrations and kept the solution overnight and centrifuged; and the P in supernatant solution was determined calorimetrically. They calculated the adsorption of P using the difference between the amount of P in supernatant and that was added in solution and plotted the adsorption data according to Langmuir and Freundlich equation.

The adsorption behavior of Cu on three solid waste material were investigated [4]: sea nodule residue (SNR), fly ash (FA), and red mud (RM). The effects of various parameters (pH of the feed solution, contact time, temperature, adsorbate and adsorbent concentrations, and particle size of the adsorbent) were studied for optimization of the process parameters. It was found that the adsorption of copper increased with increasing time, temperature, pH, and adsorbate concentration, and decreased with increasing initial copper concentration.

A laboratory experiment was conducted on synthesized hydrous stannic oxide (HSO) and Cr (VI) adsorption behavior by means of batch experiments [110], to test the equilibrium adsorption data for the Langmuir, Freundlich, Temkin and Redlich-Peterson equations. The scientists conducted batch adsorber tests by mechanical agitation (agitation speed: 120 to 130 rpm) using 0.2 g of HSO into a 100 mL polythene bottle with 50 mL of sorbate solution. Different concentrations of Cr solution were used in the range of 2.0 to 50.0 mg.L^{-1}. They finally calculated the amount of adsorbed Cr by the difference of the initial and residual amount in the solution divided by the weight of the adsorbent. It was concluded that the adsorption of Cr onto HSO took place by electrostatic interaction between adsorbent surface and species in the solution.

The phosphorus adsorption by Freundlich adsorption isotherm under rainfed conditions was examined for 10 soil series of Pothwar Plateau [29]. These soils were treated with three different P fertilizers (DAP, SSP and NP) at equilibrium solution concentrations of 10, 20, 40, 60, and 80 $\mu g.mL^{-1}$. Maximum Freundlich adsorption parameters (maximum adsorption ($\mu g\ g^{-1}$) and buffer capacity (ml g^{-1})) were observed in Chakwal soil followed by Balkassar soil. The minimum values of these two Freundlich parameters were observed in Kahuta soil. It was observed that the maximum value of KOC in Chakwal soil with DAP, SSP and NP while minimum value of KOC was observed in Bather soil with all fertilizers under investigation. A decrease in P adsorption with successive increase in equilibrium phosphorus solution concentration was recorded in all the soils under study.

A multi scale structure prediction technique was developed to study solution and adsorbed state ensembles of bio-mineralization proteins [36]. The algorithm, which employs a Metropolis Monte Carlo-plus minimization strategy, varies all torsional and rigid-body protein degrees of freedom. Authors applied this technique to fold statherin, starting from a fully extended peptide chain in solution, in the presence of hydroxyapatite (HAp) (001), (010), and (100) monoclinic crystals. Blind (unbiased) predictions capture experimentally observed macroscopic and high resolution structural features and show minimal statherin structural change upon the adsorption. The dominant structural difference between solution and adsorbed states is an experimentally observed folding event in statherin's helical binding domain. Whereas predicted statherin conformers vary slightly at three different HAp crystal faces, geometric and chemical similarities of the surfaces allow structurally promiscuous binding. Finally, they compared blind predictions with those obtained from simulation biased to satisfy all previously published solid-state NMR (ssNMR) distance and angle measurements (acquired from HAp-adsorbed statherin). Atomic clashes in these structures suggested a plausible, alternative interpretation of some ssNMR measurements as intermolecular rather than intra molecular. Finally it was revealed that a combination of ssNMR and structure prediction could effectively determine high-resolution protein structures at bio-mineral interfaces.

An experiment was carried out on ion adsorption behavior of the polyacrylic acid-polyvinylidene fluoride blended polymer [72]. Authors used polyvinylidene fluoride to remove copper from aqueous solutions. They prepared the polymer using thermally induced polymerization and phase inversion. The blended polymer was characterized by x-ray diffraction analysis, environmental scanning electron microscopy, x-ray photoelectron spectroscopy, and N_2 adsorption/desorption experiments. The sorption data was fitted to linearized adsorption isotherms of the Langmuir, Freundlich, and Dubinin-Radushkevich isotherm models. Further, they evaluated the batch sorption kinetics using pseudo-first-order, pseudo-second-order, and intra-particle diffusion kinetic reaction models. They found that ΔH° was greater than 0, ΔG° was lower than 0, and ΔS° was greater than 0, which showed that the adsorption of Cu (II) by the blended polymer was a spontaneous, endothermic process. The adsorption isotherm fitted better to the Freundlich isotherm model and the pseudo-second order kinetics model gave a better fit to the batch sorption kinetics.

The zinc adsorption was studied in 10 soils varying in texture or calcareousness in Punjab [10]. Authors executed the adsorption process by equilibrating 2.5 g soil in 25 mL of 0.01M of $CaCl_2$ solution containing 0, 5, 10, 15, 20, 25, 70, and 120 mg of Zn per liter. Sorption data were fitted to Freundlich and Langmuir adsorption models. The data were best fitted in both linearized Freundlich and Langmuir equations as evidenced by higher correlation coefficient values ranging from 0.87 to 0.98. High clay contents ranging from 8 to 32% and $CaCO_3$ 4.46 to 10.6% promoted an increase in the amount of adsorbed zinc in these soils. They found that adsorption of Zn increased with the increasing level of Zn and also increased with increase in clay content, and $CaCO_3$ contents; and the maximum adsorption of Zn was observed in the Kotli soil, whereas the minimum was in the Shahdra soil series.

The experiment was undertaken on boron adsorption at the Institute of Soil and Environmental Sciences, University of Agriculture, Faisalabad, Pakistan in five different textured calcareous soils of Punjab [109]. They executed the adsorption process by equilibrating 2.5 g of soil in 0.01 M of $CaCl_2$ solution containing different concentrations of boric acid for 24 h. They estimated the boron adsorption using Langmuir and Freundlich models. They concluded that the Freundlich model was better than the Langmuir model.

The iodide adsorption was evaluated to compare the sorption behavior of iodate and iodide [35]. They collected typical soil samples at 17 locations across China. Batch experiments of iodate and iodide adsorption were carried out by shaking soil samples equivalent to 2.5 g dry weight with 25 mL of iodine (either iodate or iodide) solution. This was performed in centrifuge tubes fitted with caps, on an end-over-end shaker (160 rpm) at 25°C and shaken for 40 h. For the sorption isotherm studies, concentrations of KIO_3 in the solution were 0, 1, 2, 4, 6, 8 mg.L^{-1} for the two soil types from Xinjiang Province and Beijing City. The results indicated that the capacity of iodate adsorption by the five soils was markedly greater than that of iodide. Furthermore, detailed comparison of sorption parameters based on the Langmuir and the Freundlich adsorption equations supported this finding showing a greater adsorption capacity for iodate than for iodide due to higher k_2 values of iodate than those of iodide.

Adsorption studies of zinc and copper ions were attempted on montmorillonite (MMT). The adsorption mechanism of the metal adsorptions was studied by the measurement of UV-VIS DRS of Zn MMT and Cu MMT [71]. They used zinc nitrate, copper sulphate, ammonium chloride and ethylenediamine. For adsorption procedure, they prepared the EDA-MMT by shaking MMT in concentrated ethylenediamine (0.9 g.cm^{-3}) for 24 h, and then filtered and dried at 105°C for 2 h. They saturated the MMT and EDA-MMT metals by shaking in the solutions of 5 mmol.L^{-1} of Cu^{2+} and 20 mmol.L^{-1} of Zn^{2+} at 170 rpm for 24 h. Then they centrifuged the suspension for 24 h and analyzed the filters for zinc and copper using atomic absorption spectrometry (AAS). They concluded that using the DRIFT method, the amount of interlayer water in Zn MMT and Cu MMT were similar.

Research was conducted on bottle brush polymers and their adsorption on surfaces and their interactions [32]. By small-angle scattering techniques, they studied the solution conformation and interactions in solution. Surfactant binding isotherm measurements, NMR, surface tension measurements, as well as SAXS, SANS and light scattering techniques were utilized for understanding the association behavior in bulk solutions. The adsorption of the bottle-brush polymers onto oppositely charged surfaces was explored using a battery of techniques: reflectometry, ellipsometry, quartz crystal microbalance, and neutron reflectivity. The combination of these techniques allowed determination of adsorbed mass, layer thickness, water content and structural changes occurring during layer formation. The adsorption onto mica was found to be very different to that on silica, and an explanation for this was sought by employing a lattice mean-field theory. The model was able to reproduce a number of salient experimental features characterizing the adsorption of the bottle-brush polymers over a wide range of compositions, spanning from uncharged bottle-brushes to linear polyelectrolyte. The interactions between bottle-brush polymers and anionic surfactants in adsorbed layers were elucidated using ellipsometry, neutron reflectivity and surface force measurements.

The adsorption characteristics of phosphorus were evaluated onto soil with the Langmuir, Freundlich, and Redlich–Peterson isotherms by both the linear and non-linear regression methods [123]. The adsorption experiment was conducted at the temperatures of 283, 288, 298, and 308°K, respectively, to choose the appropriate method and obtain the credible adsorption parameters for soil adsorption equilibrium studies. The results showed that the non-linear regression method was a better choice to compare the better fit of isotherms for the adsorption of phosphorus onto laterite. Both the two-parameter Freundlich and the three-parameter Redlich–Peterson isotherms had higher coefficients of determination for the adsorption of phosphorus onto laterite at various temperatures.

A study on nZVI particles was conducted to investigate the removal of Cd_2^+ in the concentration range of 25–450 mg.L^{-1} [20]. The effect of temperature on kinetics and equilibrium of cadmium sorption on nZVI particles was thoroughly examined. They found that the maximum adsorption capacity of nZVI for Cd^{2+} was 769.2 mg.g^{-1} at 297 K. Thermodynamic parameters were: change in the free energy (Go), the enthalpy (Ho), and the entropy (So). These results suggested that nZVI can be employed as an efficient adsorbent for the removal of cadmium from contaminated water sources.

A study was conducted on adsorption behavior of Mn from an agricultural fungicide in two south-western Nigeria soils using batch equilibrium test [91]. They applied two mathematical models described by Langmuir's and Freundlich's adsorption equations. From the isotherm analysis, they found that the sorption of Mn to the two soil types considered was best described by Freundlich model and the maximum adsorption capacities (kf) obtained from this model were 96.64 g.mL^{-1} and 30.76 g.mL^{-1} for Egbeda and Apomu soils, respectively. These maximum adsorption capacities occurred at solution pH of 5 for both soils. Finally, they found that solution pH of 3 and 4 were not significantly different as well as solution pH of 5 and 6 in their effects on the amount of Mn adsorbed.

A laboratory study was conducted at Research and Development (chemical laboratory) of product development in M/s Jain Irrigation System, Plastic Park, Jalgaon – Maharashtra – India during 2010–2011 to know the adsorption mechanism using adsorption characteristics of Langmuir and Freundlich equations for granules of PVC, LLDPE, Silicon diaphragm (rubber button) and HDPE. The results indicated that the percent of clogging was maximum for LLDPE indicating 6.94 to 11.1% from day 1 to day 15 compared to other grinded materials. The minimum percent clogging was recorded as 1.12 to 3.03% for PVC. The results demonstrated that LLDPE is more susceptible for clogging in drip irrigation system [27, 28]. For clogging test, two different types of emitter type A (2 lph) and B (4lLph) were used. The results of clogging test demonstrated that emitter type A was more susceptible for clogging compared to type B. The emitter type B was superior compared to emitter type A.

3.6 SUMMARY

The review of literature on studies on clogging mechanism in drip irrigation system indicated that the clogging of emitters can occur due to three clogging agents [57]: (i) physical clogging, (ii) chemical clogging, and (iii) biological clogging. Most of the clogging studies suggested flushing, chlorination, combination of filtration, emitter design and field practices in order to reduce emitter clogging. This is all about the temporary solution to reduce the clogging. But the information is not available on the initial adsorption mechanism of clogging of emitters in drip irrigation system for evolving strategies on reducing the clogging and such studies were very few or not done previously. Hence, an attempt was made under the present investigation to study the actual initial adsorption phenomena in emitters to evaluate the clogging mechanism.

KEYWORDS

- acidification
- adsorption
- Brasil
- China
- chlorination
- clogging
- clogging mechanism
- clogging, biological,
- clogging, chemical
- clogging, physical
- drip irrigation
- effluent
- emitter
- flushing
- Freundlich equation
- India
- Langmuir equation
- Nigeria
- Pakistan
- polymers
- saline water
- solid waste
- USA
- wastewater
- water shortage

REFERENCES

1. Aali, K. A., Liaghat, A. (2009). The effect of acidification and magnetic field on emitter clogging under saline water application. *Journal of Agriculture Science,* 1(1), 112–117.

2. Adin, A., Sacks, M. (1991). Dripper clogging factors in wastewater irrigation. *Journal of Irrigation and Drainage Engineering*, 117(6), 813–825.

3. Adin, A. (1987). Clogging in irrigation systems reusing pond effluents and its prevention. *Water Science Technology*, 19(12), 323–328.

4. Agrawal, K., Sahu, K., Pandey, B. D. (2004). A comparative adsorption study of copper on various industrial solid wastes. *AICHE Journal*, 50(10), 34–41.

5. Ahmed, B. A. O., Yamamoto, T., Fujiyama, H., Miyamoto, K. (2007). Assessment of emitter discharge in micro irrigation system as affected by polluted water. *Irrigation Drainage System*, 21, 97–107.

6. Al-Amound, A., Saeed, M. (1988). The effect of pulsed drip irrigation on water management. *Proceeding the 4th International Micro irrigation Congress*, 4b-2.

7. Alves., O., Lavorenti, A. (2003). Sulfate adsorption and its relationships with properties of representative soils of the Sao Paulo State, Brazil. *Geoderma*, 118, 89–99.

8. Alam, M., Rogers, D. H. (2002). Filteration and maintenance considerations for subsurface drip irrigation systems. Kansas State University, Agricultural Experiment Station and Cooperative Extension Service, Manhattan, Kansas.

9. ASAE, (2001). Design and Installation of Micro irrigation Systems. ASAE EP405.1 JAN 01. 2950 Niles Rd., St. Joseph, MI 49085–9659, USA.

10. Ashraf, M. S., Ranjha, A. M., Yaseen, M., Ahmad N., Hannan, A. (2008). Zinc adsorption behavior of different textured calcareous soils using Freundlich and Langmuir models. *Pakistan Journal of Agriculture Science*, 45(1), 83–89.

11. Assuncao, T. P. R; Jose E. S. P., Christiane, C. (2008). Chemical treatment to unclog dripper irrigation systems due to biological problems. *Journal of Science and Agriculture*, (Piracicaba, Braz.), 65(1), 1–9.

12. Anonymus, (2002). Micro irrigation. Jain Irrigation Systems Ltd.

13. Barges, P. J., Arbat, G., Elbana, M., Duran-Ros, M., Barragan, J., Ramirej de Cartogene, and Lamm, F. R. (2010). Effect of flushing frequency on emitter clogging in micro irrigation system with effluents. *Agricultural Water Management*, 97(6), 883–891.

14. Barges, P. J., Arbat, G., Barragan, J., and Ramirej de Cartogene, F., 2005a. Hydraulic performance of drip irrigation subunits using WWTP effluents. *Agricultural Water Management*, 77, 249–262.

15. Barges, P. J., Barragan, F., Ramirez, D. C., 2005b. Filtration of effluents for micro irrigation systems. *Transactions of ASAE*, 48(3), 969–978.

16. Boman, B. J. (1990). Clogging characteristics of various micro sprinkler designs in a mature citrus grove. *Proceedings of Florida State Horticulture Society*, 103, 327–330.

17. Boman, B. J., Bullock, R. C. (1994). Damage to micro sprinkler riser assemblies from *Selenisa sueroides* caterpillars. *Journal of Applied Engineering Agriculture*, 10(2), 221–223.

18. Boman, B. J. (2000). Insect problems in Florida citrus micro irrigation systems. In Proc. 4th National Irrigation Symposium by Evans, R. G., Benhan, B. L., Trooien, T. P. (Eds.). St. Joseph, Michigan. *American Society of Agricultural Engineers*, pp. 409–415.

19. Boman, B. J. (1995). Effects of orifice size on micro sprinkler clogging rates. *Journal of Applied Agriculture Engineering*, 11(6), 839–843.

20. Boparai, H. K., Meera, J., Denis, M., Carroll, O. (2011). Kinetics and thermody-namics of cadmium ion removal by adsorption onto nano zerovalent iron particles. *Journal of Hazardous Materials,* 186, 458–465.

21. Bresler, E. (1975). Trickle-drip irrigation: Principles and applications to soil-water management. In Brady, N. C. (Ed.), *Advances in Agronomy,* 29, 343–393. Academic Press, New York.

22. Brushwein, J. R., Matthews, C. H., Childers, C. C. (1989). *Selenisa sueroides* (Lepi-doptera: Noctuidae), A pest of sub-canopy irrigation systems in citrus in Southwest Florida. Fla. Entomol., 72(3), 511–518.

23. Bucks, D. A., Nakayama, F. S., Gilbert, R. G. (1979). Trickle irrigation water quality and preventive maintenance. *Agriculture Water Management,* 2, 149–162.

24. Capra, A., Scicolone, B. (2006). Recycling of poor quality urban wastewater by drip irrigation systems. *Journal of Cleaner Production,* 15, 1529–1534.

25. Capra and Scicolone, B. (2007). Recycling of poor quality urban wastewater by drip irrigation systems. *Journal of Cleaner Production* 16.

26. Chang, C. (2008). Chlorination for disinfection and prevention of clogging of drip lines and emitters. *Encyclopedia of Water Science,* 26

27. Chavan, V. K., Balakrishnan, P., Deshmukh, S. K., Ingle, V. K. (2013). Evaluation of adsorption mechanism in clogging of materials used in drip irrigation system. *Jour-nal of Agricultural Engineering,* 50(3).

28. Chavan V. K., Balakrishnan, P., Deshmukh, S. K., Nagdeve, M. B. (2014). Mechan-ics of clogging in micro irrigation system. Chapter 6, pages 169–182, In: *Sustainable Practices in Surface and Subsurface Micro Irrigation,* volume 2 by Megh R Goyal (ed.). Oakville, ON, Ca: Apple Academic Press Inc.,

29. Chattha, K., Yousaf, M., Javeed, S. (2007). Phosphorus adsorption as described by freundlich adsorption isotherms under rainfed conditions of Pakistan. *Pakistan Jour-nal of Agriculture Science,* 44(4), 27.

30. Childers, C. C., Futch, S. H., Stange, L. A. (1992). Insect (Neuroptera: Lepidoptera) clogging of a micro sprinkler irrigation system in Florida citrus. *Fla. Entomol.,* 75(4), 601–604.

31. Chigerwe, J., Manjengwa, N., Van der Zaag, P. (2004). Low head drip irrigation kits and treadle pumps for smallholder farmers in Zimbabwe: a technical evaluation based on laboratory tests. *Phys. Chem. Earth,* 29, 1049–1059.

32. Claesson, P. M., Makuska, R. I., Varga, R., Meszaros, S., Titmuss, P., K. Linse, (2010). Bottle brush polymers and surface adorption and their interactions. *Advances in Colloid and Interface Science,* 155, 50–57.

33. Coelho, R. D., Resende, R. S. (2001). Biological clogging of netafim's drippers and recovering process through chlorination impact treatment. ASAE Paper Number: 012231, Sacramento, California, USA.

34. Costerton, J. W., Stewart, P. S. (2001). Battling Biofilms. *Scientific American,* 285(1), 75–81.

35. Dai., M., Zhang., Hu, Q. H., Huang, Y. Z., Wang, R. Q., Zhu, Y. G. (2008). Adsorp-tion and desorption of iodine by various Chinese soils: II. Iodide and iodate. *Geo-derma,* 153, 130–135.

36. David, L. M., Jeffrey J. G. (2009). Solution and adsorbed state structural ensembles predicted for the Statherin hydroxyapatite system. *Journal of Biophysics,* 96(8), 3082–3091.

37. Davey, M. E., O Toole, G. A. (2000). Microbial biofilms: from ecology to molecular genetics. *Microbiol. Journal of Molecular Biology,* 64(4), 847–867.
38. Davies, D. G., Parsek, M. R., Pearson, J. P., Iglewski, B. H., Costerton, J. W., Greenberg, E. P. (1998). The involvement of cell-to-cell signals in the development of bacterial biofilm. *Science,* 280(5361), 295–298.
39. Dehghanisanij, H., Yamamoto, T., Ould Ahmad, B., Fujiyana, H., and Miyamoto, K. (2005). The effect of chlorine on emitter clogging induced by algae and protozoa and the performance of drip irrigation. *Transactions of American Society of Agricultural Engineers,* 48, 519–527.
40. Duran-Ros, M., Puig-Bargues, J., Arbat, G., Barragan, J., Ramirez de Cartagene, F. (2009). Effect of filter, emitter and location on clogging when using effluents. *Agriculture Water Management,* 96, 67–79.
41. De Kreij, C., Van der Burg, A. M. M., Runia, W. T. (2003). Drip irrigation emitter clogging in Dutch greenhouses as affected by methane and organic acids. *Agriculture Water Management,* 60, 73–85.
42. English SD (1985). Filtration and water treatment for micro-irrigation. In: *Drip/Trickle Irrigation In Action.* Nov. 18–21 Fresno, California. *Proceeding of the third International Drip Irrigation Congress* 1, 54–57.
43. El-Berry, A. M., Bakeer, G. A., Al-Weshali, A. M. (2007). The effect of water quality and aperture size on clogging of emitters. www.sciencedirect.com.
44. Elisa, M., and K. Suarez-Rey. (2006). Effects of chemicals on root intrusion into subsurface drip emitters. *Journal of Irrigation and Drainage,* 55, 501–509.
45. Evans, R. G. (2000). Micro irrigation. Washington State University, Irrigated Agriculture Research and Extension Center, 24106 North Bunn Road Prosser, WA 99350, USA.
46. Enciso, J., Porter, D. (2001). Maintaining subsurface drip irrigation system, Texas Cooperative Extension Service, The Texas A & M University, L-5401, 10–01.
47. Felix, B. R. (2010). Performance of drip irrigation systems under field conditions. General Manager: Operational Programs, ARC-Institute for Agricultural Engineering, Private Bag 519, Silverton (0127). South Africa.
48. Feng, F., Li, Y., Guo, Z., Li, J., Li, W. (2004). Clogging of emitter in subsurface drip irrigation system. *Trans. CSAE,* 20(1), 80–83.
49. Ford, H. W., Tucker, D. P. H. (1975). Blockage of drip irrigation filters and emitters by iron – sulfur – bacterial products. *Journal of Horticulture Science,* 10(1), 62–64.
50. Ford, H. W., Tucker, D. P. H. (1974). Water quality measurement for Drip Irrigation systems. *Journal of Florida Agriculture Experimental Stations,* 5598, 58–60.
51. Ford, H. W. (1976). Controlling slimes of sulfur bacteria in drip irrigation systems. *Journal of Horticulture Science,* 11, 133–135.
52. Gilbert, R. G., Nakayama, F. S., Bucks, D. A., French, O. F., Adamson, K. C., Johnson, R. M. (1982). Trickle irrigation: Predominant bacteria in treated Colorado River water and biologically clogged emitters. *Journal of Irrigation Science,* 2(3), 123–132.
53. Gilbert, R. G., Nakayama, F. S., Bucks, D. A., French. O. F. (1980). Trickle irrigation: emitter clogging and other flow problems. *Journal of Agricultural Water Management,* 3, 159–177.
54. Gilbert, R. G., Ford, H. W. (1986). Operational principles/emitter clogging. In: Nakayama, F. S., Bucks, D. A. (Eds.), *Trickle Irrigation for Crop Production.* Elsevier, Amsterdam, pp. 142–163.

55. Gilbert, R. G., Nakayama, F. S., Bucks, D. A., French, O. F., Adamson, K. C. (1981). Trickle irrigation: emitter clogging and flow problems. *Journal of Agricultural Water Management*, 3, 159–178.

56. Government of India, (2006). Report of the Working Group on Water Resources For the XI the Five Year Plan (2007–2012), Ministry of Water Resources. Pages 54.

57. Goyal Megh R. (Senior Editor-in-Chief), (2015). *Book Series on Research Advances in Sustainable Micro Irrigation*. Volumes 1 to 10. Oakville, ON, Canada: Apple Academic Press Inc.,

58. Granberry, D. M., Harrison, K. A. (2005). Drip chemigation; injecting fertilizer, acid and chlorine. University of Georgia College of Agricultural and Environmental Science and the US Department of Agriculture. Pages 55.

59. Harris, G. (2005). Sub surface drip irrigation system components. Queen Land Government, Department of Primary Industries and Fisheries, Australia.

60. Hills, D. J., Navar, F. M., Waller, P. M. (1989). Effects of chemical clogging on drip-tape irrigation uniformity. *Transactions of American Society of Agriculture Engineers*, 32(4), 1202–1206.

61. Hills, D. J., Brenes, M. J. (2001). Micro irrigation of wastewater effluent using drip tape. *Journal of Applied Agriculture Engineering*, 17(3), 303- 308.

62. Hills, D. J., Tajrishy, M. A., and Tchobanoglous, G. (2000). The influence of filtration on ultraviolet disinfection of secondary effluent for micro irrigation. *Transactions of American Society of Agriculture Engineers*, 43(6), 1499–1505.

63. Howell, T. A., Stevenson, D. S., Aljibury, F. K., Gitlin, H. M., Wu, I. P., Warrick, A. W., Raats, P. A.C. (1983). Design and operation of trickle (drip) systems. In: Jensen, M.E (Ed.), *Design and Operation of Farm Irrigation Systems*. ASAE, St Joseph, MI. *Monograph by American Society of Agriculture Engineers*.

64. Hussain., A., Anwar-Ul- Haq, G., and Muhammad, N. (2003). Application of the Langmuir and Freundlich equations for P adsorption in saline-sodic soils. *International Journal of Agriculture and Biology*, 5(3), 91–99.

65. International Commission on Irrigation and Drainage (ICID), (2000). Sprinkler and Micro-Irrigated Areas in Some ICID Member Countries. <http://www.icid.org/index_e.html>.

66. IARI, (2011). http://www.iari.res.in/index.php?option=com_jumi&fileid=24&Ite mid=664.

67. INCID, (2008). Indian National Committee on Irrigation and Drainage. <www.icid.org/v_india.pdf>.

68. Iyer, Ramasamy, R. (2003). *Water: Perspectives, Issues, Concerns*. Sage Publications, New Delhi.

69. James, L. G. (1988). *Principles of Farm Irrigation System Design*. John Wiley, New York, pp. 287–297.

70. Khaled, Ahmad Aali., Abdoulmajid Liaghat and Hossein Dehghanisanij, (2009). The Effect of acidification and magnetic on emitter clogging under saline water application. *Journal of Agricultural Science*, 1(1), 21–24.

71. Kozak, P. P., Vladimir, M., Zdenik, K. (2010). Adsorption of zinc and copper ions on natural and ethylenediamine modified montmorillonite. *Ceramics – Silikaty*, 54(1), 78–84.

72. Laizhou, S., Wang, J., Zheng, Q., Zhang, Z. (2008). Characterization of Cu (II) ion adsorption behavior of the polyacrylic acid-polyvinylidene fluoride blended polymer.

73. Li, J., Chen, L. (2009). Assessing emitter clogging in drip irrigation system with sewage effluent. Transactions of the *American Society of Agricultural and Biological Engineer, www.asabe.org.*

74. Li, J., Chen, L., Li, Y. (2009). Comparison of clogging in drip emitters during application of sewage effluent and groundwater. Transactions of the *American Society of Agricultural and Biological Engineers,* 52(4), 1203–1211.

75. Liu, H., Huang, G. (2009). Laboratory experiment on Drip emitter Clogging with fresh water and treated with sewage effluent. *Journal of Agriculture Water Management,* 96, 745–756.

76. Maarof, H. I., Bassim, H. H., Abdul, L. A. (2003). Adsorption isotherm of Q-cresol from aqueous solution by granular activated carbon. *Proceedings of International Conference On Chemical and Bioprocess Engineering,* 21 – 29. Universiti Malaysia Sabah. Kola Kinabalu.

77. MoWR, (1999). *Report of the Working Group on Water Availability for Use, National Commission for Integrated Water Resources Development Plan.* Ministry of Water Resources, Government of India, New Delhi.

78. Merriam, J. L., Keller, J. (1978). Farm irrigation system evaluation: A Guide for management. Agricultural Irrigation Eng. Department Utah State University Logan, Utah, pp. 271.

79. Meyer, J. L., Snyder, M. J., Valenzuela, L. H., Strohman, R. (1998). Liquid polymers keep drip irrigation lines from clogging. *Journal of California Agriculture,* 45(1), 116–120.

80. Moore, S., Dene, H. T., Lawrence, S. S., Michael, M. R., Fuhs G. W. (1981). *Journal of Applied and environmental microbiology,* 42(6), 963–975.

81. Nakayama, F. S., Bucks, D. A., French, O. F., L. Adamson. (1981). Trickle irrigation: Emitter clogging and other flow problems. *Agricultural Water Management,* 3, 159–178.

82. Nakayama, F. S. (1986). *Trickle irrigation for crop production.* Elsevier Science Publishers, Amsterdam, Netherlands, pp. 383

83. Nakayama, F. S. (1978). Water treatments in trickle irrigation systems. *Journal of Irrigation Drainage Division of American Society of Civil Engineering,* 23–24.

84. Nakayama, F. S., Bucks, D. A., 1991.Water treatments in trickle irrigation systems. *Journal of Irrigation Drainage Division of American Society of Civil Engineering,* 104S(IR1), 23–34.

85. Nakshabandi, G. A., Saqqar, M. M., Shatanawi, M. R., Fayyad, M., Al-Horani, H. (1997). Some environmental problems associated with the use of treated wastewater for irrigation in Jordan. *Agricultural Water Management,* 34, 81–94.

86. Narayanamoorthy, A. (2005). *Efficiency of Irrigation: A Case of Drip Irrigation.* Occasional Paper: 45, Department of Economic Analysis and Research, National Bank for Agriculture and Rural Development, Mumbai, India.

87. Netafim, (2009). *Drip system operation and maintenance.* Netafim irrigation USA. www.netafimusa.com.

88. Ould, A., Yamamoto T., Fujiyama H., Miyamoto K. (2007). Assessment of emitter discharge in micro irrigation system as affected by polluted water. *Journal of Irrigation and Drainage,* 21, 97–107.

89. O'Toole, G. A., Kolter. (1998). Flagellar and twitching motility are necessary for Pseudomonas aeruginosa biofilm development. *Journal of Molecular Microbiol.,* 30(2), 295–304.

90. Oron, G., Shelef, G., Truzynski, B. (1979). Trickle irrigation using treated wastewaters. *Journal of Irrigation and Drainage,* 105(IR2), 175–186.
91. Osunbitan, K. O., D. Adekalu and F. Patrick, (2011). Adsorption behavior of manganese from Dithane M– 45 to two soil types in southwestern Nigeria. *Canadian Journal on Environmental, Construction and Civil Engineering,* 2(5), 77–81.
92. Padmakumari, O., and Sivanappan, R. K. (1985). Study on clogging of emitters in drip irrigation system. *Paper presented at the 1985 Proceedings of the 3rd International Drip Irrigation System Congress* pp. 80–83.
93. Parchomchuk, P. (1976). Temperature effect on emitter discharge rates. *Transaction of American Society of Agricultural Engineers,* 19(4), 690–692.
94. Pitts, D. J., Haman, D. Z., Smajstrla, A. G. (1990). Causes and prevention of emitter plugging in micro irrigation systems. University of Florida, Florida Cooperative *Extension Service Bulletin 258.*
95. Povoa, A. F., Hills, D. J. (1994). Sensitivity of micro irrigation system pressure to emitter plugging and lateral line perforations. *Transaction of American Society of Agriculture Engineers,* 37(3), 793–799.
96. Powell, N. L., Wright, F. S. (1998). Subsurface micro irrigated corn and peanut: effect on soil pH. *Journal of Agricultural Water Management,* 36, 169–180.
97. Qingsong, Lu Gang, Liu Jie, Shi Yusheng, Dong Wenchu, and Huang Shuhuai. (2008). Evaluations of emitter clogging in drip irrigation by two-phase flow simulations and laboratory experiments. *Journal of Computers and Electronics in Agriculture,* 63(2), 294–303.
98. Raposo, M., Osvaldo, N. (1998). Adsorption mechanisms in layer-by-layer films. Universidade de Sao Paulo, CP 369 – 13560–970 Sao Carlos, SP, Brasil.
99. Rav-Acha, C., Kummel, M., Salamon, I., Adin, A. (2000). The effect of chemical oxidants on effluent constituents for drip irrigation. *Science,* 29(1), 119–129.
100. Ravina, E. P., Paj, Z., Sofer, A., Marcu, A., Shisha and Sagi, G. (1992). Control of emitter clogging in drip irrigation with reclaimed wastewater. *Journal of Irrigation Science,* pp. 129–139.
101. Ravina, E. P., Sofer, Z., Marcu, A., Shisha, A., Sagi, G., Lev, Y. (1997). Control of clogging in drip irrigation with stored treated municipal sewage effluent. *Agricultural Water Management,* 33(2–3), 127–137.
102. Ribeiro, T. A. P., Paterniani, J. E. S., Coletti, C. (2008). Chemical treatment to unclogg dripper irrigation systems due to biological problems. (Piracicaba, Braz) *Journal of Agricuture Science,* 65(1), 1–9.
103. Rowan, M., Mamel, K., Tuovinen, O. H. (2004). Clogging incidence of drip irrigation emitters distributing effluents of differing levels of treatment. *On-Site Waste Water Treatment X, Conference Proceedings,* pages 084–091.
104. Ruskin, R., Ferguson, K. R. (1998). Protection of subsurface drip irrigation systems from root intrusion. *Proceedings of Irrigation Association 19th Annual Meeting.* San Diego, CA., Nov. pp.1–3.
105. Sagi, G., Paz, E., Ravina, I., Schischa, A., Marcu, A., Yechiely, (1995). clogging of drip irrigation systems by colonial protozoa and sulfur bacteria. In: *Proceedings of 5th International Micro irrigation Congress* Orlando FL.
106. Saleth, R. (1996). *Water Institutions in India: Economics, Law and Policy.* Commonwealth Publishers, New Delhi.

107. Sahin, Ustuil, Anupali, Omer, Doumez, Mesudefigen, Sahin and Fikrettin, (2005). Biological treatment of clogged emitters in a drip irrigation system. *Journal of Environmental Management*, pp. 301–479.
108. Schwankl and Terry L Prichard. (1990). Clogging of buried drip irrigation system. *Journal of California Agriculture*, 44(1).
109. Shafiq, R., Yaseen, A. M., Mehdi, M. S. M., Hannan, A. (2008). Comparison of Freundlich and Langmuir adsorption equations for boron adsorption on calcareous soils. *Journal of Agriculture Research*, 46(2), 9.
110. Siswati, G., and Uday, G. (2005). Studies on adsorption of Cr (VI) onto synthetic hydrousstannic oxide. *Water SA*, 31(4), 21–25.
111. Smajstrla, A. G., Koo, R. C. J., Weldon, J. H., Harrison, D. S., and Zazueta, F. S. (1983). Clogging of trickle irrigation emitters under field conditions. *Proceedings of Florida State Horticulture Society*, 96, 13–17.
112. Stansly, P. A., Pitts, D. J. (1990). Pest damage to micro-irrigation tubing: Causes and prevention. *Proceedings of Florida State Horticulture Society*, 103, 137–139.
113. Stephen, D. E. (1985). Filtration and Water Treatment for Micro irrigation. Drip/Trickle Irrig. in Action. *The third International Drip/Trickle Irrigation Congress* ASAE publ. vol. (1).
114. Tajrishy, M., Hills, D., Tchobanoglous, G. (1994). Pretreatment of secondary effluent for drip irrigation. *Journal of Irrigation and Drainage*, 120(4), 716–731.
115. Taylor, H. D., Bastos, R. K. X., Pearson, H. W., Maro, D. D. (1995). Drip irrigation with waste water stabilization pond effluents solving the problem of emitter fouling. *Journal of Water Science Technology*, 31(12), 417–424.
116. Trooien, T. P., Lamm, F. R., Stone, L. R., Alam, M., Clark, G. A., Rogers, D. H., Clark, G. A., Schlegel, A. J. (2009). Subsurface drip irrigation using livestock wastewater. http://www.ksre.ksu.edu/sdi.
117. Water Global Researcher, (2008). www.globalresearcher.com.
118. Wu, F., Fan, Y., Li, H., Guo, Z., Li, J., Li, W. (2004). Clogging of emitter in subsurface drip irrigation system. *Transaction of China State Agriculture Engineering*, 20(1), 80–83.
119. Yan, Z. B., Mike, R., Likun, G. U., Ren, S., Peiling, Y. (2010). Biofilm structure and its influence on clogging in drip irrigation emitters distributing reclaimed wastewater. *Journal of Environmental Sciences*, 21, 834–841.
120. Yavuz, K. D., Okan, E., Erdem, B., Merve, D. (2010). Emitter clogging and effects on drip irrigation systems performances. *African Journal of Agricultural Research*, 5(7), 532–538.
121. Yingduo, Y., Gong, S., Di, X., Wang, J., Xiaopeng, M. (2010). Effects of Treflan injection on winter wheat growth and root clogging of subsurface drippers. *Agricultural Water Management*, 97(5), 723–730.
122. Yuan, Z., Waller, P. M., Choi, C. Y. (1998). Effect of organic acids on salt precipitation in drip emitters and soil. *Transactions of American Society of Agricultural Engineers*, 41(6), 1689–1696.
123. Zhang, J., Zhao, W., Tang, Y., Wei, Z., Lu, B. (2006). Numerical investigation of the clogging mechanism in labyrinth channel of the emitter. *International Journal for Numerical Methods in Engineering*, 70, 1598–1612.
124. Zhang, S. H., Jing, H., Fuxing, G., Yuh-Shan, H. (2010). Adsorption characteristics studies of phosphorous onto literate. *Desalinization and Water Treatment* 1944 994/1944–3986, *Desalination Publications*, pp. 98–105.

CHAPTER 4

MANAGEMENT OF EMITTER CLOGGING WITH MUNICIPAL WASTEWATER

A. CAPRA and B. SCICO-LONE

CONTENTS

4.1   INTRODUCTION

Water resources are becoming both quantitatively and qualitatively scarce. Irrigated agriculture is the biggest consumer of water in the world. In arid areas, crop irrigation requires up to 99% of total water use [13]. Sustainable water use in scarcity conditions involves treated wastewater reuse [11]. The agronomic and economic benefits of wastewater irrigation are obvious. Wastewater is used on crops, rangelands, forests, parks and golf courses in many parts of the world. Unrestricted irrigation, however, may expose the public to a variety of pathogens such as bacteria, viruses, protozoa, or helminthes. The World Health Organization (WHO) has established microbiological standards for wastewater use in irrigation [7]. In Italy, excessive restrictions imposed by past and present legislation represent one of the major obstacles to the use of wastewater irrigation.

The use of wastewater requires improvements in irrigation techniques to minimize the health and environment risks, mainly in situations where, as in developing countries or small communities, extremely stringent quality standards would lead to unsustainable costs. The irrigation method used, in particular, has to have specific characteristics which minimize the following risks: plant toxicity due to direct contact between leaves and water; salt accumulation in the root zone; health hazards related to aerosol spraying and direct contact with irrigators and product consumers; water body contamination due to excessive water loss by runoff and percolation. Drip irrigation minimizes contact with plants and operators, does not form aerosols, and water logging by runoff and deep percolation is negligible [4, 9, 10, 12]. The parts of plants growing above the soil should be practically devoid of pathogens when the drip system is buried in the soil (sub-irrigation) or covered by plastic sheets. Sprinkler and micro-sprinkler irrigation systems are not adequate for the opposite reasons.

Drip irrigation is, however, limited by emitter clogging, the largest maintenance problem with drip systems [5]. It is difficult to detect and expensive to clean, or replace, clogged emitters. Partial or complete clogging reduces emission uniformity and, as a consequence, decreases irrigation efficiency.

Filtering is an economical method against clogging. Chemical injection may be necessary to dissolve mineral precipitates or prevent the growth of slime. Settling basins can remove large volumes of sand and silt. Filtration is the simplest method in most situations. The main filters commonly used in irrigation are of the screen types that are simple, economical and easy to manage. Disk filters are also simple and economical. Gravel-sand media filters are particularly suitable for water with a high contents of suspended solids, but they are more complex and expensive and only suitable for farms with high technological and professional standards.

Clogging and mitigation procedures are closely related to the water quality. At present, analytical methods to forecast the clogging risk do not exist. Very little information is available for clean water. Nakayama and Bucks [8] classified the clogging risk for common drippers (discharge from 2 or 4 lph at an operating pressure of 101.2 kPa). Capra and Scicolone [1] classified the hazard for large-size drippers (discharge from 8 to 16 lph) and sprayers (micro sprinklers).

Field tests using wastewater showed that suspended solids and organic matter contents can cause emitter clogging, thus discouraging wastewater reuse in drip systems, mainly in areas where advanced wastewater treatment is not used [2, 6]. In these areas, farm water treatment systems are generally very elementary and irrigators prefer large-sized emitters, such as sprayers and sprinklers, which are less appropriate to minimize health risks.

To encourage the use of drip systems with wastewater, it is of fundamental importance for the costs of equipment and management to be comparable with those of a system using clean water. Farmers need emitters with assurance of good performance (uniformity) and filters that can guarantee a good trade-off between dripper protection from clogging and operating times between cleaning operations. In fact, it is obvious that water with high contents of total suspended solids (TSS) requires frequent cleaning operations.

A series of experimental trials were conducted by Capra and Scicolone [2, 3] on the behavior of different kinds of common filters and drip emitters using different types of municipal wastewater. The results showed that the performance of the emitters and filters depends on the quality of the wastewater. Wastewater quality influences the percentage of totally clogged emitters, the mean discharge emitted, the emission uniformity coefficient, and the operating time of the filters between cleaning operations. Different emitter and filter types showed different performance and management problems. According to the results of the tests, authors conducted detailed study that is presented in this chapter.

This chapter presents research results on the performance of emitters and the management problems due to filter cleaning. Researchers also compared the performance of a sub-soil drip system with that of an on-soil drip system; and identified the minimum number of wastewater characteristics to be detected for evaluation of clogging risk.

4.2   MATERIALS AND METHODS

4.2.1   EMITTERS, FILTERS AND WATER QUALITY

The tests included six trials conducted in Sicily (Italy) from 1999 to 2003 using six kinds of municipal wastewater and different kinds of emitters and filters. The emitters and filters tested in each following trial were chosen on the basis of the results of the previous trials. Table 4.1

TABLE 4.1   Summary of Parameters Used in Six Trials

Trial	Wastewater treatment	Filter type	Emitter type
1	Raw, gravity flow filtration, size 3 mm	JP Gravel media JP Disk 120 mesh JP Screen 120 mesh	Gadi Nike, labyrinth type, q = 4 lph Tirosh, vortex type, q = 4 lph
2	Secondary, dilution ratio[1] 1:1; sedimentation	JP Gravel media JP Disk 120 mesh JP Screen 120 mesh	Gadi Nike, labyrinth type, q = 4 lph Tirosh, vortex type, q = 4 lph
3	Secondary	JP Gravel media JP Disk 120 mesh JP Screen 120 mesh Arkal Disk 120 mesh	Gadi Nike, labyrinth type, q = 4 lph Siplast Mono 20, labyrinth type, q = 3.8 lph
4	Secondary, dilution ratio[1] 2.5:1	JP Gravel media JP Disk 120 mesh JP Screen 120 mesh Arkal Disk 120 mesh	Gadi Nike, labyrinth type, q = 4 lph Siplast Mono 20, labyrinth type, q = 3.8 lph
5	Secondary	JP Screen 120 mesh Arkal Disk 120 mesh Siplast Screen 120 mesh Amiad Screen 120 mesh	Gadi Nike, labyrinth type, q = 4 lph Siplast Mono 16, labyrinth type, q = 3.8 lph
6	Secondary	Siplast Disk 120 mesh Arkal Disk 120 mesh Arkal Disk 80 mesh Arkal Disk 40 mesh	Irritec DIN, long flow-path, q = 4.48 lph Siplast Mono 20, labyrinth type, q = 3.8 lph, surface and subsurface

[1]wastewater: clean water; q = discharge at an operating pressure of 101.2 kPa

summarizes the level of treatment of the wastewater used, and the emitters and filters tested.

The emitters under evaluation were:

- on-line labyrinth Gadi Nike 1, emitter discharge exponent $(x) = 0.6$;
- on-line vortex Tirosh, $x = 0.5$;
- in-line labyrinth Siplast Mono 20, $x = 0.5$, manufacturing coefficient of variation (Cvt) 5%, labyrinth length (L_1) 70 mm, minimum diameter of the labyrinth (D_{min}) 1.4 mm, in polyethylene pipe with external diameter $(D_e) = 20$ mm;
- in-line labyrinth Siplast Mono 16 (same characteristics as Siplast Mono 20) but in polyethylene pipe of $D_e = 16$ mm;
- on-line long flow-path Irritec DIN, $x = 0.66$, Cvt = 5 %, flow-path length 36 mm, $D_{min} = 0.46$ mm.

The filters tested were:

- vertical gravel media JP, 0.11 m², theoretical flow rate $(q_{Ft}) = 20$ m³ h⁻¹, crushed granite n. 8, actual size 1.5 mm, followed by a 120 nylon mesh screen filter to catch particles escaping during back washing;
- disk JP, 2″, 120 mesh (0.125 mm), filtration surface area $(A) = 0.068$ m²; $q_{Ft} = 18$ m³ h⁻¹;
- screen JP, 2″, 120 mesh, A= 0.066 m², $q_{Ft} = 18$ m³ h⁻¹; disk Arkal, 2″, 120 mesh, A= 0.095 m², $q_{Ft} = 25$ m³ h⁻¹;
- screen Siplast, 2″, 120 mesh, A= 0.049 m², $q_{Ft} = 20$ m³ h⁻¹;
- screen Amiad 2″, 120 mesh, A= 0.079 m², $q_{Ft} = 20$ m³ h⁻¹;
- disk Arkal, 2″, 80 mesh (0.18 mm), A= 0.095 m², $q_{Ft} = 25$ m³ h⁻¹;
- disk Arkal, 2″, 40 mesh (0.42 mm), A= 0.095 m², $q_{Ft} = 25$ m³ h⁻¹.

The characteristics of wastewater to evaluate the plugging potential are shown in Table 4.2. The values are calculated as the mean of two or three samples analyzed during each trial (at the start, in the middle and at the end of the trial). The water quality parameters, which show the greatest difference between the wastewater used in the different trials, are TSS and BOD5. According to the classification proposed by Nakayama and Bucks [8] for conventional water, clogging risk can be classified as low (Trial 2), to moderate (Trials 4, 5 and 6) to severe (Trials 1 and 3) for TSS, whereas all the wastewater can be classified in the same clogging risk class (moderate) for pH and electrical conductivity (EC).

TABLE 4.2   Mean Characteristics of the Wastewater For Six Trials

Characteristic	Trial					
	1	2	3	4	5	6
	Mean characteristic value					
BOD$_5$ (mg/L O$_2$)	200	15	53	38	25	25
Calcium (mg/L)	64	24	108	88	103	11
EC (dS/m)	1.7	1.5	1.0	1.2	1.0	0.9
Magnesium (mg/L)	81	90	12	44	108	9
pH	7.8	7.5	7.7	7.6	7.8	7.1
TSS (mg/L)	376	3	146	96	78	47

4.2.2   EXPERIMENTAL SET-UP

The experimental set-up consisted of sub-units whose number equaled the filter types tested in each trial. Each sub-unit (Fig. 4.1) included: a flow-meter, two filters of the same type, two pressure gauges (installed before and after each filter), and four polyethylene lateral lines (2 for each emitter type). Each lateral line was 50 m long and had 100 emitters connected at a spacing of 0.5 m. The external diameters of the laterals were 32 mm for on-line emitters, 20 mm (trials 3 and 4) and 16 mm (trial 5) for in-line emitters.

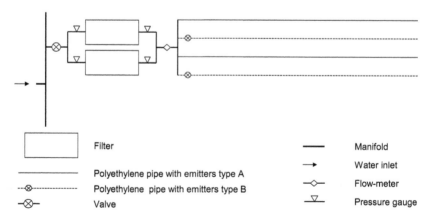

FIGURE 4.1   Schematic of the experimental sub-units.

Each filter type was used in pairs to allow it to continue operating during cleaning process (one filter was operating while the other was being cleaned). At the end of each trial, all the equipment and materials were substituted with new materials.

4.2.3   EXPERIMENTAL CONDITIONS

Because the head loss was very small, the pressure along the lateral can be considered essentially constant, thus the discharge variations are only due to Cvt and clogging. During each trial, the system was in operation for about 60 hours (4–6 hours of operation twice a week). On each day of operation and for each kind of filter, data were recorded for: the exact time of operation, the total flow volume, the number of totally clogged emitters and the number of filter cleaning operations. The filters were cleaned by back flushing, whenever the pressure drop caused by partial clogging of the filter increased by 20.24 kPa. Disk and screen filters were also manually cleaned, by pulling out the filter basket and washing it, when the pressure drop did not diminish after two minutes of back flushing. During trials 5 and 6, only manual cleaning was used.

4.2.4   DATA ANALYSIS

The performance of the emitters was evaluated based on the following parameters: the percentage of totally clogged emitters; the reduction in the mean discharge of each sub-unit with respect to the theoretical discharge (discharge of 400 new unclogged emitters at the same operating pressure as in the test); and the field emission uniformity coefficient EU [5]. The performance and management problems of the filters were discussed in terms of emitter protection from clogging; and discharge and time of operation between cleaning. The influence of the water quality on clogging was evaluated by the correlation among wastewater characteristics and emitter/filter performance indices.

4.3   RESULTS AND DISCUSSION

4.3.1   EMITTER PERFORMANCE

Table 4.3 summarizes the results for the performance of emitter indices at the end of the trials. Vortex emitters were not suitable for the use of wastewater with the exception of very good quality water (Trial 2) filtered by gravel media filters. Long-path flow emitters (Irritec DIN) assured good performance with 120 mesh disk filters only (ratio between filtration dimension and diameter of flow passage of the emitter = 1:4). When the Siplast-Mono-labyrinth-type emitter was used with

TABLE 4.3   Totally Clogged Emitters and Field Emission Uniformity Coefficients (%)

Emitter	Filter	Trials						
		1	2	3	4	5	6on	6sub
Gadi	Arkal disk 120			1 (54)	3 (76)	4 (72)		
Nike 1	JP gravel media	1	1	0 (64)	0 (69)			
	Amiad screen 120					10 (25)		
	JP screen 120	72	0	65 (0)	91 (0)	12 (34)		
	Siplast screen 120					23 (0)		
Irritec	Arkal disk 120						1 (92)	
DIN	Arkal disk 80						55 (0)	
	Arkal disk 40						30 (44)	
	Siplast disk 120						1 (63)	
Siplast	Arkal disk 120			0 (67)	0 (77)		0 (87)	0 (98)
Mono	Arkal disk 80						1 (95)	
20	Arkal disk 40						1 (98)	
	JP gravel media			1 (61)	0 (69)			
	JP screen 120			24 (54)	19(72)			
	Siplast disk 120						1 (93)	
Tirosh	JP gravel media	19	1					
vortex	JP screen 120 mesh	69	5					

[a] = field emission uniformity in brackets; 6on = emitters on soil surface; 6sub = emitters buried in soil; 120, 80, 40 = mesh.

gravel media or disk filtration, practically no totally clogged emitters were observed, and the distribution uniformity ranged from 67 to 98%. The ratio between the disk filtration dimension and the diameter of the flow passage of the labyrinth ranged from almost 1:10 (Trials 3 and 4) to almost 1:4 (Trial 6). These differences are probably due to the different composition of TSS (i.e., granulometric size). No significant difference was observed between the same kind of emitter placed on soil surface or at sub-soil. Screen filters were not capable of assuring good emitter performance for wastewater, which had not undergone previous advanced treatment.

4.3.2 FILTER PERFORMANCE

Table 4.4 shows that the reduction in mean discharge, for each type of wastewater, was maximum for a locally manufactured disk filter (JP) and for screen filters. Gravel filters and 120 mesh disk filters of good quality assured better performance. The differences between the disk filters tested show the importance of the manufacturing technology. The differences between the disk filters and the screen filters with the same dimension of filtration are due to the different mechanisms of filtration. The differences between the screen filters are not significant if analyzed in terms of discharge per filtering area unit.

TABLE 4.4   Reduction in Mean Emitted Discharge (%)

Filter	Trials					
	1	2	3	4	5	6
Amiad screen 120 mesh					45	
Arkal disk 120 mesh			31	24	43	16
Arkal disk 40 mesh						23
Arkal disk 80 mesh						28
JP disk 120 mesh	90	15	62	70		
JP gravel media	50	8	13	18		
JP screen 120 mesh	83	7	51	60	52	
Siplast disk 120 mesh						11
Siplast screen 120 mesh					65	

Table 4.5 shows the mean operating time of the filters (duration between cleaning operations). The operating time is clearly dependent on both the water quality and the type of filter. For a filtration dimension corresponding to 120 mesh, with dilute and settled wastewater only (Trial 2), all the filters needed a number of cleaning operations comparable to that of a drip system using conventional water. For wastewater with a TSS of about 50 mg/L (Trial 6), the 80 and 40 mesh disk filters also had acceptable operating times. In the other cases, the short operating times caused management problems.

4.3.3 INFLUENCE OF WASTEWATER QUALITY ON PERFORMANCE OF EMITTER AND FILTER

On the basis of the results discussed in the previous sections, the influence of wastewater characteristics on emitter and filter performance indices is evident. Table 4.6 shows the correlation coefficients (r) among water quality parameters and emitter and filter performance indices, processed for the filters tested in a minimum of four trials. The water quality parameters were best correlated with both the reduction in mean discharge. The mean duration of filter operation is generally affected by pH, TSS and BOD_5. Due to high correlation between TSS and BOD_5 (r = 0.99), pH and TSS are sufficient to define the clogging risk. The great importance of suspended solids (TSS) is confirmed by the good correlation (r = 0.71 to 0.96)

TABLE 4.5   Mean Duration of Operating Time (h) Between Cleanings

Filter	Trials					
	1	2	3	4	5	6
Amiad screen 120 mesh					0.3	
Arkal disk 120 mesh			0.6	0.4	0.7	1.0
Arkal disk 40 mesh						6.1
Arkal disk 80 mesh						4.3
JP gravel media	0.3	4.5	0.7	0.9		
JP screen 120 mesh	0.4	4.5	1.5	0.4	0.2	
Siplast disk 120 mesh						1.0
Siplast screen 120 mesh					0.2	

TABLE 4.6 Correlation Coefficients Between Water Quality Parameters and Emitter and Filter Performance Indices

Filter type	Water quality parameters					
	pH	TSS	EC	BOD$_s$	Ca	Mg
Reduction in mean discharge						
Arkal disk 120 mesh	0.657	0.068	-0.591	-0.242	0.204	-0.677
JP gravel media	0.825	0.957	0.648	0.984	0.030	0.327
JP screen 120 mesh	0.776	0.840	0.080	0.743	0.538	-0.211
Mean duration of operation						
Arkal disk 120 mesh	-0.513	-0.666	0.564	-0.666	0.797	-0.639
JP gravel media	-0.847	-0.734	0.213	-0.597	-0.811	0.537
JP screen 120 mesh	-0.980	-0.613	0.148	-0.517	-0.554	0.433

between frequency of cleaning operation and solid discharge rate (Eq. (1): product between TSS and water discharge):

$$Fc = k \, (Q \times TSS)^a \tag{1}$$

where: Fc = frequency of cleaning operation, h^{-1}; Q = wastewater discharge rate, lps; TSS, mg/L; k, a = non-linear regression coefficients for each type of filter. The value of the coefficients a, was almost equal to 1 for all the filters tested. Value of regression coefficient, k, can be determined by a simple regression analysis between Fc and solid discharge, Q x TSS. Therefore, there is a linear dependence of the frequency of the cleaning operations on the wastewater discharge.

4.4 CONCLUSIONS

The results confirm that wastewater reuse in drip irrigation can lead to management problems. In fact, the performance of the system depends on wastewater quality and emitter and filter type. Of the emitters with a similar discharge (almost 4 lph), the vortex type was more sensitive to clogging than the long-path and labyrinth types. A ratio between the disk filtration dimension and the diameter of the flow passage of the labyrinth, no higher than 1:4, seems necessary to obtain acceptable emitter performance. The optimal performance of subsoil emitters shows the feasibility of subsurface drip irrigation, which reduces health risks related to wastewater reuse.

Among the different filters tested, the gravel media filter guaranteed the best emitter performance. Disk filters, as long as they are of good quality, are cheaper and simpler to manage, and assured performance similar to that of the gravel media filter. Screen filters were not suitable for use with wastewater, with the exception of diluted and settled wastewater. With wastewater of very good quality only, all the 120 mesh filters needed a number of cleaning operations comparable to those of a drip system using conventional water. For wastewater with a maximum TSS of about 50 mg/L, the 80 and 40 mesh disk filters also had acceptable operating times. In the other cases, the shorter operating times caused management problems. The extremely shorter operating times (less than one hour in most trials) suggest the use of automatic cleaning systems. These systems can contribute towards solving the problem, and are also preferable to avoid contact between the wastewater and the irrigator, but they are expensive. The theoretical discharge of filters, suggested by the manufacturers for clean water, is not adequate for types of wastewater used here. In fact, the operating times were very short with a discharge of only 1/15–1/20 of the theoretical discharge. The frequency of cleaning operations depends on the solid discharge (product between TSS and water discharge), and can be estimated by an empirical equations.

Among the water characteristics, pH and total suspended solids (TSS) content are sufficient to define the clogging risk. The use of wastewater with suspended solids content greater than 50 mg/L, did not permit optimal emission uniformity ($\geq$85–90%).

4.5 SUMMARY

Drip systems, mainly if buried in soil, are suitable for wastewater reuse because they reduce the health risks to farmers and product consumers. Drip irrigation is, however, limited by emitter clogging. Clogging and mitigation procedures are closely related to the quality of the water used. Field tests using wastewater showed that suspended solids and organic matter content can cause emitter clogging, discouraging wastewater reuse in drip systems, mainly in areas where advanced wastewater treatment is not used.

The paper discusses the performance of emitters and filters shown in trials conducted in Sicily (Italy) using six kinds of municipal wastewater that had not undergone previous advanced treatment. Nine kinds of filters (gravel media, disk and screen) and five types of drip emitters (vortex, labyrinth and long-path) were tested.

The performance of the emitters and filters depended on the quality parameters of the wastewater: pH and total suspended solids, organic matter, calcium and magnesium content influenced the percentage of totally clogged emitters, the emission uniformity coefficient, the reduction in mean discharge, and the operating time of the filters between cleaning operations. Total suspended solids and pH were the water characteristics best correlated to emitter and filter performance.

Vortex emitters were more sensitive to clogging than labyrinth emitters. To obtain acceptable performance, it was necessary to remove any particles larger than almost one-quarter the diameter of the flow passages of the long-path and labyrinth-type emitters, using disk filters. The optimal performance of emitters buried in soil showed the feasibility of drip sub-irrigation, which reduces health risks related to wastewater reuse.

The gravel media filter guaranteed the best performance, but the disk filter, which is cheaper and simpler to manage, assured performance similar to that of the gravel media filter. Screen filters were shown to be unsuitable for use with wastewater of low quality. The short operating times of most filters cause management problems. An empirical equation is proposed to estimate the frequency of cleaning operations depending on the solid discharge (calculated as the product between TSS and water discharge).

KEYWORDS

- **backwash**
- **clay**
- **clogging**
- **clogging risk**
- **disk filter**

- distribution uniformity
- drip irrigation
- effluent quality
- flow passage
- flushing
- gravel media filter
- health risk
- labyrinth emitter
- long-path emitter
- organic matter
- sand
- sand media filter
- screen filter
- silt
- subsurface drip irrigation
- surface drip irrigation
- trickle irrigation
- TSS
- vortex emitter
- wastewater
- water quality

REFERENCES

1. Capra, A., Scicolone, B. (1998). Water quality and distribution uniformity in drip/trickle irrigation systems. *J. Agricultural Engineering Research*, 70:355–365.
2. Capra, A., Scicolone, B. (2001). Wastewater reuse by drip irrigation. *Proceedings of First International Conference on Environmental Health Risk*, Cardiff.
3. Capra, A., Scicolone, B. (2004). Emitter and filter tests for wastewater reuse by drip irrigation. *Agricultural Water Management*, 68(2), 135–149.
4. Capra, A., Li Destri Nicosia, O., Tamburino, V. (1984). Experience subirrigation using wastewater (French). *Proc. of 10th Int. Congress of Agricultural Engineering, Budapest*, September, 1.3(6), 18–28.

5. Keller, J., Bliesner, R. D. (1990). *Sprinkler and Trickle Irrigation.* AVI Book, New-York.
6. Li Destri Nicosia, O., Capra, A. (1989). Emitter and irrigation system performance with water with high suspended organic matter content (French). *J. Irrigation and Drainage,* 4, 238–243.
7. Mara, D., Cairncross, S. (1989). *Guidelines For the Safe Use of Wastewater and Excreta in Agriculture and Aquaculture.* World Health Organization, Geneva.
8. Nakayama, F. S., Bucks, D. A. (1991). *Water Quality in Drip/Trickle Irrigation: A Review. Irrigation Science,* 12(4), 187–192.
9. Oron, G., DeMalach, J., Hoffman, Z., Manor, Y. (1996). Effect of effluent quality and application method on agricultural productivity and environmental control. *Water Sci. Tech.,* 26(7–8), 1593–1601.
10. Papayannopoulou, A., Parissopoulos, G., Panoras, A., Kampeli, S., Papadopoulos, F., Papadopoulos, A., Ilias, A. (1998). Emitter performance in conditions of treated municipal wastewater. *Proc. of the 2nd Int. Conf. AWT Advanced Wastewater Treatment, Recycling and Reuse,* Milan, pp. 1011–1014.
11. Pereira, L. S., Oweis, T., Zairi, A. (2002). Irrigation management under water scarcity. *Agricultural Water Management,* 57:175–206.
12. Pescod, M. B. (1992). *Wastewater Treatment and Use in Agriculture.* FAO Irrigation and Drainage Paper 47.
13. Ragab, R. (2001). Policies and strategies on water resource in the European Mediterranean region. *Proc. of the Water and Irrigation Development International Conference,* Editoriale Sometti, Mantova, Italy, pp. 37–55.

CHAPTER 5

MANAGEMENT OF MICRO IRRIGATION USING WASTEWATER TO REDUCE EMITTER CLOGGING: A REVIEW

ANTONINA CAPRA

CONTENTS

5.1 INTRODUCTION

Irrigation is the most important factor for increasing crop production in water scarcity regions. In many areas, agriculture is impossible without irrigation. In these areas, agriculture sector is the biggest consumer of water and consumes 50–85% of total water use [19, 36]. At the same time, water resources are becoming both quantitatively and qualitatively scarce.

Due to well-known agronomic, environmental and economic benefits, sustainable water use in scarcity conditions involves wastewater (WW): mainly urban, reuse [30].

Wastewater is utilized to irrigate crops, rangelands, forests, parks and golf courses in many parts of the world [37]. Unrestricted irrigation may expose the public to a variety of pathogens such as bacteria, viruses, protozoa, or helminthes. Disease transmission occurs through contact with wastewater by farmers, aerosol spraying, and the consumption of products irrigated with effluent and contaminated surface or groundwater. Therefore, the use of WW requires improvements in irrigation techniques to minimize the health and environmental risks in situations, in developing countries or small communities for example, where extremely stringent quality standards would lead to unsustainable costs. The factors influencing the transmission of disease by irrigation are: the degree of WW treatment, the crop type (e.g., human consumption or not, consumption after cooking or not, animal consumption fresh or sun-dried, etc.) and the harvesting practices used, the degree of contact with WW, and the irrigation method.

The irrigation method used with wastewater should minimize risk of plant toxicity due to direct contact between leaves and water; salt accumulation in the root zone; health hazards related to aerosol spraying and direct contact with irrigators and product consumers; water body contamination due to excessive water loss by runoff and deep percolation [30]. Drip irrigation (DI) is suitable for WW reuse as highlighted by different researchers [9–11, 29, 35, 47]. Microbial contamination of the products of plants growing above the soil should be reduced when the drip system is buried in the soil (SDI) or covered by plastic sheets. Sprinkler and micro-sprinkler systems are less adequate for the opposite reasons [30]. The advantages of DI and SDI systems [9] are: minimum contact with plants and operators, do not form aerosols, restrict and control water logging by runoff and deep percolation, the negligible percolation minimizes nitrates migration, and the fact that the soil surface remains dry and minimize weed growth and the pollution hazard due to the use of herbicides. Oron et al. [29] verified that DI technology, primarily SDI, meets environmental contamination control standards.

Micro irrigation has limitations of emitter clogging. It is a challenging management problem with drip systems [23]. Partial or complete clogging causes poor distribution of the water along the laterals. Plants can be severely damaged when emitters are clogged for long time. Emission uniformity and irrigation efficiency decrease in many cases. To assure all irrigated plants receive their water needs, it is necessary to reduce water losses due to over irrigation. It causes deep percolation and consequent disadvantages due to water and energy costs, fertilizer leaching, drainage needs and groundwater contamination. Clogging adversely affect the longevity of drip system. It depends on the emitter characteristics, the original water quality, the level of WW treatment, the degree of dilution and the distance from the polluting source in the case of indirect reuse, and the high temporal variability (time of day, season, etc.). It is difficult and expensive to detect and to clean or replace clogged drippers. At farm level, filtering is the main defense against emitter clogging caused by mineral and organic particles. Chemical treatment is necessary to dissolve mineral precipitate and to control the growth of bacterial slimes.

WW treatment technologies are to achieve desired quality of water. To control emitter clogging caused by bacteria and algae tertiary treatment and chlorination is required. Farmers are not trained on the knowledge to maintain trickle irrigation systems [48] and filtering equipment's are generally very elementary. Under these conditions, suspended solids and organic matter content of WW also cause emitter clogging [9, 14, 24, 42]. As a consequence, irrigators are discouraging the use of WW with drip systems and prefer large-sized emitters such as sprayers and sprinklers. Therefore, a good trade-off between farm water treatment costs and micro irrigation system performance has to be investigated. Filters do not prevent clogging completely and low-quality water clog filters consequently system management failures.

This chapter discusses management problems of micro irrigation using wastewater that should be considered in the design and management of a micro irrigation system to reduce the risk of failure due to emitter and filter clogging.

5.2 WATER QUALITY AND EMITTER CLOGGING

5.2.1 WATER CHARACTERISTICS AND CLASSIFICATION OF CLOGGING RISKS

Clogging occurs as a result of physical, biological and chemical contaminants [1, 6, 20, 26]. Physical clogging is caused by suspended inorganic particles (such as sand, silt, clay, plastics), organic materials (animal residues, snails, seeds and other parts of plants, etc.), and microbiological debris (algae, bacteria, etc.). Physical materials can be combined with bacterial slimes. Inorganic particles deposit near the end of laterals and/or along the wall of labyrinth type emitter, where flow is slow. Inorganic particles are classified according to size by passing them through a standard series of screens. Table 5.1 depict the classification of soil particles by size and the corresponding screen mesh number (the number of openings per inch of standard screen).

Chemical problems are due to dissolved solids when they interact with each other to form precipitates of calcium carbonate under calcium and bicarbonates rich water. Precipitates are formed inside the pipe; and emitters cause clogging due to change in pH and temperature. Mineral deposits outside the emitter outlets due to evaporation are also a problem. Some of dissolved solids (iron, sulphur, manganese) are a source of nutrients for algae and bacterial slimes.

Biological clogging is due to algae, iron, sulphur and manganese slimes. Algae are a family of plants commonly found in almost all surface water. Most algae need sunlight to grow; therefore black emitters and pipe are used above ground. Slime is used as a generic term for filaments of microorganisms mainly produced by bacteria. The slime creates a gelatinous substance (biofilms) from the particulate matter that can clog the emitters. Use of WW showed the clogging composition included biological species and clogging process was usually initiated by bacterial biofilms [14, 45]. Some of the bacterial species also cause emitter clogging through the precipitation of iron, manganese and sulphur dissolved in the water [14, 41]. Precipitated iron forms the red crust inside of tube and clogs the drippers. Some filamentous bacteria, such as *Gallionella*, *Crenothrix* and *Leptothrix*, oxidize Fe^{+2} into Fe^{+3}, which can precipitate and cause clogging.

Water analysis prior to system design is a preventive maintenance program and field evaluation of clogging and uniformity are strongly recommended [1, 4, 5, 22, 23]. Clogging also depends on local conditions, such as water temperature and emitter features. No water quality evaluation method is capable of representing potential emitter clogging. Little information on clogging risk classification is available even with fresh water. Nakayama and Bucks [26] and Boswell [4] classified the potential clogging risk (Table 5.2) for common drippers (discharge 2–4 lph at an

TABLE 5.1 Approximate Particle Size of Soils and Corresponding Standard Screen Mesh Number and Recommended Minimum Diameter of Emitter Flow Passage

Soil classification		Very coarse to coarse sand	Coarse to medium sand	Fine sand				Very fine sand	
Standard pore [1] filtration size	(mm)	0.711	0.42	0.18	0.152	0.125	0.105	0.089	0.074
	(mesh)	20	40	80	100	120	150	180	200
Minimum emitter flow passage	(mm)	5	3	1.2	1	0.9	0.7	0.6	0.5

[1] almost 1/7 of the minimum dimension of the emitter flow passage.

TABLE 5.2 Potential Clogging of Irrigation Water For Common Drippers[1] [26]

Clogging factor	Risk level		
	minor	moderate	severe
Bacteria populations, mL^{-1}	<10,000	10,000–50,000	>50,000
Calcium carbonate[3], $mg\ L^{-1}$	<150	150–300	>300
Calcium[3], $mg\ L^{-1}$	<40	40–80	>80
Electrical conductivity, $dS\ m^{-1}$	< 0.8	0.8–3.1	>3.1)
Hydrogen sulfide, $mg\ L^{-1}$	<0.2 (<0.1[2])	0.2–2.0	>2.0
Iron, $mg\ L^{-1}$	<0.2 (<0.1[2])	0.2–1.5	>1.5
Manganese, $mg\ L^{-1}$	<0.1 (<0.2[2])	0.1–1.5	>1.5
pH	<7	7–8	>8
Total dissolved solids, $mg\ L^{-1}$	<500	500–2000	>2000
Total suspended solids, $mg\ L^{-1}$	<50	50–100	>100

[1] discharge = 4 lph at an operating pressure of 100 kPa.

[2] according to Ref. [4];

[3] according to Ref. [28].

operating pressure of 101.2 kPa). Capra and Scicolone [8] classified the hazard (Table 5.3) for large-size drippers (discharge from 8 to 16 lph) and sprayers. These authors took into consideration the water quality parameters, such as: total suspended solids (TSS), total dissolved solids (TDS or EC), calcium (Ca), hydrogen sulphide (H_2S), magnesium (Mg), manganese (Mn), pH, total iron (Fe), and number of bacteria. All researchers reported clogging hazard in three classes: minor, moderate and severe.

5.2.2 EXPERIENCE WITH FRESH WATER

The results of several laboratory and field studies in different irrigated areas using fresh water showed the main characteristics of the water responsible of the emitter clogging. In an investigation performed in the field on a number of irrigation systems in Southern Italy, Capra and Scicolone [8] found that suspended solids, Fe, S, Ca and Mn are responsible of clogging.

TABLE 5.3   Potential Clogging of Irrigation Water For Large Size Drippers[1] and Sprayers [8]

Clogging factor	Risk level		
	minor	moderate	severe
Calcium, mg L^{-1}	<250	250–450	>450
Electrical conductivity, dS m^{-1}	< 1.0	1.0- 4.5	>4.5
Iron, mg L^{-1}	<0.5	0.5–1.2	>1.2
Magnesium, mg L^{-1}	<25	25–90	>90
Manganese, mg L^{-1}	<0.7	0.7–1.0	>1.0
Total suspended solids, mg L^{-1}	< 200	200–400	>400

[1] discharge >8 lph at an operating pressure of 100 kPa

The water quality parameter phosphorus creates greater problem. Although there is no information about phosphorus restriction in irrigation water for micro irrigation system. When this element is present in springs at concentrations of more than 0.01 mg L^{-1}, it determines a mass proliferation of algae, which in turn causes serious problems for water use. The water source, which contributed to algae proliferation was from a lake, showed phosphorus concentration of 1 mg L^{-1}, practically 100 times greater than the acceptable limit. Yavuz et al. [47] concluded that precipitation of the Ca and Mg ions existing in water (32 to 59 mg L^{-1} of Ca and 7 to 16 mg L^{-1} of Mg) played a vital role in the dripper clogging due to the influence of high temperature and pH (7.2 to 8.1). Results of a field study involving trickle irrigation systems in Iran [48] showed high percentage of totally clogged emitters and low emission uniformity mainly due to clogging problems. The tested irrigation waters, based on properties (pH, TDS, TSS, Fe, Mn), were classified, in general, as minor hazardous to severe hazardous. Some cases bicarbonate concentrations of more than 305 mg L^{-1} were observed. Large formations of biological biofilms were also observed in one region.

5.2.3 EXPERIENCE WITH WASTEWATER

Wastewater used in irrigation is mainly urban WW. It is 99% water, which however contains chemical, physical, and microbiological contaminants from domestic effluent, water from commercial establishments and institutions, industrial effluent and storm water and other urban runoff [37]. It is expected that the characteristics relevant to the clogging risk are, in some way, the same described for fresh water. However, the WW quality is extremely complex, and there is are series of physical, chemical, and biological reactions between suspended solids and other materials contained such as microorganisms, which complicates the characteristics of suspended particulates in WW [24]. Ravina et al. [39] reported emitter and filter clogging problems mainly due to mucous products of microbial activity using treaded effluents stored in two surface reservoirs (see Table 5.6 for effluent characteristics). Mucous biomass was formed mainly by colonial protozoa and, to a lesser extent, *bryozoa* and *sulphur bacteria beggiatoa* for effluents containing sulphur.

The experimental trials [9–11] were conducted to evaluate the behavior of several kinds of drip emitters using municipal WW of different quality not submitted to advanced treatment. The results showed the influence of WW quality on the clogging. Organic suspended solids had an influence on clogging so high to mask that of the other water contaminants. In fact, of the water characteristics in Tables 5.2 and 5.3, TSS and organic matter content, expressed by Biochemical Oxygen Demand at 5 days (BOD_5), were sufficient to classify clogging risk. The authors highlighted that the use of WW with a TSS>80 mg L^{-1} and a BOD_5>25 mg L^{-1} did not permit optimal emission uniformity. They proposed to classify the clogging risks due to organic matter content according to BOD_5: minor when BOD_5 is <15 mg L^{-1}, moderate between 15 and 40 mg L^{-1}, and severe >40 mg L^{-1}.

Depositions of brown color on drip line walls were observed by Duran-Ros et al. [15] during field tests with two kind of treated WW (See (a) and (b) in Table 5.6). These depositions were composed of biofilm, calcium, aluminum silicate, manganese, sand and algae. External views of integrated emitters at the end of the trial with effluent (a) showed an organic film, which could clog emitters, over the drip line and emitter surface. Algae (among them different diatomea genera such as *Nitzchia, Navicula, Hantzschia*), bacteria (among them spirochete), ciliated protozoa (*thecamoebian* and from the genus *Vorticella*), organic and inorganic particle and vegetal remains were detected in the solid deposit at the end of lateral with both the effluents.

Liu and Huang [25] carried out a laboratory experiment to study the emitter performance of commonly used emitter types with the application of freshwater (FW) and treated WW. Both the FW and WW were practically free of suspended solids (TSS not detectable and equal to 0.02 mg L^{-1}, for FW and WW, respectively), but they were at moderate or high risk for pH (equal to 7.5 and 8.6 for FW and WW, respectively). WW was also at moderate risk for TDS and Mg, which content was equal to 845 and 34 mg L^{-1}, respectively. Results showed that all performance indices were affected by water quality and clogging was higher for WW. Analyses of the precipitation components inside and at the outlet of emitters revealed that chemical precipitation was the main reason for emitter clogging due to high pH and ions concentration, especially in the WW. The main components of the precipitated materials were carbonate and sulphate.

Cirelli et al. [13], compared the results of irrigation tests on drip system lateral performance using FW and tertiary treated WW of good quality (TSS lesser than almost 50 mg L^{-1} and BOD$_5$< 14 mg L^{-1}). They showed worst emission uniformity and discharge reductions (due to emitter clogging) for FW due to high calcium carbonate content (95 to 286 mg L^{-1}). The authors found that clogging was more pronounced on the surface laterals than in the subsurface pipes due to higher temperature exposure. Li et al. [24] showed that the biofilms grew and detached in irrigation system and deposited continuously at the inlet and outlet of emitter labyrinth path was the major reason for the emitter clogging in two WW treated with fluidized-bed reactor (FBR) and biological aerated filter (BAF) processes, respectively. Differences were not significant between WW characteristics treated by FBR and BAF methods, except for TSS, equal to 78 and 24 mg L^{-1}, respectively. The Mn content was 10 times in FBR treated WW (0.11 mg L^{-1}) compared to BAF method (0.1 mg L^{-1}). Using the scanning electron microscope technology, they showed that the suspended particulates were flocculent and porous; and the pore system mainly consisted of solid suspended particulates (mainly clay), and most areas between the particulates were filled with microbes (such as *cocci*, *bacillus*, and *sheathed bacterium*) and extracellular polymers. The degree of clogging was more obvious for FBR treatment compared to BAF treatment.

5.3   PREVENTION OF CLOGGING

5.3.1   FILTRATION

The main defense against clogging consists of: filtering and keeping contaminants entry to the emitters, periodic or continuous chemical treatments and system flushing [23, 26]. Filtration is mandatory for all micro irrigation systems, but the careful selection of filters must be considered for WW reuse [39]. Usually, the main bank of filters, that may serve more than one unit, is located at the pumping plant. The primary filter can be gravel media, screen or disk filter. In addition, small screen or disk filters should be installed at the head of unit manifold as a safety precaution, to stop any debris due to the filter cleaning or the pipeline repairs. Filtration of water entering before the pump is called pre-filtration. It is recommended to protect the pump and other equipment such as counters, pressure gages, etc. Capra and Scicolone [9] and Capra et al. [12] showed considerably improved system performance after installing a purpose-built wire netting cage with a large filtering surface. Taylor et al. [45] used a 50 mm filter grit, equipped with flushing system, flow meters and frequently cleaned 250 mm cartridge filters.

Good maintenance requires periodical cleaning of filters. Filters can be cleaned automatically (mainly by back flushing) or manually. Where frequent back flushing is required, automatic self-cleaning system is recommended. Researchers have recommended cleaning of filters under 35–70 kPa head loss [7, 23, 39]. Puig-Bargués et al. [32] developed several equations for calculating the head loss in disk, screen and sand filters when using effluents. Using WW high filtration levels are recommended, but filters clog frequently. Consequently, the main problem of micro-irrigation system operation is with clogged filters [11, 38, 39], which involves both failure of the irrigation system and the continuous presence in the field of an irrigator (for manual filters), or the use of automatic cleaning filters which, in turn, contribute towards an increase in equipment costs.

Furthermore, frequent cleaning of the filters involves contact between the irrigator and WW and the production of large amounts of dirty water to eliminate. To encourage the use of micro irrigation systems with WW, it is important to consider the costs of equipment and management to be comparable with those of a system using clean water. Farmers need filters that guarantee a good trade-off between system performance (uniformity), cost and filter cleaning interval [9–11].

5.3.1.1 Screen Filters

The most common filter is screen filter, which is economical compared to other types of filters. It is easy to clean and maintain. The screens may be made of steel, plastic or synthetic material and kept in pressurized chamber. The method of operation varies with the manufacturer design and relates to the way of water entry, circulation, and exit. Screen filters are adequate for removing mineral suspended particles from water. They are not efficient at removing organic materials such as algae, bacteria, slime. These non-solid materials tend to embed themselves into the screen material and they are difficult to remove. In other cases, they simply slide through the holes in the screen by temporarily deforming their shape [22]. In WW, organic particles acting as a bridge, cause the formation of biofilms, which adhere to the screen clogging it. A screen filter may be used as a primary filter, or it may be secondary to a vortex sand separator, or to a media filters, which removed organic matter. In the latter case, it prevents washed out media from entering the irrigation system [17]. Table 5.1 shows standard classification of the screens used in micro-irrigation systems. It should be noted that the smallest hole size (200 mesh screen) does not filter the smallest portion of the very fine sand, but the only part larger than 0.074 mm. Due to the aggregation of fine particles, it is recommended to remove all particles larger than 1/7–1/10 the small diameter of the orifice or flow passage of the emitter by water [4, 23].

Screen filters can be cleaned manually by removing the screen and washing it with clean water. However, dismantling is not very practical for use with waters containing high levels of contaminants, since it must be done often. Self-cleaning greatly simplify the cleaning process. Filters are cleaned with the water diverted through the screen by opening a gate valve [17, 18]. This operation does not require disassembling. Several methods of flushing are common. The most effective is the backwash method, but these filters are typically more expensive. In this method, the water is forced backwards through the screen for a very effective cleaning [22]. Periodically manual cleaning to remove particles not removed by self-flushing should be necessary depending on the flush method used.

5.3.1.2   Disk Filters

Disk filters also are simple and economical. They are relatively new types of filters, which can replace screen filters in trickle irrigation systems. Disk filters are a cross between a screen filter and a media filter, with many of the advantages of both. Disk filters are good at removing both particulates, like sand, and organic matter [22]. The filtration element in disk filter consists of a number of plastic or plastic-coated metal disks that are placed side-by-side on a telescopic circular shaft inside the housing. When these disks are stacked tightly together, they form a cylindrical filtering body, which resembles a deep tubular screen [27]. The face of each disk is covered with various sized small bumps. Because of the bumps, the disks have tiny spaces between them when stacked together. Water flows through the disks from the outside inwards along the radii of the disks. Particles suspended in the water are trapped in the grooves of the disks, and clean water is collected in the center of the disks. The organics are snagged by the sharp points on the bumps [17].

These filters have more surface area than screen filters of similar sizes and like screen filters must be cleaned periodically [7]. Disk elements can be manually or automatically cleaned. During manual cleaning, the housing is removed, the telescopic shaft is expanded, and the compressed disks separated for easy cleaning. They are normally cleaned by rinsing with a water hose. Automatic back flushing is triggered by the preset pressure differential. This pressure differential opens the exhaust valve and water flows backwards through the disks, removing trapped particles from the grooves [17].

5.3.1.3   Media Filters

Gravel-sand media filters are particularly suitable for water with high inorganic and organic suspended solids. Media filters are best for removing organic material from the water, but they are more complex and expensive and thus only suitable for farms with high technological and professional standards. Media filters usually consist on sand or gravel of selected sizes placed in a pressurized tank. Nevertheless, many materials in different forms are used as filter media. Pressure-type, high-flow sand or mix-bed media filters are more popular ones to filter irrigation water for micro irrigation systems. They are almost always in groups of two or more. The water passes through the small spaces between the media grains and the debris are entrapped when it can't fit through these spaces. Due to their three dimensional nature, the sharp edges media increase the ability to entrap large amounts of pollutants. Sand media filters have additional advantages like the capability of removing microbial contaminants and heavy metals. Battilani et al. [3] reported that gravel media filter can remove up to 60% of *Escherichia coli*, 41% of arsenic, 36% of cadmium and lead, 48% of chromium and 46% of copper. Aronino et al. [2] showed gravel filter was able to remove viruses in the size range of 200 nm.

Uniformity coefficient and mean effective size are the two factors describing the media used in the filter [18]. Uniformity coefficient reflects the range of sand sizes. It is the ratio between the size of the screen opening, which will pass 60% of the filter sand to the screen opening, which will pass 10% of the same sand. The uniform size of filtering media assures better control of the filtration, since only particles large enough to clog the emitters should be retained by the filter. Instead of, grading in size from fine to coarse causes premature clogging of the filter, since even small particles, which can be tolerated by the emitters are retained by the non-uniform pores of the media [18]. For irrigation purposes a uniformity coefficient of 1.5 is considered adequate. The mean effective sand size is the size of the screen opening, which will pass 10% of the sand sample. It is an indicator of the particle size removed by the media. Table 5.4 shows some examples of standard sand media.

TABLE 5.4 Materials and Sizes of Media

Parameter	Crushed granite		Crushed silica		
Designation number	8	11	16	20	30
Mean effective size (mm)	1.5	0.78	0.66	0.46	0.34
Filtration quality (mesh)	100–140	140–200	140–200	200–230	230–400

Media filters are cleaned by back flushing that is achieved by reversing the direction of water flow in the tank. Clean water is usually supplied from the second tank. The upward flow fluidizes the media and flushes out the contaminants. The back-flush water is discharged and does not enter the irrigation system. In order to provide sufficient cleaning without accidental removal of the media, the backwash flow must be carefully adjusted according to recommended flow rates for a given media [18].

5.3.1.4   Selection of Filters

There is no specific answer to the question "which filter to use?" even for clean water. The choice between filter types depends on water quality (origin, quality and quantity of contaminants, type of treatment of WW), the emitter type, the size of the irrigation system and the budget; availability of the filters and parts must also be considered. It is for this reason that several researchers have studied micro irrigation system performance using different effluents, filters and emitters [9–11, 14, 15, 24, 25, 32–34, 38, 39]. In general, when WW is used, a combination of more than one type of filter is needed. Screen filters are generally the least expensive. For a smaller size system, a disk filter should be a good trade-off because it would remove both sand and organics, but it might need to frequently clean it. Disk filters have larger filtration capacity and a bigger dirt-holding capacity than screen filters, and are less expensive of media filters for low flow rates [7, 27].

5.3.2   *CHEMICAL TREATMENTS*

Filters may not be sufficient in some cases, where water contains dissolved minerals or organic matter. In such cases chemical treatment is required to prevent biological growth and precipitation reactions, or dissolve existing deposits in the irrigation system and, in particular, on the inside surfaces of emitters. Acid injection reduces or eliminate mineral precipitation, can remove deposits, and create an environment unsuitable for microbial growth [16]. Chlorination is an effective measure against microbial activity. Preventing water treatment can be continuous or intermittent, and organic or inorganic chemicals can be used [40]. Table 5.5 summarizes problems and possible solutions for micro irrigation system maintenance.

TABLE 5.5   General Summary of Micro-Irrigation Problems and Possible Solutions [7]

Problem	Trouble shooting
Small slimy bacteria	1. Continuously apply chlorine at low dosage. Free chlorine at the far end of the system should measure 1 ppm.
	2. Superchlorinate to 200–500 ppm free chlorine at the far end of the system: thoroughly flush system, chlorine, allow to sit overnight, flush system the next day.
Iron and manganese bacteria	1. Pump well water into a reservoir and aerate it before pumping it to the irrigation system. The oxidation cause precipitation of iron and manganese.
	2. Inject a long chain linear polyphosphate or polymaleic acid into the irrigation water to sequester iron and manganese, keeping them suspended while they move through the irrigation system.
	3. Inject chlorine gas (1.4 parts chlorine for each 1.0 part iron in the water) prior to a fine media or disk filter to precipitate the iron from the water and trap it in the filter. This technique does not work for manganese.
Calcium and magnesium carbonate precipitation	1. Neutralize the carbonates by injecting phosphoric or sulfuric acid into the irrigation water such that the pH decreases below 6.5.
	2. Inject a very long chain linear polyphosphate into the irrigation water at a rate of 1 to 2 ppm to sequester calcium and magnesium, keeping them suspended while they move through the irrigation system.

5.3.2.1 Acidification

Acidification helps to prevent emitter plugging by lowering the pH. The pH-lowering can prevent precipitation of solid compounds, particularly calcium carbonate, and can enhance the effectiveness of chlorine. Sulfuric, hydrochloric, and phosphoric acids are used for this purpose. Phosphoric acid should not be used if irrigation water contains more than 50 ppm of calcium because calcium phosphate will likely precipitate [16, 40]. Acid can be injected in the same way as fertilizer; however, extreme caution is required. Always acid must be added to water; do not add water to acid. Adding water to acid can cause a violent reaction, and the acid can splash on the person treating the water. Irrigators working with acids should wear protective clothing and eyewear, and safety devices should be provided, including a shower and eyewash [40]. Check and vacuum relief valves (anti-siphon devices) are necessary safety devices. They prevent mixtures of water and chemicals from draining or siphoning back into the water source. Both valves must be located between the pump and the point where chemicals are injected into the pipeline [21].

Neutralization of 80% of the bases (carbonates and bicarbonates) is suggested to eliminate carbonate precipitation. The amount of acid to inject is determined by a titration curve, which is unique for each water source and type of acid. Acid can be injected on a continuous basis to prevent calcium and magnesium precipitates. The injection rate should be adjusted until the pH of the irrigation water is 6.5 (just below 7.0). Periodic injection of a greater volume of acid is a control method to remove deposits as they are formed by. Acid should be injected continuously for 45 to 60 minutes in quantity enough to reduce the water pH to 4.0 [21]. The release of water into the soil should be minimized during this process since plant root damage is possible. The acid injected into the irrigation system should remain in the system for several hours, after which the system should be flushed with irrigation water. Concentrations that may be harmful to emitters and other system components should be avoided.

5.3.2.2   Chlorination

Chlorination is a common method to prevent biological emitter plugging. Chlorine injected into irrigation water kills microorganisms like algae and bacteria. These organisms are most commonly found in surface water and in WW. Since bacteria can grow within filters, chlorine injection should occur prior to filtration. The easiest form of chlorine to handle is liquid sodium hypochlorite (laundry bleach) available at several chlorine concentrations. Powdered calcium hypochlorite is not recommended for injection into micro irrigation systems, since it can produce precipitates that can plug emitters, especially at high pH levels [16]. Chlorine gas is the most effective and economical source (if legal in the local area), but its use is limited by the high toxicity [40].

Several possible chlorine injection schemes [16, 26] can be followed: a. inject continuously at a low level to obtain 1 to 2 ppm of free chlorine at the ends of the laterals; b. inject at intervals (once at the end of each irrigation cycle) at concentrations of 20 ppm and for a duration long enough to reach the last emitter in the system; c. inject a slug treatment in high concentrations (50 ppm) weekly at the end of an irrigation cycle and for a duration sufficient to distribute the chlorine through the entire piping system. At these concentrations, chlorine kills microbes and oxidizes iron. Super-chlorination with up to 500 ppm is used to reclaim a system that has bacterially-plugged emitters. At higher concentration (100–500 ppm), chlorine can oxidize organic matter, and can be used to disintegrate organic materials that have accumulated in emitters. Super-chlorination may damage sensitive plants and irrigation system components. Emitter parts may be made of silicon or other materials that chlorine will degrade [40].

The choice of the method depends on the growth potential of microbial organisms, the injection method and equipment, and the scheduling of injection of other chemicals. Similarly to acidification, precautions must be used when injecting chlorine [28]: acid and chlorine must be injected at two different injection points; mixing acid and liquid chlorine together will produce highly toxic chlorine gas; always chlorine have to be added to water, not vice versa; chlorine should be injected upstream of the filter.

The amount of liquid sodium hypochlorite required for injection depends on the desired chlorine concentration in the irrigation water, the system flow rate and the percentage chlorine in the product used [16, 26]. The chlorine activity depends on the pH of the water; the lower the water pH, the more the chlorine exists as $HOCl$. It is about 60 times more powerful as a biocide than OCl^-. For a more economical chlorine treatment, acidification of alkaline water is suggested so that $HOCl$ predominates.

5.3.3 ALTERNATE MAINTENANCE OPERATIONS

Micro irrigation systems require significant maintenance to run at maximum possible efficiency. Evaluating and monitoring system performance, cleaning or replacing plugged emitters, and periodic line flushing are the main practices, apart from chemical analysis of irrigation water source, the filter backwashing and cleaning and the chemical treatments described in this chapter.

5.3.3.1 Evaluating and monitoring system performance

Partial or complete emitter clogging causes poor distribution of the water along the laterals and can severely reduce water application uniformity. Solomon [44] and Capra and Scicolone [8] studied the effects of various factors known to influence emission uniformity. They found that the most important issue was plugging. Evaluations of application uniformity should be done at least once in each irrigation season to determine the effects of emitter plugging or changes in other components on system performance. More frequent evaluations may be required to diagnose and treat emitter plugging problems.

The most common coefficient used to estimate application uniformity is the emission uniformity coefficient (EU) based on emitter flow measurements [26]:

$$EU = 100 \left(\frac{q_{1/4\min}}{q_m} \right) \qquad (1)$$

where: $q_{1/4\min}$ = mean of the low quarter of the emitter flow rates, lph; and q_m = mean of all flow rates, lph. Using this method, two contiguous drippers are selected at four locations in each emitter line (at the beginning, at 1/3 of the length, at 2/3 of the length and at the end of the emitter line). In addition, the average discharge of all the emitters, q_m, should be considered to evaluate overall system performance.

The coefficient EU is simple and easy to determine, but it is only able to evaluate global uniformity. In other words, it does not permit to separate the effects on emitter flow rate of lack of uniformity due to clogging and cleaning actions, pressure and temperature variability, manufacturers' specifications. It is only useful when EU of the same irrigation system with new and unclogged emitters is known.

Bralts [5] evaluated the uniformity distribution by two coefficients. The statistical uniformity coefficient, U_s, based on emitter flow measurements, which allows to evaluate the global uniformity and the coefficient of variation due to emitter performance in the field, V_{pf} based on pressure measurements, which allows the effects of pressure differences on emitter flow rate to be separated from those due to other causes. Statistical uniformity coefficient, U_s, is determined by the Eq. (2):

$$U_s = 100(1-V_q) = 100\left(1 - \frac{SD_q}{q_{af}}\right)$$
(2)

where: V_q is the field flow rate variation; SD_q the standard deviation of field emitter flow rate, lph; and q_{af} the average emitter flow rate in the field, lph.

The coefficient of variation due to emitter performance in the field, V_{pf} can be estimated by the EQ. (3):

$$V_{pf} = 100(V_q^2 - x^2 V_h^2)^{1/2} = 100\left[V_q^2 - x^2\left(\frac{SD_h}{h_{af}}\right)^2\right]^{1/2}$$
(3)

where: x is the emitter discharge exponent, V_h the coefficient of variation of the field pressure head; SD_h is the standard deviation of pressure measured in the field; and h_{af} is the mean pressure in the field. V_{pf} includes manufacturer's variation, V_m, variation due temperature and emitter plugging. If variation due temperature is negligible (when, $x \cong 0.5$), a comparison between V_m and V_{pf} provides an indication of the extent of clogging.

The three separate tests are useful to evaluate the performance of a micro irrigation system: (a) overall water application uniformity (e.g., EU or U_s); (b) hydraulic uniformity or pressure variation (e.g., SD_h/h_{af}), and (c) emitter performance variation, V_m. If the overall water application uniformity is high, there is no need to perform further tests [43]. If the water application uniformity is low, then hydraulic uniformity tests should be conducted in order to determine the cause of the low uniformity. The hydraulic uniformity test will indicate whether the cause of low water application uniformity is excessive pressure variation in the system or emitter performance problems such as plugging.

A graduated cylinder and a stop watch can be used to measure the volume collected for a given time to calculate emitter flow rates. To accurately determine uniformity, measurements should be made at a minimum of 18 points located throughout each irrigated unit. Care should be taken to distribute the measurement points throughout the irrigation system. Some points should be located near the inlet, some near the center, and some at the distant end. The specific emitters to be tested should be randomly selected at each location, without visually inspect the emitters to select those with certain flow characteristics before making measurements [43].

5.3.3.2   Cleaning or Replacing Plugged Emitters

When emitters become severely plugged, it is necessary to replace them with new ones. The material plugging the emitters should be identified and the irrigation water re-analyzed to identify the plugging source before new emitters are reinstalled.

If the plugged material is calcium carbonate, the chance of reclamation without removing emitters from the field is good. If the primary cation is iron, removal and cleaning or replacement with new emitters is the only solution because iron compounds are very difficult to clean in the field. Reclaiming plugged emitters by chemical treatment is not always success-ful because most of the injected chemical flows through the open emitters and not through the plugged ones. Iron-fouled emitters can be cleaned by soaking in a strong (0.5–1.0%) citric acid solution for 24 to 48 hours [40].

Cleaning plugged emitters by scraping or reaming with a small wire dis-tort the emitter orifice and can introduce another source of non-unifo mity in irrigation water application.

5.3.3.2   Lateral Flushing

Agricultural filters do not remove all suspended materials from the water because of high cost of removing very small particles (clay and silt size particles). Although these particles are small enough to be dis-charged through the emitters but these cause clogging problems when present in large quantities [27]. Fine inorganic particles usually settle out at the ends of manifolds and laterals, where flow velocities are slow. Chemical precipitation may occur inside pipelines after the irrigation system shuts down.

Flushing of the micro irrigation pipelines is an essential part of the maintenance program needed to remove particles that accumulate in the lines before they build up to sizes and amounts that cause clogging problems [43]. Annual flushing is often sufficient, but some combinations of water and emitters require almost daily flushing to prevent clogging. Where frequent flushing is required, semiautomatic or automatic flushing valves at the ends of lateral are recommended [23]. Irrigation laterals are flushed by opening the ends of the lines during operation and allowing water to freely discharge along with carrying particulate matter. This operation usually requires only a minute or two [27]. Flushing must be done at a suitable velocity to dislodge and transport the accumulated sediments [27, 38]. Flow velocity of 0.3–0.5 m.s^{-1} is recommended to assure that all particles are removed [28]. This must be taken into account in the irrigation system design and where semiautomatic or automatic flushing valves are to be utilized. When flushing is manual, only few lateral lines are opened at a time.

5.4   FIELD EVALUATION OF EMITTER AND FILTER PERFORMANCE USING WASTEWATER

The micro irrigation system performance using WW depends on a complex combination of WW quality, emitter and filter type, chemical water treatments and other system maintenance operations, which interact each to other. For this reason, at present it is not possible to exactly foresee the system performance for a specific set of conditions. The knowledge of the results of experimental tests on the examined topics can enhance the choices in the design and maintenance of a micro irrigation system, which use WW. This section shows an analysis of the main results of some recent papers (Table 5.6) dealing with emitter and filter performance using WW of different quality.

An attempt to compare the results obtained by the different researchers was made despite the different objectives and the different methodologies used. A minimum number of emitter and filter performance indices are discussed. Further details can be found in the literature cited. The most common emitter performance indices, all referred to the end of the experiments, are: the percent of totally clogged emitters, two discharge reduction

TABLE 5.6 Literature Review on Field Emitter and Filter Tests Using WW

WW type, treatment and quality [1]	Emitter types [2]	Filter types and others maintenance operations
Ravina et al. [39]		
Municipal WW (a) and (b), secondary treatment, stored in two reservoirs: (a) mix of secondary sewage effluents, fresh water and winter storm runoff: pH=8–9 (S), TSS=40–300 (m, M, S), H_2S= 0 (m), BOD<20 (M); Chlorophyll (g l^{-1})=0.05–0.6; (b) only secondary sewage effluent: pH=7.2–8 (M); TSS=15–40 (m); H_2S=0–8 (m, M, S); BOD=70–130 (S); Chlorophyll (g l^{-1})=0.2–0.7	12 dripper types, online and inline, regulated and not regulated, q=1.55–3.8 $l\,h^{-1}$	4 filter types: gravel, screen, and disk filters 120, 80 and 40 mesh with automatic backwashing and screen and disk filters 120, 80 and 40 mesh with manual cleaning; daily chlorination to control the mucous biomass development with 3–5 mg of Cl l^{-1} (effluent (a)) and reaching a level of 80 mg Cl l^{-1} (effluent (b))
Capra and Scicolone [9–11]		
6 series of test (see below) with municipal WW (see Table 5.7 for the characteristics)	5 dripper types (G, T, S20, S16 and DIN, see below for the description)	8 filter types (JP g, JP d, AR d 120, AR d80, AR D40, JP s, Si s, A s, see below for the description)
Tests 1 and 2 Effluent (a) raw, gravity flow size 3 mm prefiltration; effluent (b) secondary treatment, dilution ratio 1:1, sedimentation	2 dripper types; online labyrinth (G) and orifice (T) types, q = 4 lh^{-1}	3 filter types: gravel media (JP g), effective sand size=1.5 mm, followed by a 120 nylon mesh screen filter; disk (JP d) and screen (JP s) 120 mesh;
Tests 3 and 4 Effluent (c) secondary treatment; effluent (d) secondary treatment, dilution ratio 2.5:1	2 dripper types; online (G) and embedded in 20 mm tube wall (S20) labyrinth type, q = 4 $l\,h^{-1}$ (G) and 3.8 $l\,h^{-1}$ (S20)	4 filter types: gravel media (JP g), effective sand size=1.5 mm, followed by a 120 nylon mesh screen filter; 2 disk types (JP d and AR d120) 120 mesh; and 1 screen type (JP s) 120 mesh;

TABLE 5.6   Continued

WW type, treatment and quality [1]	Emitter types [2]	Filter types and others maintenance operations
Test 5 Effluent (e), secondary treatment	2 dripper types; online (G) and embedded in 16 mm tube wall (S16) labyrinth type, q = 4 l h⁻¹ (G) and 3.8 l h⁻¹ (S16)	4 filter types: 3 screen types 120 mesh (JP s, SI s and A s); 1 disk type 120 mesh (AR d120)
Test 6 Effluent (f), secondary treatment	2 dripper types: long flow-path (DIN), q = 4.48 l h⁻¹; and embedded in 20 mm tube wall (S20) labyrinth type, q = 4 l h⁻¹, positioned in surface (DI) and subsurface (SDI) drip systems	4 disk types: 2 of 120 mesh (SI d) and (AR d120), 80 mesh (AR d80) and 40 mesh (AR d 40)

Puig-Barguès [32, 33]

(a) Meat industry wastewater, preliminary treatment; pH=7; TSS= 176; EC=2.6; (b) Municipal WW, secondary treatment through a sludge process (from Girona, ES); pH=7.25 (M); TSS=24 (m); EC=1.15 (M); (c) Effluent (b) filtered through a sand filter; pH=7.4 (M);TSS=9 (m); EC=1.12 (M); (d) Municipal WW, secondary treatment through a sludge process (from Castell-Platja d'Aro, ES); pH=7.63 (M); TSS=11 (m); EC=1.63 (M); (e) effluent (d) filtered through sand and disinfected by ultraviolet light and chlorination with 5 mg NaClO l⁻¹; TSS=5 (m)		7 filter types, 3 screen filter types with a nylon screen filtration surface of 946 cm² and openings of 0.09 mm (S98), 0.115 mm (S115), and 0.178 mm (S178); 3 disk filter types, with a filtration surface of 953 cm² and openings of 0.115 mm (D115), 0.130 mm (D130), and 0.2 mm (D200); one sand filter effective sand size= 0.65 mm

Dazhuang et al. [14]

Reclaimed WW:	6 dripper types with	80-mesh weave wire
pH=8 (M); Fe= 0.78–1.04 (M); HCO$_3^-$= 204–222; SO$_4^{2-}$ = 52.2–62.4; TDS= 542–567 (M); TSS=33.6–101.4 (m-M-S); BOD$_5$= 16.7–27.1 (M, S)	different flow path structure; w= 0.8–1.46 mm, d=0.8–1.03 mm; l=102–360 mm	nylon screen filter and 120-mesh disk filter

TABLE 5.6 Continued

WW type, treatment and quality [1]	Emitter types [2]	Filter types and others maintenance operations
Duran-Ros et al. [15]		
Municipal WW (a) and (b): (a) biological treatment, pH=7.48–7.50 (M), TSS=10.02–10.59 (m); 3×10^4 particles ml^{-1}; (b) effluent (a) treated by 120 mesh disk filter and ultraviolet radiation, pH=7.34 (M), TSS=6.26–6.96 (m); and 4×10^4 particles ml^{-1}	6 dripper types; q=2–8.5 l h^{-1}; compensating and not compensating; w=0.76–1.26 mm; d=0.95–1.08 mm; l=17–75 mm; 8.5 l h^{-1} emitter not tested with effluent (b)	4 filter types: 2 media filters in parallel, effective sand size=0.40 mm for effluent (a) and 0.27 mm for effluent (b); two 120 mesh disk filters in parallel; one 120 mesh screen filter; one screen filter followed by two disk filters in parallel, both 120 mesh
Liu and Huang [25]		
Reclaimed WW (a) and fresh water FW (b): (a) treated by active sludge technology: pH=8.6 (S); TSS=0.02 (m); EC=1.469 (M); Ca=95.2 (S); Fe=0.23 (M); Mn=0.033 (m); H_2S=0.175 (m); Bacteria number=4032 (m); (b) pH=7.49 (M); TSS=n.d. (m); EC=0.468 (m); Ca=48.4 (M); Fe=0.14 (m); Mn=0.00039 (m); H_2S=n.d.; Bacteria number=69 (m)	3 dripper types; (a) inline labyrinth at turbulent flow, q=2.83 l h^{-1}; minimum flow passage=1.53 mm; (b) inline labyrinth at laminar flow, q=1.88 l h^{-1}; minimum flow passage=0.8 mm; (c) online pressure compensating, q=2.29 l h^{-1}, minimum flow passage= n.a.	WW filtered with a 80 mesh screen filter cleaned daily; fresh water not filtered; 2 flushing treatments: (i) flushed 5 times for 4 min each, separated by 5 min each, using FW at a pressure of 75 kPa and flow velocity of 0.6 m s^{-1}, after 1680 h of experiment; (ii) not flushed.
Puig-Bargués et al. [34]		
Reclaimed WW treated by tertiary treatment: pH=7–8 (M); turbidity=8–10.3 (m); mesophilic aerobic bacteria=1×10^4 cfu ml^{-1} (M)	2 emitter types both welded onto the interior drip-line wall and positioned in surface (DI) and subsurface (SDI) drip systems;dripper (a): q=2.3 l h^{-1}; pressure compensating; w=1.15 mm; d=0.95 mm; l=22 mm;dripper (b): q= 2 l h^{-1}; turbulent flow; w=0.76 mm; d=1.08 mm; l=75 mm.	2 sand filters in parallel with effective sand size=0.40 mm (1st and 2nd season) and 0.27 mm (3rd season); 3 flushing frequency treatments: (i) no flushing, (ii) one flushing at the end of each irrigation period and (iii) a monthly flushing during the irrigation period

TABLE 5.6   Continued

WW type, treatment and quality [1]	Emitter types [2]	Filter types and others maintenance operations
Li et al. [24]		
Reclaimed WW (a) and (b): (a) treated with fluidized-bed reactor (FBR); pH=7.26 (M); Ca= 68.7 (M); Fe=0.15 (m); Mn=0.01 (m); TSS=24 (m); (b) treated with biological aerated filter (BAF); pH=7.31 (M); Ca=65.8 (M); Fe=0.41 (M); Mn=0.11 (M); TSS=78 (M)	4 dripper types; q= 2.3–3.2 l h^{-1}; w=0.52–1.32 mm; d=0.65–1.12 mm; l=6.15–121.72 mm; 2 types of tube inline and two tape embedded in wall; flow path types in Figure 2.	120 mesh screen filter cleaned every 4 days or less.

[1] Fe, HCO_3^-, SO_4^{2-}, H_2S, TDS, TSS, BOD_5 in mg L^{-1}; EC in dS.m^{-1}; Bacteria number in n.ml^{-1}; n.d.= not detectable; in parenthesis, the clogging risk level classified according to Table 5.2: m = minor, M = moderate, S = severe; BOD classified according to Capra and Scicolone [2004].

[2] q = nominal discharge; w = flow path width; d = flow path depth and l = flow path length; n.a.= not available.

TABLE 5.7   Mean Characteristics of the WW tested by Capra and Scicolone [9–11]

Parameter	Tests 1 and 2				Tests 3 and 4				Test 5		Test 6	
	effluent (1)		effluent (2)		effluent (3)		effluent (4)		effluent (5)		effluent (6)	
		cr		cr		cr		cr		cr		cr
BOD_5 (mg L^{-1})[2]	200	S	15	M	53	S	38	M	25	M	25	M
Calcium (mg L^{-1}) [1]	64	M	24	m	108	S	88	S	103	S	11	m
EC (dS m^{-1}) [1]	1.7	M	1.5	M	1	M	1.2	M	1	M	0.9	M
Magnesium (mg L^{-1})[3]	81	M	90	M	12	m	44	M	108	S	9	m
pH [1]	7.8	M	7.5	M	7.7	M	7.6	M	7.8	M	7.1	M
TSS (mg L^{-1}) [1]	376	S	3	m	146	S	96	M	78	M	47	m

cr = clogging risk; m= minor; M= moderate; S= severe;

[1] according to Table 5.2;

[2] according to Capra et al. [9];

[3] according to Table 5.3.

indices (R_d, and P_0), and the emission uniformity coefficient EU (estimated according to the Eq. (1)).

The percent of totally clogged emitters (T_{clog}, %) is estimated as:

$$T_{c\log} = 100 \left(\frac{N_{c\log}}{N_t} \right) \qquad (4)$$

where: N_{clog} = number of totally clogged emitters; and N_t = total number of tested emitters.

The reduction in the system mean discharge (R_d, %) is calculated as:

$$R_d = 100 \left(1 - \frac{Q_m}{Q_0} \right) \qquad (5)$$

where: Q_m = system mean discharge measured during last operation ($l\,h^{-1}$); and Q_t = discharge (lph) of the same number of new, unclogged emitters, at the same operating pressure measured during last operation. The Average emitter flow reduction (P_0, %) is defined as:

$$P_0 = 100 \left(\frac{\sum_{i=1}^{t} (q_f / q_0)}{N_t} \right) \qquad (6)$$

where: q_f = mean emitter discharge during last operation (lph); and q_0 = mean emitter discharge at the beginning of the trials (lph). The filter performance is usually discussed in terms of mean time of operation, for example, the mean interval between cleanings. Some authors also used the percent of removal efficiency (E_f, %):

$$EU = 100 \left(\frac{q_{1/4\min}}{q_m} \right) \qquad (7)$$

where: N_0 = value of every WW quality parameter at the filter inlet; and N = value of the same parameter at the filter outlet.

Ravina et al. [39] reported a synthesis of the results of an experimental program, initiated in 1987 and concluded in 1992, to test in field conditions filters and emitters widely used in Israel. The trials were carried out successively at two different sites using WW supplied from Kfar Barukh reservoir and Burgata reservoir (effluents (a) and (b) in Table 5.6). Twelve types of emitters were tested both online and inline, regulated and not regulated, with discharge ranking between 1.55 and 3.8 lph. The filters tested were gravel, screen, and disk filters 120, 80 and 40 mesh with automatic backwashing and screen and disks filters with manual cleaning. The effects on dripper and filter performance of different chlorination and lateral flushing schemes were also tested. The main results highlighted by the authors are briefly described here. Emitter and filter clogging hazards were associated mainly with mucous products of microbial activity. Chlorination was required for emitter clogging abatement and to maintain satisfactory operation of the drip laterals with most of the tested emitter types. The effects of daily chlorination and once every 3 days was almost the same and much better than once every 10 days. Emitters of a higher discharge rate clogged less than similar ones of a lower discharge rate. Non uniformity in the discharge rate of emitters along the lateral increased with the increase of clogging level. More clogged emitters were found towards the end of the drip lateral, than at its beginning section. Clogging was influenced more by the velocity than by the supply pipelines length. Biological growth accumulates on the wall of the pipes and carried into the drip system under the velocity less than 0.5 m s^{-1}. Media filters were advantageous in reducing the levels of emitter clogging, particularly for the more sensitive emitters. Emitter clogging after 80 and 120 mesh filtration was almost the same in most types of emitters that were tested, and less than after 40 mesh. There was no difference in emitter clogging when flushing the drip laterals daily or every two weeks. Rapid clogging and malfunction of the filters have occurred at peak periods of the mucous biomass growth, at both sites.

Capra and Scicolone [9–11] conducted field experimental trials using five kinds of municipal WW not undergone previous advanced treatment conducted in Sicily (Italy), from 1999 to 2003, (Table 5.6). They tested the behavior of eight kinds of filters (gravel media, disk and screen with different orifice sizes and constructed by different manufacturers) and five online and inline types of drip emitters (orifice, long-path and labyrinth) commonly used in the area. The emitter, filter and effluent characteristics are presented in Tables 5.6 and 5.7. Figures 5.1 and 5.2 show some experimental system details.

FIGURE 5.1   Screen filters tested with WW by Capra and Scicolone [9–11].

FIGURE 5.2   Media filters (followed by a screen filter) tested with WW by Capra and Scicolone [9–11].

The performance of the emitters and filters under study depended on the interactions between WW quality, emitter and filter types. The orifice (vortex) emitters (T in Table 5.6) were not suitable for the use of WW due to both the high percentage of totally clogged emitters (P_{clog}), the low emission uniformity (EU) and the high reduction in the mean discharge (R_d), the only exception being very good quality water (effluent (2) in Tables 5.6 and 5.7) filtered by gravel media filter. Long-path flow emitters (I in Table 5.6), tested only during Trial 6, assured good performance with 120 mesh disk filters only (ratio between filtration dimension and diameter of flow passage of the emitter almost equal to 1:4). When the labyrinth-type emitter (S) was used with gravel media or disk filtration (AR d and SI d in Tables 5.6 and 5.7), no totally clogged emitters were observed, and EU varied from 67% to 98%. The ratio between the disk filtration dimension and the diameter of the flow passage of the labyrinth ranged from almost 1:10 (Trials 3 and 4) to almost 1:4 (Trial 6); these differences were probably due to the different composition of TSS (i.e., granulometric size). Between 40, 80 and 120 mesh disk filters produced by the same manufacturer, the best emitter performance was assured by 120 mesh filtration. All the screen filters tested were not capable of assuring good emitter performance, with the only exception of the good quality effluent (effluent (2) in Table 5.7). With the same effluent: emitters with a higher discharge rate (e.g., 4 or 8 l h⁻¹) clogged less than similar ones with a lower discharge rate (e.g., 2 L h⁻¹); gravel media filters guaranteed the best emitter performance; the disk filters, as long as they were of good quality, assured performance similar to that of the gravel media filter using WW with TSS content not higher than 100–150 mg L⁻¹; the screen filters, irrespective of the commercial type, were not suitable for use with WW, with the exception of WW with very low values of TSS (3 mg L⁻¹). No significant difference was observed during Trial 6 between the same kinds of emitters placed on soil (DI) or sub-soil (SDI): no totally clogged emitters were observed and EU coefficients of 87% and 98% were shown, respectively for emitters on soil and subsoil. All filters tested showed clogging problems. With WW of very good quality only (e.g., effluent (2)), all the 120 mesh filters needed a number of cleaning operations comparable to those of a drip system using conventional water; in the other cases (TSS higher than about 50 mg L⁻¹) the operating times of the filters were very short (less than 1 h in most cases). The 80 and 40 mesh disk filters had acceptable operating times, but they were not capable to assure emitter optimal performance. The design discharge of filters, suggested by the manufacturers for clean water, was not adequate for WW of poor quality; in fact, the frequency of cleaning operations depends on the solid discharge (product between water discharge and TSS); in the tests a discharge of 1/15–1/20 of the design discharge for clean water was used.

Puig-Bargués et al. [32, 33] studied the quality of filtration and the time required between cleanings of sand, disk and screen filters with different opening sizes (Table 5.6) used in micro irrigation with different effluents (Table 5.6). They concluded that physical and chemical parameters have the greatest influence on clogging. In the trials, the filters were considered to be completely clogged when the head loss across the filter was greater than 49 kPa. Related to the quality of the filtration, the best filtration quality using the worst quality effluent (the meat industry effluent (a) in Table 5.6) was achieved by the sand filter. There were no difference among screen, disk, and sand filters using the Girona municipal treated secondary effluent (effluent (b) in Table 5.6). There was a slightly greater efficiency in removing particles when using the screen and disk filters with smaller openings. No significant differences were observed using disk or screen filters with the same effluent filtered in a sand filter with an effective size of 0.65 mm (effluent (c) in Table 5.6). Despite no significant differences being observed between the 0.130 mm disk and screen filters with the Castell-Platja d'Aro municipal treated secondary effluent (effluent (d) in Table 5.6), the filtration quality was slightly higher using the screen filter.

Related to the clogging of filters, measured through the time of operation: the disk and sand filter required earlier backwashing than the screen filters when using the poorest quality effluent (a); the sand filter had a tendency to need more cleanings than the disk and screen filter with the effluent (b); the disk filters tended to need less time between cleanings than the screen filters with the Girona effluent, whether filtered with sand (effluent (c)) or not (effluent (b)), but more time when using the better quality effluents (d) and (e). The relative importance of each WW parameter varied with effluent, but those with the highest incidence were total suspended solids and electrical conductivity.

In a study on the microbial characteristics of mature biofilms present in the emitters, Dazhuang et al. [14] examined the performance of six types of drippers with different characteristics (Table 5.6 and Fig. 5.3) using reclaimed WW classified at medium/high clogging risk for different parameters, filtered by a 80-mesh weave wire nylon screen filter and

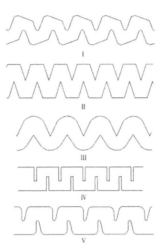

FIGURE 5.3   Flow path type of the labyrinth emitters tested by Dazhuang et al. [14].

120-mesh disk filter. The results showed that suspended particles combining with microbial slime were the main cause of the slowly progressive emitter clogging occurred during the irrigation season (360 h). Between the emitter characteristics, the flow path type had the major role in explaining the emitter performance. The worst performance, in terms of both average emitter flow reduction (DR_a=72.2%) and emission uniformity (EU= 57.2%), was obtained in emitter with flow path type II (Fig. 5.1). Emitters with flow path type I and IV (Fig. 5.1) resisted clogging and performed better with only 14.4% and 23.4% of flow reduction and EU equals to 95% and 86.1%, respectively. The emitter with the unsymmetrical dentate structure (Type I in Fig. 5.3) had the best anti-clogging capability.

Duran-Ros et al. [15] tested four filtration systems (sand, screen, disk and a combination of screen and disk filters) and six emitter types for 1000 h, using two different effluents ((a) and (b) in Table 5.6) both with low suspended solid levels (<11 mg L^{-1}) from a WW treatment plant. Emitter clogging, mainly due to an organic film, was affected mainly by emitter type, location along the lateral and the interaction between these two factors for both effluents. The emitter with the smallest water passage section (width = 0.78 mm; depth = 0.67 mm) experienced an important decrease in flow rate (the final flow rate was 53% of the initial). Instead of, emitters with higher water passage sections, showed smaller flow reductions and a small number of totally clogged emitters. The tests also showed the importance of the lateral flow, an emitter with characteristics similar to that which showed good performance, but installed in the lateral at higher flow (q = 0.14 lps, almost double respect to the other laterals tested) showed the greatest flow reduction (final flow rate was 41% of the initial) due to the high WW volume and, thus, to the high particles entering and clogging the emitters. Differences among emitters with larger clogging were only observed at the end of drip-line attributable to the lower velocity at this point, which favors particle settling. Only with the effluent that had a higher number of particles the filter and the interaction of filter and emitter location had a significant effect. Emitters placed after screen and sand filters showed the largest flow rates at the lateral ending, even though only sand filtration significantly reduced turbidity and suspended solids. Emitters protected by a disk filter experienced the largest flow rate reductions. However, the operating time of the filters (interval between cleaning operations) was different for the different filters tested. Table 5.8, obtained re-elaborating the data published by Duran-Ros et al. [15], shows mean operating time lesser than 1 h (0.3–0.4 h) for screen filters with effluent (a); that means screen filters can be used only if cleaned automatically. The short operating times cause management problems if manual systems are used.

Liu and Huang [25] conducted a laboratory experiment to investigate the emitter-clogging pattern for three emitter types with the application of FW and treated WW (see Table 5.6 for the water characteristics). The three-emitter types were the inline-labyrinth type of emitters both with a turbulent flow (a) and with a laminar flow (b), and the online pressure-compensation type of emitter (c). At the end of the tests, after 1680 h of irrigation (in two periods separated by three months), completely clogged emitters were not found for emitter type (c) with the application of both FW and WW, and for emitter types (a) and (b) with the application of FW. At the same time, more than 13% of emitters of type (a) and 31% of emitters of type (b) were completely clogged for the WW treatment. The calculated EU for emitter types (a) and (c) for the FW treatment was equal to almost 98% in the entire experimental period. At the end of the experiment, the EU of emitter type (b) was 0 and 38.7% respectively for the WW and FW treatment. With the WW treatment, the EU of emitter types (a) and (b) decreased with the increase of operational time and almost 15% and 81.4%, respectively, at the end of the experiment. The worst performance of the emitter (b) may

TABLE 5.8 Operational Characteristics of Filters Tested by Duran-Ros et al. ([15] re elaborated).

Effluent	Filter	Number of filtration cycles	Mean number of cleaning per h [1]	Mean operating time (mean interval between cleaning) (h) [1]	Backwashing water consumption (%)
(a)	Sand	219	0.22	4.57	1.14
			Combined		
	Disk	96	0.10	10.42	0.14
	Screen	2689	2.69	0.37	2.08
	Disk	719	0.72	1.39	1.48
	Screen	3257	3.26	0.31	2.77
	Sand	645	0.65	1.55	4.99
(b)			Combined		
	Disk	60	0.06	16.67	0.11
	Screen	189	0.19	5.29	0.17
	Disk	190	0.19	5.26	0.37
	Screen	184	0.18	5.43	0.16

[1] elaborated in this chapter, considering 1000 hours of operation.

be partially due to the relatively smaller flow-path dimension and discharge and to the laminar flow. Compared to emitter types (a) and (b), emitter type (c) had relatively low reduction in emitter discharge and high EU for the WW treatment. This implies that the pressure-compensation type of emitters showed a better anti-clogging potential than the labyrinth types of emitters. Chemical precipitation (carbonate and sulphate) was found to be the primary source of emitter clogging for both the FW and WW treatments, which was mainly due to their relatively higher total dissolved solids and alkalinity. During the test the effects of lateral flushing was also evaluated. Laterals were flushed 5 times for 4 min each, separated by 5 min each, using FW at a pressure of 75 kPa and flow velocity of 0.6 m.s^{-1}, after 1680 h of experiment. Results showed that flushing did not increase the discharge of the completely clogged emitters for emitter types (a) and (b), but did increase the discharge of partially clogged emitters, especially for emitter type (c) which may be due to the elastic membrane. The EU for emitter types (a) and (b) were not affected by flushing. For emitter type (c), the EU only increases by 5% after flushing. The clogging materials (i.e., chemical precipitation) sticking to the inside and outside walls of emitters could not easily be flushed out; therefore, flushing using FW is not practical for cleaning emitters which are completely or severely clogged by chemical precipitation.

Puig-Bargués et al. [34] studied the clogging of two different emitters (Table 5.6), one pressure compensating (a) and the other not (b) working during three irrigation seasons of 540 h each with a WW treatment plant effluent that presented a minor physical clogging hazard, but moderate chemical and biological clogging hazards. Both the dripper types were placed in both a surface (DI) and subsurface (SDI) irrigation systems. WW was filtered by two sand filters in parallel with effective sand size=0.40 mm (1st and 2nd season) and 0.27 mm (3rd season). Three flushing frequency treatments were tested: (i) no flushing; (ii) one flushing at the end of each irrigation season; and (iii) monthly flushing during the irrigation period. At the end of the three irrigation seasons, there were only a small number of completely clogged emitters with the greatest amounts (3.7–4.5%) for emitter (b) with SDI. Emitter (b) with SDI had significantly more totally clogged emitters than any other combination of emitter and irrigation system. Additionally, emitter (b) had an approximately 13% smaller nominal discharge than emitter (a), and this may have led to greater clogging. Distribution uniformity tended to decrease through the experiment in the surface system, especially in the not flushed treatment (EU=76% and 68% for pressure compensating (a) and at turbulent flow (b) emitters, respectively). EU was equal to 54% and 24% for the subsurface system in the same conditions. Greater EU was achieved with monthly flushing (EU equal to 94% and 90% for DI emitter type (a) and (b), respectively, and 88% and 69% for the corresponding emitters in SDI system). As showed, at the end of the experiment, EU was smaller in SDI units for both emitter types. It could be partly related with soil heterogeneity, which was the condition of the present experiment. The mean operating time of sand filter was 3.8 h during the first irrigation period, 1.5 h during the second one and 0.9 h in the third irrigation period. This increase was primarily due to the sand media clogging by trapped particles that were observed at the end of the second irrigation period and to the smaller effective sand media size in the third period. Biofilm formation was the primary problem in the DI emitters. However, this biofilm problem was exacerbated in the SDI emitters because external soil particles became stuck in the biofilm at the emitter outlet leading to increased clogging. The intrusion of soil in the drippers used in SDI was also observed by other authors (Fig. 5.4).

Li et al. [24] evaluated the effects on clogging in four types of labyrinth drippers (Table 5.6) operating for 432 h with two WW effluent ((a) and (b) in Table 5.6) treated with fluidized-bed reactor (FBR) and biological aerated filter (BAF). The results showed that the average discharge of emitters decreased continuously with some fluctuations. The clogging degree of emitters, similar for the two types of WW was low during the first 256 h. Then, clogging increased greatly, and its effect on emitter discharge was more obvious in the FBR treatment than BAF. It was closely related to the growth, detachments and sediment of biofilms in the irrigation system. The tests showed the importance of dripper type. At the end of the experiment, the relative mean discharges of the two drip irrigation tapes (emitters (a) and (b) in Fig. 5.5, inserted inline in the tube) were larger than two tubes (emitters (c) and (d) in Fig. 5.5, embedded in the tube wall),. It shows that anti-clogging abilities of the two tubes were higher.

FIGURE 5.4 Intrusion of soil in a dripper used in SDI (photo by Capra A).

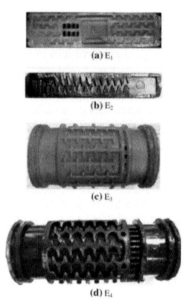

FIGURE 5.5   Flow path types of emitters tested by Li et al. [24].

In comparison, the emitters clogging degrees of drip irrigation with BAF WW were lower than that of FBR, and the Relative mean discharge (100* measured discharge/ discharge of new and unclogged emitters at 20°C) of FBR was lower 7.8–15.2% than that of BAF. So using BAF reclaimed WW for drip irrigation is more suitable.

5.5   RECOMMENDATIONS USING WASTEWATER

The results of laboratory and field tests on micro irrigation systems using WW (municipal WW, in most cases) showed the performance of the emitters depends on the complex interaction between the emitter type, the filtration type and the quality of the WW. Total suspended solids and organic matter, contained in the water or developed in the system, are the main factors affecting the percentage of totally and partially clogged emitters, the mean discharge emitted by the system, the emission uniformity coefficient, and the operating time of the filters between cleaning operations. Only for WW practically free of suspended solids, emitter clogging due to chemical precipitates becomes evident. Filtering is the main defense against clogging. In some cases, acidification treatment may be used to prevent chemical precipitation.

Chlorination can be used to prevent and control microorganism development. The quality of the effluent, which depends on the type and quality of the original water, and the WW treatment and dilution rate, has a fundamental role in the clogging process. Unfortunately, the WW quality is extremely complex, and little information on clogging risk classification is available even for clean water. Clogging risk due to organic matter content according to BOD_5 should be classified as minor when BOD_5 is <15 mg L^{-1}, moderate between 15 and 40 mg L^{-1}, and severe >40 mg L^{-1}. Between the tested dripper types, the main results of the experimental trials showed:

- orifice (vortex) emitters were more sensitive to clogging than labyrinth emitters;
- for labyrinth emitters, clogging depended on the flow path dimension and structure; the emitter with small water passage section (or discharge, for example, <2 l h^{-1}) experienced decrease in flow rate more important respect to emitters with higher water passage sections; the flow path type had the major role in explaining the labyrinth emitter performance when biological clogging was observed; the emitters with the unsymmetrical dentate structure (Type I in Fig. 5.3) had the best anti-clogging capability;
- the pressure-compensation emitters showed a better anti clogging potential than the non-compensating emitters; however, it should be noted that the pressure-compensation emitters are generally more expensive, and their performance may be affected by material fatigue caused by temperature, extended chlorination, microbial activity, and acid injection; therefore, only under proper micro-irrigation system design and with sufficient financial support, pressure-compensation emitters are recommended for drip irrigation with WW;
- emitters at laminar flow had worst performance respect to emitters at turbulent flow;
- drippers inserted inline in the tube (e.g., (a) and (b) types in Fig. 5.5) showed anti clogging abilities higher respect to drippers embedded in the tube wall (e.g., (c) and (d) types in Fig. 5.5);
- drippers installed in lateral at higher flow rate (almost 0.15 L s^{-1} in the trials conducted by Duran-Ros et al. [15]) showed the greatest flow reduction respect to emitters with characteristics similar installed in lateral at lower flow due to the high WW volume and, thus, to the high particles entering in the pipes;

- more clogged emitters were found towards the end of the drip lateral, than at its beginning section;
- the effects on emitter clogging of the position of the laterals on surface (DI) or sub surface (SDI) are not clear; in some tests [10, 12] no totally clogged emitters and *EU* similar to that of emitters on soil were observed for emitters on subsoil; other experiments [34] showed more totally clogged emitters with SDI; the authors commented that emitter discharge of SDI could also decrease as a result of positive pressure in the soil water matrix creating a backpressure at the emitter orifice; this backpressure phenomenon may explain the reduction in emitter discharge for emitters used in SDI rather than in DI; other authors [24] observed biofilm problems was exacerbated in the SDI emitters because external soil particles became stuck in the biofilm at the emitter outlet leading to increased clogging.

Filtering efficiency should be examined according two criteria: the level of emitter protection and the time of operation of the filters between cleanings. The two criteria can be in contrast each to other depending on WW quality. Generally WW use requires a compromise between degree of filtration and level of uniformity (acceptable although not optimal). Clogging of the filters involves management problems due to the need to frequent cleaning operations, for example, a continuous presence in the field of the irrigator (for manual filters), or the use of automatic cleaning equipment, which, in turn, contribute an increase in equipment costs. Farmers need filters that guarantee a good trade-off between system performance (uniformity), costs and operating times between cleaning operations. The main results of the experimental trials on filter performance are:

- most of the tested filter types (screen, disk and gravel media), size and dimension of filtration showed clogging problems variables as a function of the WW quality;
- the time of operation between cleanings could be very low (<1 h) for all the filter types depending on the WW quality; consequently, automatic flushing equipment's are suggested;
- between screen, disk and gravel media filters, media filters guaranteed the best emitter performance, but they are relatively more complex and expensive; disk filters, as long as they were of good quality,

assured performance similar to that of the gravel media filters using WW with TSS content not higher than 100–150 mg L^{-1};
- the differences between screen and disk filters were not clear; in general, it seems that disk filters assured better performance in terms of both emitter protection and time of operation using WW of low quality (e.g., classified at medium or sever clogging risk for the total suspended solid content); instead of, no important differences among screen and disk filters were observed using WW of better quality (e.g., classified as at minor clogging risk for TSS content); however, some results on the mean operating time are in contrast each to other; it was observed that the time of operation of disk filters could be both higher, similar or lower respect to screen filters;
- disk filters of good quality (e.g., some commercial brands internationally distributed) assured better performance respect to disk filters with the same size and filtration dimension locally manufactured and available;
- the ratio between the disk filtration dimension and the minimum dimension of the flow passage of the emitter labyrinth ranged from almost 1:10 to almost 1:4;
- screen filters, the most widespread, cheaper and simple to manage, irrespective of the commercial brand, were able to prevent clogging only for WW of very good quality for TSS and organic matter content (classified at minor risk for clogging);
- between 40, 80 and 120 mesh screen and disk filters, the best emitter performance was assured by 120 mesh filtration; even contrasting results were obtained for the 40 and 80 mesh filters; but, sometimes, 40 or 80 mesh filters could be preferred to 120 mesh to elongate the operating time although the sub-optimal values of uniformity;
- the theoretical discharge of the filters suggested by the manufacturers for clean water was not adequate for WW classified at medium or severe clogging risk for TSS and BOD$_5$; some authors [10–12] proposed to choose filter sizes having a discharge of 1/15–1/20 of the design discharge for clean water when WW was used.

Pre-filtration before the water is pumped in the system was recommended to protect the pump and other equipment.

Chlorination is required to avoid biological emitter clogging and to maintain satisfactory operation of the drip laterals when WW was stored in open reservoirs. The effect of daily chlorination and once every 3 days was almost the same and much better than once every 10 days.

Evaluating and monitoring system performance (mainly emission uniformity), cleaning or replacing plugged emitters, and periodic line flushing are accompanying measures apart the test of irrigation water source. In the experiments described, flushing did not increase the discharge of the completely clogged emitters, but increased the discharge of partially clogged emitters, especially for compensating emitters.

5.6 SUMMARY

Irrigated agriculture is the biggest consumer of water in comparison to other sectors. In arid and semi-arid regions of the world competition for fresh-water resources with domestic and industrial sectors is very strong. At the same time, increasing urbanization is producing large volumes of wastewater that have become a serious environmental problem in most of the countries. Sustainable water use in these conditions involves wastewater reuse for irrigation. Municipal wastewater (WW) is the most suitable for irrigation reuse but requires irrigation methods, which minimize health and environmental risks. Micro irrigation with surface (SI) and subsurface drip irrigation (SDI) is the suitable techniques for WW reuse because edible parts of plants grow mainly above the soil (tomatoes, oranges, etc.) surface.

Wastewater contains physical, chemical and biological particles. It could lead to filtering difficulties and high risk of system failure due to clogging of the emitters. The clogging of emitters is the largest maintenance problem with micro irrigation systems. It is difficult to detect and expensive to clean, or replace, clogged emitters. Partial or complete clogging reduces emission uniformity and irrigation efficiency. Clogging risk depends on the level of treatment with the WW. There are treatment technologies to achieve any desired water quality, but extremely stringent quality standards would lead to unsustainable costs especially in developing countries and in small communities, where the development of sewage treatment plants does not follow that of water use. In these conditions, a good trade-off between treatment costs and micro irrigation system performance has to be found.

Filtering is the prime defense against emitter clogging, but does not prevent it completely and, in many cases of low-quality water the filters clog and causes system management problems. The main filter types used in drip irrigation systems are screen, disk and gravel-sand media filters. The most common are screen and disk filters; they are simple, economical and easy to maintain. Gravel-sand media filters are particularly suitable for water with high inorganic and organic suspended solids but they are relatively more complex and expensive. Chemical injection (with chlorine or acid compounds) may be necessary to dissolve mineral precipitates or prevent the growth of slime. Gravel-sand media filters and chemical injection are more suitable for farms with high technological and professional standards. In many areas, farm water treatment systems are generally very elementary and irrigators prefer large-sized emitters, such as sprayers and sprinklers, which are less appropriate to minimize health and environmental risks.

Field tests using WW showed that a complete protection of the emitters from clogging may require expensive filtration and chemical treatment equipment's. The filters require frequent cleaning, which involves both failure of the irrigation system and the continuous presence of the irrigator in the field. Frequent manual cleaning of the filters involves contact between the irrigator and WW and an high risk of contamination in the case of WW not subjected to disinfection treatment. The use of automatic cleaning filters increases the costs of equipment. Emitters with good performance and filters which warrant good trade-off between cost and operating times between cleaning operations are required for farmers. Identification of water quality parameters is essential to foresee the emitter clogging risk and to reduce the cost of water analysis.

The micro irrigation system performance using WW depends on a complex combination of WW quality, emitter and filter type, chemical water treatments and other system maintenance operations, which interact each to other. For this reason, at present it is not possible to exactly foresee the system performance for a specific set of conditions. The results of experimental tests, analyzed in the final section of the chapter, can enhance the choices in the design and maintenance of a micro irrigation system, which use WW.

KEYWORDS

- acidification
- algae
- alkaline water
- anti-siphon device
- bacteria
- bacterial slimes
- biochemical oxygen demand, BOD
- biocide
- biological aerated filter, BAF
- biological biofilms
- biological contaminants
- calcium, Ca
- chemical contaminants
- chlorination
- clogging
- complete clogging
- deep percolation
- disk filter
- domestic effluent
- drainage
- drip irrigation (DI)
- effluent quality
- emission uniformity
- emitter clogging
- emitter flow
- emitter performance
- emitters
- environmental contamination
- fertilizer leaching
- filter

- physical contaminants
- plant toxicity
- pollution hazard
- polymers
- runoff
- salt accumulation
- sand separator
- scanning electron microscope
- screen filter
- self-cleaning
- sodium hypochlorite
- sprayers
- statistical uniformity coefficient
- suspended solids
- temporal variability
- titration
- total dissolved solids, TDS
- total iron, Fe
- total suspended solids, TSS
- uniformity coefficient
- urban WW
- vacuum relief valve
- vortex filter
- wastewater effluent
- wastewater reuse
- wastewater, WW
- water application uniformity
- water logging
- water quality
- weed growth

REFERENCES

1. Adin, A., Sacks, M. (1991). Dripper clogging factors in wastewater irrigation. *Journal of Irrigation and Drainage Division, ASCE*, 117(6), 813–825.
2. Aronino, R., Dlugy, C., Arkhangelsky, E. (2009). Removal of viruses from surface water and secondary effluents by sand filtration. *Water Research*, 43(1), 87–96.
3. Battilani, A., Steiner, M., Andersen, M. (2010). Decentralized water and wastewater treatment technologies to produce functional water for irrigation. *Agricultural Water Management*, 98, 385–402.
4. Boswell, M. J., 1990. *Hardie Irrigation, micro-irrigation design manual*. 4ᵗʰ ed., James Hardie Irrigation, El Cajon, California.
5. Bralts, V. F. (1986). Field performance and evaluation. In: *Trickle Irrigation for Crop Production. Design, Operation and Management* (Nakayama F. S., Bucks S. A., Eds.). Amsterdam, Elsevier.
6. Bucks, D. A., Nakayama, F. S., Gilbert, R. G. (1979). Trickle irrigation water quality and preventive maintenance. *Agricultural Water Management*, 2, 149–162.
7. Burt, C. M., Styles, S. W. (2000). *Drip and micro irrigation for trees, vines, and row crops, with special section on buried drip*. Irrigation Training and Research Centre, Department of Agricultural Engineering, California Polytechnic State University, San Luis Obispo, California, USA, pages 166.
8. Capra, A., Scicolone, B. (1998). Water quality and distribution uniformity in drip/trickle irrigation systems. *Journal Agricultural Engineering Research*, 70, 355–365.
9. Capra, A., Scicolone, B. (2004). Emitter and filter tests for wastewater reuse by drip irrigation. *Agricultural Water Management*, 68, 135–149.
10. Capra, A., Scicolone, B. (2005). Assessing dripper clogging and filtering performance using municipal wastewater. *J. Irrigation and Drainage*, 54, S71-S79.
11. Capra, A., Scicolone, B. (2007). Recycling of poor quality urban wastewater by drip irrigation systems. *Journal of Cleaner Production*, 15, 1529–1534.
12. Capra, A., Tamburino, V., Zimbone, S. M. (2010). Irrigation systems for land spreading of olive oil mill wastewater. *Terrestrial and Aquatic Environmental Toxicology*, 4(Special Issue I), 65–74.
13. Cirelli, G. L., Consoli, S., Licciardello, F. (2012). Treated municipal wastewater reuse in vegetable production. *Agricultural Water Management*, 104, 163–170.
14. Dazhuang, Y., Zhihui, B., Rowan, M. (2009). Biofilm structure and its influence on clogging in drip irrigation emitters distributing reclaimed wastewater. *Journal of Environmental Sciences*, 21, 834–841.
15. Duran-Ros, M., J. Puig-Bargués, J. G., Arbat G. (2009). Effect of filter, emitter and location on clogging when using effluents. *Agricultural Water Management*, 96, 67–79.
16. Haman, D. Z. (2011). Causes and prevention of emitter plugging in micro irrigation systems. http://edis.ifas.ufl.edu/pdffiles/AE/AE03200.pdf. University of Florida, IFAS Extension.
17. Haman, D. Z., Zazueta, F. S., (2011a). Screen filters in trickle irrigation systems. http://edis.ifas.ufl.edu/wi009. University of Florida, IFAS Extension.

18. Haman, D. Z., Zazueta, F. S., (2011b). Media filters for trickle irrigation in Florida. https://edis.ifas.ufl.edu/pdffiles/WI/WI00800.pdf. University of Florida, IFAS Extension.

19. Hamdy, A. (2001). Agricultural water demand management: a must for water saving. In: *Advanced short course on Water saving and increasing water productivity: challenges and options.* University of Jordan, Faculty of Agriculture, March 2001, at Amman, Jordan, 2001, b18.1 to b18.30 pages.

20. Gilbert, R. G., Nakayama, F. S., Bucks, D. A. (1981). Trickle irrigation: emitter clogging and flow problems. *Agricultural Water Management,* 3, 159–178.

21. Granberry, D. M., Harrison, K. A., Kelley, W. T. (2012). Drip chemigation: injecting fertilizer, acid and chlorine. University of Georgia Cooperative Extension. http://www.caes.uga.edu/ applications/publications/files/pdf/B%201130_2.

22. Irrigation tutorial, (2014). Irrigation water filtration and filter recommendations. http://www.irrigationtutorials.com/filters.htm.

23. Keller, J., Bliesner, R. D. (1990). *Sprinkler and Trickle Irrigation.* AVI Book, New York, USA, 652 pages.

24. Li, Y. K., Liu, Y. Z., Li, G. B. (2012). Surface topographic characteristics of suspended particulates in reclaimed wastewater and effects on clogging in labyrinth drip irrigation emitters. *Irrigation Science,* 30, 43–56.

25. Liu, H., Huang, G. (2009). Laboratory experiment on drip emitter clogging with fresh water and treated sewage effluent. *Agricultural Water Management,* 96, 745 756.

26. Nakayama, F. S., Bucks, D. A. (1991). Water quality in drip/trickle irrigation: a review. *Irrigation Science,* 12(4), 187–192.

27. Nakayama, F. S., Boman, B. J., Pitts, D. J. (2007). Maintenance. In *Micro irrigation for Crop Production. Design, Operation, and Management* (Lamm, F. R., Ayars, J. E., Nakayama, F. S. Eds.). Elsevier, Amsterdam, pages 389–430.

28. Obreza, T. (2004). Maintenance guide for Florida micro irrigation systems. Soil and Water Science Department, University of Florida, Gainesville, CIR 1449, pages 29. http://edis.ifas.ufl.edu.

29. Oron, G., Campos, C., Gillerman, L., Salgot, M. (1999). Wastewater treatment, renovation and reuse for agricultural irrigation in small communities. *Agricultural Water Management,* 38, 223–234.

30. Pereira, L. S., Oweis, T., Zairi, A. (2002). Irrigation management under water scarcity. *Agricultural Water Management,* 57, 175–206.

31. Pires – Ribeiro, T. A., Stipp-Paterniani, J. E., Coletti, C. (2008). Chemical treatment to unclog dripper irrigation systems due to biological problems. *Scientia Agricola* (Piracicaba, Brazil), 65(1), 1–9.

32. Puig-Bargués, J., Barragán, J., Ramírez de Cartagena, F., 2005a. Filtration of effluents for micro irrigation systems. *Transactions of the ASAE,* 48(3), 969–978.

33. Puig-Bargués, J., Barragán, J., Ramírez de Cartagena, F., 2005b. Development of equations for calculating the head loss in effluent filtration in micro irrigation systems using dimensional analysis. *Biosystems Engineering,* 92(3), 383–390.

34. Puig-Bargués, J., Arbat, G., Elbana, M. (2010). Effect of flushing frequency on emitter clogging in micro irrigation with effluents. *Agricultural Water Management,* 97, 883–891.

35. Qadir, M., Wichelns, D., Raschid-Sally, L. (2010). The challenges of wastewater irrigation in developing countries. *Agricultural Water Management*, 97, 561–568.
36. Ragab, R. (2001). Policies and strategies on water resources in the European Mediterranean region. Proceedings of the Water and Irrigation Development Int. Conf., 2001, Mantova, Italy, Editoriale Sometti, Mantova, Italy, pages 37–55.
37. Raschid-Sally, L. (2010). The role and place of global surveys for assessing wastewater irrigation. *Irrigation and Drainage Systems*, 24, 5–21.
38. Ravina, I., Paz, E., Sofer, Z. (1992). Control of emitter clogging in drip irrigation with reclaimed wastewater. *Irrigation Science*, 13(3), 129–139.
39. Ravina, I., Paz, E., Sofer, Z. (1997). Control of clogging in drip irrigation with stored treated municipal sewage effluent. *Agricultural Water Management*, 33(2–3), 127 137.
40. Runyan, C., Obreza, T., Tyson, T. (2007). Maintenance guide for micro irrigation systems in the Southern Region. Southern Regional Water Program. http://fawn.ifas.ufl. edu/tools/irrigation/citrus/maintenance/FAWN%20Irrigation%20Maintenance%20 Guide.pdf.
41. Sahin, Ü., Anapalı, Ö., Donmez, M. F. (2005). Biological treatment of clogged emitters in a drip irrigation system. *Journal of Environmental Management*, 76, 338–341.
42. Scischa, A., Ravina, I., Sagi, G. (1996). Clogging control in drip irrigation systems using reclaimed wastewater, the platform trials. Proceedings of the 7th Int. Conf. on Water and irrigation, at Tel.: Aviv, Israel, pages 104–114.
43. Smajstrla, A. G., Boman, B. J., Haman, D. Z. (2012). Field evaluation of micro irrigation water application uniformity. University of Florida, IFAS Extension, http:// edis.ifas.ufl.edu/pdffiles/AE/AE09400.pdf.
44. Solomon, K. H. (1985). Global uniformity of trickle irrigation systems. *ASAE Transactions*, 28(4), 1151–1158.
45. Taylor, H. D., Bastos, R. K., Pearson, H. W. (1995). Drip irrigation with waste stabilization pond effluents: Solving the problem of emitter fouling. *Water Resources*, 29(4), 1069–1078.
46. Vivaldi, G. A., Camposeo, S., Rubino P. (2013). Microbial impact of different types of municipal wastewaters used to irrigate nectarines in Southern Italy. *Agriculture Ecosystems and Environment*, 181, 50–57.
47. Yavuz, M. Y., Demirel, K., Erken, O. (2010). Emitter clogging and effects on drip irrigation systems performances. *African Journal of Agricultural Research*, 5(7), 532–538.
48. Zamaniyan, M., Fatahi, R., Boroomand-Nasab, S. (2013). Evaluation of emitters and water quality in trickle irrigation systems under Iranian conditions. *International Journal of Agriculture and Crop Sciences*, Online at www.ijagcs.com.

PART II

CURRENT STATUS OF WASTEWATER MANAGEMENT

CHAPTER 6

REUSE OF MUNICIPAL WASTEWATER FOR IRRIGATED AGRICULTURE: A REVIEW

JUAN C. DURÁN–ÁLVAREZ and BLANCA JIMÉNEZ–CISNEROS

CONTENTS

Modified from Juan C. Durán–Álvarez and Blanca Jiménez–Cisneros, 2014. Beneficial and negative impacts on soil by the reuse of treated/untreated municipal wastewater for irrigation—a review of the current knowledge and future perspectives. Chapter 5, 137–197, In: *Environmental Risk Assessment of Soil Contamination by M. C. Hernandez (ed).* <http://www.intechopen.com/books/environmental-risk-assessment-of-soil contamination/beneficial-and-negative-impacts-on-soil-by-the-reuse-of-treated-untreated-municipal-WW-for-a>

6.1   INTRODUCTION

The scarcity of water for human use, such as food and energy production, manufacturing, drinking water and ecosystem conservation is a global problem for which the solution goes beyond merely the preservation of freshwater sources [60, 170]. Although three quarters of the Earth's surface is covered by water, most of this water is either contained in oceans or confined in glaciers [199]. The volume of freshwater (FW) available for human activities (less than 1%) is unequally distributed throughout the globe; in some cases this water is confined to the deep sub-soil or is polluted [182]. Furthermore, the desertification of large areas caused by climate change has intensified the lack of water sources in cities and rural areas throughout the world [70]. Water scarcity results in food scarcity, since 70% of the water withdrawn for human activities goes to agriculture [58]. In zones where rain-fed agriculture is practiced, decay in crop yields is observed when droughts occur, which results not only in the scarcity of food but also the decrease in incomes due to falling crop sales [160]. The use of freshwater for irrigation limits the volume of freshwater available for human consumption; therefore, recycling of water becomes necessary for irrigation in dry zones [56, 90]. The idea of reusing wastewater (WW) to irrigate is not new; it actually originated around 3000 B.C. People in these ancient civilizations knew that WW contained both water and compounds that benefited the soil and thus they used it in a planned way to increase crop yields [52].

Commonly, reusing WW in agriculture is considered a deleterious practice since it may introduce pollutants to the environment, spread waterborne diseases, generate odor problems and result in aversion to the crops. Nevertheless, this kind of reuse may result in some benefits for soils, crops and farmers. Nowadays, the reuse of WW in agriculture is seen in some countries as a convenient environmental strategy [8, 131]; municipal WW is therefore considered an appropriate option for reuse. This kind of WW contains a significant load of biodegradable organic material (carbon and nitrogen) as well as most of the mineral macronutrients (e.g., phosphorous, potassium, magnesium and boron) and micronutrients (e.g., molybdenum, selenium and copper), which are necessary for the growth of crops. Accumulation of organic matter in soil by irrigation with WW can be beneficial as it may result in the enhancement of the physical structure of the soil, the increase in the soil microbial activity and the improvement of soil performance as a filter and degrading media for pollutants. Conversely, a fraction of the organic matter contained in WW is due to the occurrence of organic pollutants (e.g., polyaromatic hydrocarbons and polychlorinated biphenyls) and pathogenic microbial agents [13, 103]. Because of the presence of organic, inorganic and microbial pollutants in WW, a prior step of depuration is necessary before reuse in irrigation in order to avoid the pollution of soil, crops and the nearby water sources, and thus the dissemination of waterborne diseases or the degradation of soil. The extent at which WW has to be treated prior to irrigation depends on the restrictions established in local or international water quality criteria for irrigation [56, 90, 124].

Primary treatment schemes (coagulation-flocculation with sedimentation or aerobic/anaerobic stabilization pounds) are used for treating WW to irrigate crops that are not intended for human consumption (e.g., fodder), while secondary treatment of WW (biological treatment followed by disinfection) is recommended when unrestricted crops are irrigated [180, 209]. In developing countries, most or the whole volume of WW produced in cities is treated prior to irrigation, while in low income countries WW treatment is not a priority, and thus untreated or partially treated WW or a mixture of treated and untreated WW is commonly used for agricultural purposes [103, 108]. In Mexico, China, India and Pakistan, for instance, large areas exist where untreated WW has been reused in irrigation for a considerable time [110]. The World Health Organization estimates that nearly 20 million hectares throughout the world are irrigated using untreated WW [208]. It is also reported that in some cities up to 80% of the vegetables locally consumed are produced using WW for irrigation [62]. The application of WW to soil, particularly untreated WW, followed by its infiltration poses a significant risk of pollution, not only to soil and crops but also to the surface and subterranean water sources surrounding the irrigated area [117, 198].

Pollution by pathogenic agents is the main cause of concern regarding the application of treated/untreated WW to soil. Due to the variety of microorganisms entering the soil via the WW there is a high risk of enteric disease outbreaks for farmers and consumers [44, 142]. This chapter addresses the contamination of WW irrigated soils by helminthes (intestinal worms) and pathogenic bacteria common in developing countries (where untreated WW is used to a greater extent), as well as the risk of outbreaks of parasitic diseases for both farmers and consumers in agricultural areas where untreated WW is reused. The occurrence of antibiotic resistance in indigenous organisms of soil and pathogens reaching soil via WW is gaining the attention of scientists and health organizations around the world [15, 120], thus a review of what it is known and the research opportunities in this field are presented in this chapter.

A review on the presence of some organic contaminants of emerging concern, such as pharmaceutical substances, personal care products and industrial additives, in WW–irrigated agricultural soils is presented in this chapter along with some of the known potential effects caused to soil organisms, plants and consumers. With regard to organic pollution, a current topic of interest is the entry to the soil and potential risks within crops of so-called "contaminants of emerging concern." These pollutants are substances that have not previously been considered as pollutants since they are part of everyday products; however, due to the subtle but harmful effects that these substances may cause in a variety of aquatic and terrestrial organisms, concerns have risen due to their continuous entry into the environment via WW [19]. Such effects have just begun to be elucidated, and only for some groups of contaminants of emerging concern [45, 177], even though it is now known that up to 7 million commercially available chemicals are routinely disposed of in sewage after use [51]. In this regard, this chapter makes some suggestions regarding the next steps in the toxicity studies for this class of pollutants, such as testing the synergistic effects of mixtures of contaminants of emerging concern in soil organisms.

In spite of the variety and quantity of contaminants that soil regularly receives through WW irrigation, this ecosystem possesses self-purification processes that maintain homeostasis within the system. Such self-purification processes may either inactivate or reduce the population of pathogenic microorganisms reaching the soil via WW through predation by the indigenous microbiota within the soil [71, 169], the production of antibiotics by some organisms in the rhizosphere [21] and by retention of microorganisms in the surface layers of the soil profile through physical and chemical processes. For organic pollutants, mechanisms such as photolysis and biodegradation promote the dissipation of contaminants in the soil, while adsorption onto the soil particles lead to the retention—and the potential confinement—of organics within the solid matrix [215].

In this chapter, current knowledge concerning the environmental fate of pathogen and organic contaminants of emerging concerns in WW reuse in irrigated soils is discussed. The chapter highlights the laboratory approaches that show the best results in simulation of the field conditions. Knowledge of the environmental fate of contaminants in irrigated soils is important in order to perform more accurate risk assessment studies on contamination of water sources, soil and crops in WW irrigated areas. Furthermore, it provides information to policy makers to make proper legislation aimed at promoting environmentally responsible management of treated/untreated WW in agricultural irrigation.

Depuration of WW prior to its reuse is the most plausible option to prevent soil pollution by WW reuse. However, since WW represents a cheap source of water and fertilizer for farmers [65], it is necessary to consider the needs of users before planning schemes of WW treatment. The use of WW treatment systems aimed at removing carbon, nitrogen, phosphorous and minerals in WW leads to the reduction in quality of effluents as fertilizers, impacting crop yields and thus in the livelihood of farmers. In this sense, the use of advanced primary treatment systems can be a feasible option to: (a) remove suspended solids, pathogens and heavy metals in WW without significantly impacting the content of nutrients in effluent; (b) preserve the quality of agricultural soils to properly perform ecosystem services such as the production of food; and (c) fulfill the needs of farmers that use WW as a source of water and nutrients. Treating WW by these kinds of systems may be an opportunity to couple sanitation with reuse within a program of comprehensive management of WW, the recycling of nutrients and the use of soil as a food producer and purification system.

This chapter aims to describe what it is known and what it is unknown regarding the positive and negative impacts of the reuse of treated/untreated WW in irrigation. It is shown in details how this practice can benefit soil/water and farmers, while at the same time posing a risk of contamination to the ecosystem. Emphasis is given to the purification processes occurring in the soil and how soil manages the continuous entrance of pollutants via WW. Lastly, some perspectives for further studies on the presence and environmental fate of pollutants in WW irrigated soils are proposed.

6.2   IMPACTS OF WASTEWATER REUSE IN AGRICULTURE

The reuse of WW results in both beneficial and negative impacts on soil, some of which are explained in this section. The aim is to identify both and to understand their origins in order to assist scientists and policy makers to balance them and even to greater advantage of the benefits compared to the drawbacks in certain situations.

6.2.1   BENEFITS OF WW REUSE IN IRRIGATION

Figure 6.1 summarizes the positive impacts of reusing WW in irrigation. The extent of the positive impacts depends on local conditions of the specific project.

FIGURE 6.1   Beneficial impacts of reusing WW for irrigation [14, 18, 21, 36, 37].

6.2.2   BENEFITS IN CROPS

Since WW is produced constantly and thus is always available, it is possible to select a wider range of crops to be sown year-round, specifically those of high profitability which normally have higher and more stringent water demands in terms of quantity and timing. The consistent use of WW in irrigation may stabilize the content of nutrients in the soil, even when growing crops with high nutritional requirements; this is because the continuous withdrawal of nutrients by plants is compensated by the constant input of organic and mineral components into the soil via WW. Examples of how the reuse of WW has led to increases in crop yields in arid zones can be found worldwide. Studies conducted in Hubli–Dharwad, India, showed that irrigation with treated and untreated WW made it possible to produce vegetables during the dry season; yields and selling prices increased by 3–5 times compared to the monsoon season [132]. In Pakistan, Ghana and Senegal the reliability and flexibility of WW supply allows rural and urban farmers to cultivate profitable crops in a shorter time, resulting in 3 to 6 harvests per year [82, 200]. Treated/ untreated WW is a source of organic matter and the same large diversity of nutrients contained in any formulated fertilizer. It is estimated that 1,000 m^3 of municipal WW applied to one hectare can contribute 16–62 kg of organic nitrogen, 4–24 kg of phosphorus, 269 kg of potassium, 18–208 kg of calcium and 9–110 kg of magnesium each year [108]. Table 6.1 shows the contribution of water and nutrients that untreated WW make to several crops.

Nitrogen is a plant macronutrient, which can be found in the form of nitrate ions (N–NO_3), mostly in treated WW, or as ammoniacal nitrogen (N–$NH^{4+)}$ and organic nitrogen in untreated WW. The sum of all these forms is known as total nitrogen (TN). Most crops absorb nitrates to the greatest extent (85% of the nitrate contained in WW); whereas 50% of ammoniacal and less than 30% of organic nitrogen contained in WW can be assimilated by plants. The remaining nitrogen is taken up by soil microorganisms and transformed into nitrates or volatilized as N_2. In WW irrigated soils, organic nitrogen is transformed into nitrates by soil microorganisms to a greater extent than that observed in non-irrigated agricultural soils [38]. Problems related to high inputs of nitrate ions are due to their high solubility in water, and thus their rapid percolation through the soil to the aquifer.

TABLE 6.1   Contribution of Nutrients and Sodium From Untreated WW and Water Requirements of Selected Crops [108]

Crop	Water requirements	Nutrients and sodium contribution by WW (kg.ha/year)					
	(mm/year)	N_{total}	P_{total}	K	Ca	Mg	Na
Alfalfa	1360	218–843	54–326	27–938	245–2829	122–843	367–2475
Barley	516	83–320	21–124	10–356	93–1074	46–568	139–939
Beans	370	59–229	7–89	7–255	67–770	33–407	100–673
Chili	601	96–373	24–144	12–415	108–1250	54–661	162–1094
Green tomatoes	653	104–405	26–157	13–451	118–1358	59–718	176–1188
Maize	673	108–418	27–162	13–465	121–1401	61–741	182–1226
Marrow	364	58–226	15–87	7–251	66–757	33–400	98–662
Oats	353.6	57–219	14–85	7–244	64–735	32–389	95–644
Wheat	520	83–322	21–125	10–359	94–1082	47–572	140–946

A significant quantity of nitrate leaching through soil subsequently becomes unavailable for plants; this does not necessarily represent a problem, as nitrate is continuously supplied to soil via WW. More important, the presence of nitrates in subterranean water is related to occurrence of methemoglobinemia disease in infants ingesting nitrate at levels higher than 45 mg/L via drinking water [64, 168]. The quantity of nitrogen washed out from soil depends on the irrigation rate, the frequency of rain events, the type of crops sown and the characteristics of the soil [87]. The amount of nitrogen that can be applied to soil to produce minimal nitrate leaching rates depends on the demand of crops, which usually varies between 50 and 350 kg of nitrogen per hectare [87]. Such demand is within or slightly above the amount of nitrate supplied by treated WW. In this sense, the limited removal of nitrogen by WW treatment would not significantly affect the input of this macronutrient to agricultural soils.

Phosphorous is another plant macronutrient, which is very scarce in soil, at the point it needs to be added through the application of fertilizers. Due to its stability and low solubility, this nutrient can be accumulated in soil. WW normally contains small amounts of phosphorous, so its use for irrigation is beneficial to plants and it does not impact negatively upon the environment, even if applied consistently for long periods of time [53, 87]. The recycling of phosphorous and nitrogen in WW–irrigated soils is important because it allows closure the P-cycle rather than its breakage. Breakage of the cycle occurs when phosphorous is removed from WW during treatment, becoming trapped in sludge and dumped to confinement sites or landfills. An advantage of the availability of phosphorus in WW is that it is partly bound to organic components and thus it cannot form complexes with iron or aluminum ions upon its entry to soils [108]. In contrast to phosphorous, potassium is contained in soil at high concentrations (around 3% of the lithosphere) but in chemical forms that impede its bioavailability. As a result it is necessary to add potassium to soils via fertilizers. Approximately 185 kg of potassium per hectare are required to cultivate some crops [108]. Sewage contains low concentrations of potassium, insufficient to cover the theoretical demand in most cases. Meeting the demand for potassium in irrigated soils will depend on the amount of WW supplied at each irrigation event, the WW quality and the frequency of irrigation. Fertilization with potassium has not resulted in adverse impacts to the environment [147]. Recycling nutrients by the reuse of WW promotes savings in energy, which would otherwise be consumed in the production of fertilizers [173]. In particular, the recycling of phosphorus is important since the world's phosphorus reserves are becoming scarce [189]. Fertilizing agricultural soils by the reuse of WW invariably leads to the increase of crop yields. An example of this can be found in Mezquital Valley, Mexico [111].

Table 6.2 shows the differences in the agricultural production in croplands of Mezquital Valley when either untreated WW or groundwater is used for agricultural irrigation. The use of WW in Mezquital Valley has also contributed to changing the landscape of the zone, transforming barren soils into productive and green vibrant soils, as shown in Fig. 6.2.

6.2.3   BENEFITS IN SOIL QUALITY

In order to define the improvements in soil quality produced by the application of treated/ untreated WW, it is necessary to establish the use of the irrigated soil. It is known that soil complies with five ecological

TABLE 6.2 Comparison of Crop Yields For Some Vegetables in Plots Where WW and Groundwater Are Used For Irrigation in Mezquital Valley, Central Mexico [108, 144]

Crop	Crop yield (tons/ha)		Increment (%)
	Untreated WW	**Groundwater**	
Alfalfa	120.0	70.0	71
Barley	4.0	2.0	100
Chili	12.0	7.0	70
Corn	5.0	2.0	150
Oats for forage	22.0	12.0	83
Tomato	35.0	18.0	94
Wheat	3.0	1.8	67

FIGURE 6.2 Comparison of untreated WW irrigated (right) and rain-fed (left) croplands in Mezquital Valley, Central Mexico [108, 144].

functions: (a) a medium for plant growth (including agriculture); (b) a biodiversity pool and habitat for plants and (micro and micro) fauna; (c) a carbon sink; (d) a storage, filter and transforming medium for nutrients, pollutants and water; and, (e) a landscaping and engineering medium [46]. This chapter focuses on the functions of soil as a medium for plant growth as well as in its role as a transforming medium for nutrients and pollutants.

In addition to the continuous supply of nutrients to the soil, irrigation with treated/untreated WW confers significant improvements in soil quality. Favorable changes reported in irrigated soils comprise: (a) an improvement in the physical structure of soil; (b) an increase in soil microbial activity; and (c) the improvement of the soil performance as a WW treatment system.

6.2.3.1   Improvement of the Physical Structure of Soil

The physical structure of soil is defined as the arrangement of the solid particles and the size, shape and interconnection of pores and voids. Soil structure is closely related to its capacity to store and transport gases and water (and thus dissolved substances) [128]. Gas exchange between the soil and the atmosphere determines whether aerobic, anoxic or anaerobic conditions prevail within the soil. This in turn regulates the metabolism of soil microorganisms and impacts, inter alia, upon the nitrogen fixation, the transformation of soil organic matter and the degradation of pollutants. Additionally, the physical structure of soil affects the plant growth by influencing root distribution and thus the ability to take up water and nutrients [157]. Improvements in the physical structure of soil are related to the increase in both the stability of the soil aggregates and soil porosity. The enhancement of the physical structure of soil results in a rise in agronomic productivity, the augmentation of water infiltration through soil to the aquifer and a decrease in erodibility [26]. The hierarchical theory of aggregation proposes that micro-aggregates (particle size below 250 µm) in the soil are formed initially by the attachment of organic material to some inorganic components of soils (e.g., clay and hydroxides); in turn these micro-aggregates join together to form macro-aggregates (particle size above 250 µm). Alternatively, macro-aggregates can form around the particulate organic matter, while exudates produced by soil microorganisms serve as cementing agents, making micro and macro-aggregates more stable [59]. Micro-aggregates can be also formed from bacterial colony clusters, which use bacterial polysaccharide exudates to bind with clay particles. The clay particles act as a protective shell for clusters and macro-aggregate formation continues as here [33].

Since the formation of aggregates in the soil is related to the presence of organic matter, and in some cases microorganisms, it might be expected that the continuous supply of these two elements via WW would result in the increased formation and stability of soil aggregates and thus an improvement in the physical structure of soil. For example, the study referred in Ref. [144] establishes that increased soil microbial activity due to the augmentation of organic carbon content by the application of WW impacts positively upon the stability of soil aggregates.

Furthermore, there are substances contained in WW other than organic matter and microorganisms that may contribute to the formation and stability of soil aggregates. Calcium and magnesium cations, which are abundant in WW, increase the formation of micro-aggregates through cationic bridging between clay and organic matter, resulting in aggregation. In arid soils and soils with low organic matter contents, insoluble calcium and magnesium carbonates can trigger the formation of soil micro and macro-aggregates [33]. Additionally, calcium can inhibit clay dispersion, and thus the breakup of aggregates, when sodium concentration increases in soil [11]. Dissolved organic matter in WW can form complexes with iron and aluminum in soil forming mobile organo-metallic compounds which can further precipitate and act as cores for micro-aggregates formation. Particulate organic matter (i.e., suspended solids in WW) may enhance the binding of micro-aggregates to subsequently form macro-aggregates; for instance, extracellular polysaccharides of microorganisms in the surface of suspended solids can act as binding agents in the formation of macro-aggregates [106]. In the case of phosphorous, the formation of insoluble aluminum and calcium phosphates in the soil can induce the formation of micro-aggregates and additionally it may act as a macro-aggregate binding agent [98]. The entry of certain chemicals to the soil via WW increases the stability of soil aggregates. For example, hydrophobic substances (e.g., surfactants, lipids and hydrocarbons) decrease the wettability of aggregates by inducing water repellency, which in turn leads to increased cohesiveness and low decomposition rates of soil aggregates [33]. Agricultural activities in WW irrigated soils may also contribute to the improvement of the physical structure of soil. Previous studies have found that some crops (i.e., maize, alfalfa and leguminous plants) have beneficial effects on the conservation of the physical structure of soil. Aggregation of soil particles tends to increase when planting crops characterized by high density and long length of roots; this is because chemicals released by roots (i.e., mucilage) enhance the stability of soil aggregates in the rhizosphere by increasing the bond strength and decreasing the wetting rate [49]. According to the study reported in reference [97], roots of leguminous crops increase the aggregation of soil particles. Corn (Zea mays) residues (leaves and shoots) also increase aggregation of soil particles compared with other crops; this is attributable to the liberation of phenolic compounds from plant tissues, since phenols favor the agglutination of particles and prevent wetting [97, 167]. Municipal and industrial WW may also be a source of phenolic compounds to soil through irrigation, producing similar effects to those of corn wastes [34]. The study referred to in Ref. [167] demonstrated that the stability of soil aggregates is high for continuous cultivation of alfalfa (Medicago sativa), while the opposite effect was observed for soybean. This is attributable to the low concentration of phenols in the latter [143]. Some studies have addressed the changes in the physical structure of agricultural soils caused by long-term irrigation with WW. The results of these studies show a decrease in soil porosity caused either by occlusion of pores by the suspended solids contained in WW or by the augmentation of micro-pores (radius < 0.01 µm) in the soil matrix [47, 48]. Depending on the method of water application during irrigation, an increase in the compaction of soil may be observed in the plot after an irrigation event [205]. Soils irrigated by flooding exhibit high compaction while water-dropping effects (erosion) may be observed in soils irrigated by spraying. In any of both cases, WW irrigated soils exhibit large populations of earth-worms, which may assist in the formation and connection of pores within the soil matrix. Undoubtedly WW contains agents that improve the physical structure of soil. However, studies performed so far show contrasting results, either an increase in the soil micro-porosity or soil compaction. It is therefore necessary to carry out studies aimed at measuring changes in the physical structure of soil throughout several irrigation cycles and for longer periods (months or years); additionally, it is of interest to assess changes in the physical structure of soil at landscape level (piedmont or catena), as it may be useful for evaluating the horizontal displacement of soil particles and nutrients.

6.2.3.2   Increase of Soil Microbial Activity

Either due to the extra supply of organic carbon or because of the addition of microorganisms via WW, microbial activity in WW irrigated soils tends to be higher than that found in non-irrigated soils [74, 194]. This increase in the microbial activity of the soil brings benefits to both agriculture and the development of flora and fauna in the soil ecosystem. The C/N ratio in soils irrigated with WW for long periods tends to decrease by up to 45%, which implies an improvement in the nutritional conditions for soil micro-organisms [69]. The authors report an increase in the population of copi-notrofic and oligotrophic bacteria (234 and 217%, respectively), and in the populations of actinomycetes (234%) and fungi (206%) in soils irrigated with WW for 100 years compared with those populations found in non-irrigated soils. Rises in the metabolic activity of soil, measured as the pro-duction of ATP and enzymatic activity have been also reported [74, 194]. According to reference [69], soil enzymatic activity remained unchanged 20 years after WW irrigation ceased. In contrast, elevated microbial activ-ity in soils irrigated with treated WW decreases after few days without irrigation [116]. Due to the augmentation of the populations of bacteria, actinomycetes and fungi in the irrigated soil, a rise in the rhizospheric activity is experienced, resulting in:

a. the increase in the growth and development of plants;
b. high rates in stabilization of organic matter entering the soil through WW;
c. higher performance of the depuration of WW and degradation of the pollutants fixed in the soil in comparison with non-irrigated soils; and,
d. the improvement in the formation and stability of soil aggregates.

The latter may be explained by the role of polysaccharides exuded by bacteria as transient binding agents, which initialize aggregation of soil micro-aggregates [116]. The transformation of carbon and nitrogen by soil microorganisms supports the proliferation of soil (micro and macro) fauna, which is essential for soil formation as well as for the develop-ment of plants. The use of treated WW to irrigate an agricultural soil over 20 years has resulted in the improvement of the metabolic efficiency of soil microflora to transform carbonaceous and phosphorous substances into nutrients readily available to plants and macrofauna [9].

Soil biomass has proven to be capable of adsorbing a certain proportion of heavy metals contained in the WW. The biosorption rates for cadmium and nickel within the range of 5–55 mg/g of biomass have been reported in an irrigated soil with WW for two decades [10]. In that soil, the predomi-nant bacteria after irrigation were Enterobacteriaceae and Pseudomonas.

The effects of WW irrigation on soil nitrogen fixing organisms have not been investigated thoroughly. An increase in soil nitrifying activity accompanied by a low rate of denitrification has been observed in WW irrigated forest soils [119]. However, a peak in N_2O production in a soil irrigated with treated WW was reported, followed by an immediate drop in gas production [172]. So far, the metabolic processes performed by different soil microbial species in WW irrigated soils have been explored very little. However, it is important to keep in mind the important role that soil microorganisms play in both the development of the soil and plants as well as in the purification of WW, when planning agricultural systems based on the reuse of WW. Even when soil microbial populations show some kind of resilience to a wide variety of contaminants, some other chemicals can cause not only toxic effects to soil microorganisms but the proliferation of pathogenic organisms and the occurrence of antibiotic resistance within the agricultural soils.

6.1.1.3   Improvement of Soil Performance as a WW Treatment System

The WW can be purified through its application into soils and infiltration of WW accelerates the process. In practice, specific WW treatment systems are based on soil infiltration, which have been demonstrated to improve water quality to levels obtained using tertiary treatment systems [30, 31]. Purification of WW is one of the ecological functions of soil; through this mechanism, soil maintains, at least partially, the quality of surface and groundwater bodies. The extent at which this natural system works is highly variable, from almost nonexistent to very high, depending on local conditions and types of pollutants.

Table 6.3 shows the extent to which pollutants in WW are removed by infiltration through the soil. The application of WW to soil reduces the content of pathogenic microorganisms by 6–7 log units for bacteria and 100% for helminthes and other protozoa. Total organic carbon can be reduced by up to 90%, while levels of recalcitrant compounds in WW, such as phosphorus (20–90%), nitrogen (20–70%), and metals (70–95%) are also reduced dramatically. In sewage, organic phosphorus (5–50 mg/L) is biologically converted to phosphate; subsequently, in alkaline or calcareous

TABLE 6.3   Processes in Soil That Improve the Quality of the WW

Parameter	Effects
Heavy metals	Heavy metals can be removed by the formation of complexes with soil organic matter, precipitation or methylation at efficiencies of 70–95%.
Micro organisms	*Helminth eggs* and protozoa are easily removed by straining in the soil surface; bacteria and viruses can also be adsorbed onto the soil particles and then desiccated or killed by indigenous soil microorganisms. The performance of these processes depends on the texture, physical structure and organic matter content of soil.
Nitrogen	Nitrogen is removed from water at a level similar to tertiary treatment systems by transformation in soil as well as by assimilation by soil microorganisms and plants.
Organic matter	Biodegradable material is reduced by more than 90%, while less readily biodegradable material is adsorbed and later biodegraded or volatilized.
Phosphorus	Phosphorous is reduced to levels of 1 mg/L or less by assimilation by plants.
Toxic organic compounds	Most are retained in soil and then biodegraded at different rates.

soils, phosphate precipitates with calcium to form calcium phosphate and remains available for plants. In contrast, in acidic soils phosphate reacts with iron and aluminum oxides to form insoluble compounds, which are unavailable to plants. Sometimes soluble phosphate is initially immobilized by adsorption onto soil particles and then slowly returns to insoluble forms, allowing for further adsorption of mobile phosphate. This process is generally known as phosphate aging [30].

Most of the organic compounds (natural and synthetic) in sewage are rapidly transformed in soil to stable, and in some cases non-toxic, organic compounds (e.g., humic and fulvic acids). Actually, soil biodegrades a greater amount and variety of organic pollutants than that reported for water streams. WW application to soil under controlled conditions (e.g., limited irrigation rate and intermittent flooding) permits the biodegradation of hundreds of kilograms of carbonaceous substances per hectare per day, with no impact on the environment [30]. Total organic carbon levels in WW are dramatically reduced from levels of 80–200 mg/L to 1–5 mg/L in the infiltrated water [166]. Heavy metals can be removed from WW during soil infiltration and confined within the organic domain of the soil for several hundred years. Metals are retained in the surface layer of the soil either by complexation with soil organic matter or by precipitation at high pH values. Only a small fraction of metals infiltrates to lower layers of the soil profile and even less can be assimilated by crops. For instance, around 80–94% of cadmium, copper, nickel, and zinc can be removed in the first 5–15 cm of the soil profile, 5–15% is leached to lower layers and only 1–8% can be absorbed by grass [162]. A similar process occurs with fluorine [14]. This phytoremediation process is used to treat WW in planned natural treatment systems such as wetlands. However, it is necessary to be aware that some edible crops are able to take up heavy metals to a greater degree than grasses [12].

The capability of soil to act as a filter and transforming medium for WW pollutants can be observed in both long-term and newly WW irrigated soils [30, 105]. The operation of this natural purification system is closely related to the physical and chemical properties of the soil and thus modifications in soil characteristics caused by irrigation with WW may either improve or worsen the performance of this natural WW treatment system. The increase in the soil organic matter content is the main factor resulting in an improvement in the removal of biological, organic and inorganic pollutants as WW leaches through the soil. This is because soil organic matter promotes the immobilization of pollutants either by adsorption or formation of complexes, while at the same time stimulating the proliferation of degrading microorganisms [105, 136]. Regularly, heavy metals are fixed in the upper layers of the soil profile by complexation with organic matter [74], thus organic matter enrichment in WW irrigated soils results in greater retention of heavy metals by the solid matrix. Heavy metals cannot be biodegraded but they may be modified by soil microorganisms. Biological methylation of metals and metalloids, such as selenium, arsenic and mercury, has been reported in WW irrigated soils. It is expected that this process is elevated in WW irrigated soils, where microbial biomass occurs at higher levels than in non-irrigated soils. Methylation of heavy metals leads either to reduced toxicity or increased loss of metals in soil through volatilization [75, 127].

Alternate process, observed in long-term WW irrigated soils, related to those aforementioned, is the potential of soil microorganisms to develop resistance to the harmful effects caused by the presence of heavy metals in the solid matrix [5, 10]. Such resistance is similar to that developed to antibiotics and has been reported for cadmium, chromium, zinc and nickel in soils irrigated with WW over the long term [6, 10]. It is plausible that the expression of these resistances results in an increase of heavy metal methylation in the soil, which allows soil microorganisms to survive and to continue with those metabolic functions that increase agricultural productivity and purify WW. With regard to organic contaminants, the increase in the soil organic matter content produces, in most cases, an incremental boost in the adsorption of solutes onto soil particles. Increased adsorption of organic contaminants (i.e., pesticides, pharmaceuticals and estrogenic hormones) in long-term WW irrigated soils has been reported compared to rain-fed soils from the same agricultural area [150]. Organic compounds displaying high hydrophobicity are adsorbed by soil not only faster and to a greater extent but with greater strength than is observed for semi-polar and polar compounds [202]. The increase in the hydrophobicity of soil due to the application of WW increases the capacity of such soils to strongly retain non-polar organic contaminants within the solid matrix. The increase in the adsorption of organic pollutants by soil results in an extended retention time in the solid matrix, encouraging biodegradation processes. Similar to the results reported for adsorption, higher rates of biodegradation of organic pollutants have been observed in treated/untreated WW irrigated soils compared to non-irrigated ones [151]. This may be caused, on the one hand, by the continuous supply of organic matter to the soil via WW, which can be used by soil microorganisms as co-substrate in the biodegradation of target organic pollutants, and on the other hand, by the prolonged exposure of soil organisms to pollutants. The latter case can be understood as the acclimation of the degrading organisms to the occurrence of organic pollutants in the soil followed in the short term by the acquisition of the capability for using organic contaminants as a carbon source. The increase in the soil organic matter content caused by WW irrigation has a positive impact not only in the adsorption of organic compounds but also on the retention by soil of WW-borne pathogens. This is due to the high affinity of the cell membranes to the organic domain of soil. The study referred to in Ref. [129] reports a higher adsorption of enteric bacteria *Escherichia coli* and the enteric protozoa *Giardia lamblia* in long-term WW irrigated soils compared with long-term groundwater irrigated soils from the same agricultural zone.

In general terms, an increase in pH values has been observed in agricultural soils irrigated with treated/untreated WW; although in less cases soil pH tended to decrease following the application of WW [127]. The first phenomenon is attributed to the continuous addition of salts (carbonates, calcium, magnesium, sodium) in WW. The second case is explained by the high mineralization rate of organic matter in the irrigated soil, which is highly dependent on the soil type, the climatic conditions of the site, and the quality of WW, among other reasons. The increase in soil pH, in combination with the continuous supply of organic matter, results in the buffering of soil pH, which prevents the drop of soil pH values during rain events (including acid rain). Stabilization of soil pH values also contributes to the retention of heavy metals in the surface layers of the soil by the formation of insoluble basic salts. Furthermore, basic values of soil pH can facilitate the adsorption of neutral and basic organic contaminants; as these compounds tends to be better adsorbed to neutral and basic soils than to acidic ones.

Since WW irrigation improves the physical structure of soil (i.e., increased formation and greater stability of aggregates), aerobic conditions may be maintained within the soil matrix; which in turns contributes to an increase in the aerobic biodegradation rate of organic pollutants. Additionally, an increase in the adsorption of pollutants can be achieved in better-structured soils due to the increase in the specific surface area of soil particles. Moreover, higher biodegradation of the adsorbed contaminants can be expected as long as they remain available to microorganisms after adsorption. In irrigated soils where occlusion of the pores by the suspended solids in WW occurs, anoxic conditions may be achieved. Under such conditions, toxic species of heavy metals are chemically reduced into non-toxic species (e.g., Cr^{+6} into C^{+3} and As^{+5} into As^{+3}), then they may be immobilized by the formation of insoluble hydroxides. The extent to which WW irrigation contributes to the function of the soil as filter and degradation medium for pollutants is just beginning to be studied. The potential of soil to act as an efficient WW depuration system is a powerful argument to convince policy makers that irrigation with treated/untreated WW can be an appropriate strategy to simultaneously solve problems of water stress and low agricultural productivity with no negative impacts in the quality of water sources surrounding the irrigation site. This, of course, is achieved when all of the appropriate precautions to avoid contamination are taken at each site.

6.3 NEGATIVE IMPACTS OF WASTEWATER REUSE IN IRRIGATION

The main drawback of reusing treated/untreated WW in agriculture is the pollution of soil, the potential contamination of crops and water sources, and the inherent risk of harmful effects that contamination poses to the exposed organisms. Even when soil acts as an efficient living filter to remove, inactivate and transform the pollutants contained in WW, it is not fully effective at eliminating some of them. Moreover, as a result of the increasing industrial development, WW irrigated soils continuously receive newly synthesized substances, which may negatively impact the effectiveness of soil as a treatment system by poisoning the degrading microorganisms, destroying the physical structure of soil or damaging the natural cycles occurring within soil. The pollutants received by soil via WW may be different in developing and developed countries. Examples of this include pathogenic microbial agents. In developed countries most WW is treated prior to reuse and thus pathogens are not present in irrigation water, while in developing countries untreated WW is used in most of cases. Pathogens vary for different zones; for instance, the enteric protozoa Giardia is commonly found in WW of developing countries (Latin American and African countries), while the parasitic protozoa Cryptosporidium occurs in developed countries (United States and western European countries). Similar to microorganisms, some organic pollutants can be found in WW from developing countries and not in developed countries. Examples include some herbicides (e.g., DDT and atrazine) whose use is restricted in developed countries; on the other hand, nanomaterials and new-generation antibiotics, all of which are much more likely to occur in WW of developed countries. The determination of pollutants in soil initially requires specific sampling methods, which take into consideration the heterogeneity of the soil matrix. In addition, specialized extraction techniques able to efficiently isolate analytes (or microorganisms) from soil are necessary prior to analysis. Specialized analytical methods have been developed and validated for the determination of trace contaminants and microorganisms in soil. However, in most cases, these methods are time-consuming, expensive and require the use of specialized reagents and personnel.

It is therefore necessary to continue research towards the development of simpler and environmentally-friendly analytical techniques. Determining the occurrence and concentration of contaminants in soil is a task that requires a significant effort. However, this is only a part of the job. The study and understanding of the environmental fate of contaminants in soil is also a priority task to accomplish truly useful environmental risk assessment studies comprising soil, water sources, crops, farmers and consumers. Knowing the environmental fate of contaminants in the soil is necessary to understand the potentialities and limitations of each soil as a natural purification system of WW and an effective tool to define the capacity of each site to support WW irrigation in agriculture. Since soil is a complex and heterogeneous matrix, the fate of contaminants can vary significantly from one site to another. In this sense, it is worth defining which parameters are determinant in the fate of contaminants within soil and, on the basis of this knowledge, elucidating the fate of contaminants in other sites using mathematical tools to achieve such extrapolations. In this section, attention will be focused on pathogenic microbial agents, heavy metals and organic pollutants contained in municipal WW. The occurrence of such pollutants in WW–irrigated soil as well as their environmental fate in soil is addressed; additionally the most significant effects of these contaminants will be treated in some detail. Lastly, perspectives for further studies on the occurrence and fate of the studied pollutants in soil are presented.

6.3.1   SOIL POLLUTION BY PATHOGENIC MICROBIAL AGENTS

Contamination of soil and crops by pathogenic agents is the effect of WW reuse in agriculture that receives most attention from environmentalists and scientists. Municipal WW contains a huge quantity and variety of bacteria, protozoa and viruses passed from human and animal feces and urine; therefore this water is a vector for intestinal infections (although some other diseases can spread from the environment via WW). Exposure may be direct through contact or ingestion of WW and soil, or indirect through contact with sick people or by ingestion of polluted crops, meat or milk. There are four groups at risk:

a. farmers and their families,
b. crop handlers,

c. product consumers, and

d. people living nearby to irrigated fields.

For any of these groups children and elderly are the most vulnerable, especially when they are undernourished. The most affected group is agricultural workers due to high exposure to WW and contaminated soils [208]. Table 6.4 shows the risk of infection of water-borne diseases for vulnerable groups in irrigated areas using treated/untreated WW.

6.3.1.1 Effects Caused by Microbial Pollution in Soil

Several diarrheal outbreaks have been associated with the use of WW to irrigate [184, 208]. However, since this occurs in places where sanitation/hygiene practices and drinking water are of low quality, it is always difficult to define the specific contribution to the total burden of diseases. Cholera, caused by the bacterium Vibrio cholera, is one infection closely linked to WW irrigation in poor countries. Other intestinal diseases related to the use of WW to irrigate are traveler's diarrhea caused by *Escherichia coli*, shigellosis caused by Shigella spp., gastric ulcers caused by Helicobacter pylori, giardiasis caused by the parasitic protozoan Giardia intestinalis and amebiasis caused by Entamoeba histolytica. Additionally, viral enteritis (caused by rotaviruses) and Hepatitis A are the most reported viral infections caused by consumption of polluted vegetables [179]. Some studies [29] report skin diseases, such as dermatitis (eczema), in farmers that come into contact with untreated WW and WW irrigated soil. Nail problems in farmers, such as koilonychias (spoonformed nails), have also been reported as related to the presence of fungi in WW irrigated soils [112].

Health and growth problems have been observed in cattle that consume forage produced by WW irrigation. Furthermore, in low-income areas where water is scarce, cattle are not only fed with fodder grown using WW but also they are allowed to drink the WW used for irrigation. Some protozoa can survive in the surface layers of soil or even in aerial parts of crops; animals can be infected after eating these crops, although this is a remote way of transmission. There is some evidence indicating that beef tapeworm (Taenia saginata) can be transmitted from livestock fed with WW-irrigated forage to meat consumers. Furthermore strong evidence

TABLE 6.4   Summary of Health Risks Associated with the Use of WW in Agriculture

Group exposed	Helminth infections	Bacterial/viral infections	Protozoan infections
Consumers	Significant risk of Ascaris infection for both adults and children consuming vegetables contaminated with helminth ova.	Cholera, typhoid and shigellosis outbreaks reported due to the consumption of polluted crops. Helicobacter pylori in crop consumers. Increase in risks of suffering non-specific diarrhea when concentration of thermotolerant bacteria in WW used for irrigation exceeds 104 CFU/100 mL.	Evidence of parasitic protozoa found on the surface of found on the surface of vegetables, but no direct evidence of disease transmission.
Farm workers and their families	Significant risks of Ascaris infection for both adults and children in contact with untreated WW and irrigated soils. Risk remains, especially for children, when WW presents more than one nematode egg per liter. Increased risk of *hookworm* infection in farmers.	Increased risk of diarrheal diseases for children in contact with WW when it exceeds 104 CFU/100 mL for thermotolerant coliforms. Elevated risk of *Salmonella* infection in children exposed to untreated WW and WW irrigated soils. Elevated seropositive responses to norovirus in adults exposed to partially treated WW and WW irrigated soil.	Risk of Giardia intestinalis infection insignificant for contact with both treated/untreated WW and soil. Increased risk of amoebiasis observed due to contact with untreated WW and WW irrigated soils.
Nearby communities	High risk of infections when flood and furrow irrigation is used.	Sprinkler irrigation with untreated WW and high aerosol exposure associated with increased rates of bacterial infections due to the use of partially treated WW (10^4–10^5 CFU/100 mL or less). No risks of viral infection associated with sprinkler irrigation.	No data of protozoan infections transmission during irrigation with WW.

indicates that cattle grazing on fields freshly irrigated with raw WW or drinking from raw WW canals or ponds can become heavily infected by Cysticercus bovis, the early stage of the Taenia saginata life cycle [184].

6.3.1.2 Microbial Agents in WW Irrigated Soils

The study of microbial contamination by the use of treated/untreated WW in irrigation is focused in the pollution of crops rather than the soils receiving WW. This is because, on the one hand, a greater number of people are exposed to pathogenic microorganisms through consumption of contaminated crops, meat and milk than by direct contact with irrigated soils. On the other hand, the difficulties have been encountered in the analysis of microorganisms in soil; for instance the inherent problems of extracting microorganisms from such a complex soil matrix. Ascaris transmission associated with increased irrigation with WW was studied for sprinkler irrigation. Rates of bacterial infections due to the use of partially treated WW (104–105 CFU/100 mL or less) indicated no risks of viral infection associated with sprinkler irrigation.

Mezquital Valley, Central Mexico, found the occurrence of fecal contamination indicators (*Escherichia coli*). *Giardia lamblia* cysts and *helminth eggs* (*Ascaris lumbricoides*) at different depths of long-term WW irrigated soils. Figure 6.3 indicates the evidence the accumulation of the three microorganisms in the first few centimeters of the soil profile, indicating that infectious agents are removed from WW at the beginning

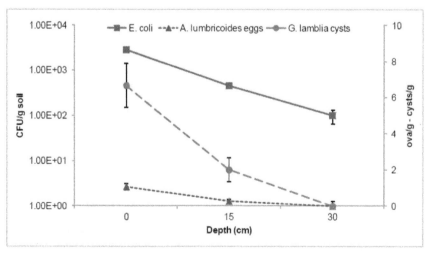

FIGURE 6.3 Abundance of three pathogenic microorganisms in a long-term WW irrigated soil at different depths.

of percolation through soil; such removal can be achieved by several physical and chemical phenomena. In this chapter, the content of pathogenic microorganisms in soils with different time under irrigation was also reviewed. Results showed that the accumulation of microorganisms in the tested soils is not related to the irrigation timing, suggesting that soils have mechanisms to inactivate and/or destroy these microorganisms after irrigation.

As mentioned earlier, different types of microorganisms can be found in WW irrigated soils depending on the zone where reuse is taking place. For example, studies have reported a higher prevalence of Cryptosporidium spp. compared with Giardia spp. in WW irrigated and manure amended soils of dairy farms in southeastern New York [18]. Cryptosporidium is a protozoan commonly found in developed countries, while different species of Giardia are widespread in developing countries. The occurrence of *Ascaris lumbricoides, hookworm* and *Trichiuris trichiura* has been observed in 69% of the soil samples taken in an untreated WW irrigated area in West Bengal, India [94].

The entry of antibiotic-resistant pathogens (ARPs) and antibiotic resistance genes (ARGs) into the soil via WW is an emerging issue. Since municipal WW contains both sub-therapeutic amounts of antibiotics, ARPs and ARGs, which occur to a greater extent when sewer systems combine municipal and hospital WW [171], these substances can reach the soil, modifying the dynamic of soil microbial populations. Antibiotic resistance may occur naturally in the soil, and to a greater extent in the rhizosphere, which functions as a hotspot for both antibiotic-resistant bacteria and ARGs [21]. Previous studies have reported presence of opportunistic pathogens (*Stenotrophomonas maltophilia*, responsible for respiratory tract infections and endocarditis) in the rhizosphere of Brassicacaea type plants [22]. The transfer of ARGs from these opportunistic bacteria to human pathogens reaching the soil through WW has not yet been demonstrated. The ARPs reaching the soil through WW may survive on the soil surface and, if conditions are appropriate, reproduce or migrate to surface and groundwater sources. ARGs may be mobilized into aquifers by infiltration of WW or into surface water sources by runoff. So far a relationship between the presence of traces of antibiotics in WW and the occurrence of antibiotic resistance in the irrigated soil has not been categorically established. In earlier studies, the incidence of two sulfonamide resistance genes (sul1 and sul2) was determined in the Mexico City in agricultural soils irrigated with such WW over different time periods and rain-fed soils [50]. The authors found the presence of ARGs in the three analyzed matrices; the concentration of resistance genes was 150 to 1500 times higher in irrigated soils than in non-irrigated ones. The occurrence of ARGs was positively related to the time under irrigation, with a higher content of resistance genes occurring in Enterococci bacteria living in soils irrigated for longer periods of time [50]. Such behavior may indicate that prolonged irrigation with WW promotes both the proliferation of indigenous ARPs in soil, due to the high and constant supply of nutrients via WW, and the increase in the assimilation of resistance genes due to the higher biomass content in old WW irrigated soils.

pesticides adsorbed contaminants suspended solids water stress living filter developing countries developed countries viruses amoebiasis livestock ponds microbial contamination sprinkler irrigation antibiotic-resistant pathogens (ARPs) and antibiotic resistance genes (ARGs) runoff Zucchini.

Conversely, the abundance of isolates resistant to tetracycline, ciprofloxacin, sulfonamides and erythromycin were identical in WW irrigated soils and freshwater irrigated soils despite the high load of ARGs and ARPs in the WW used for irrigation [154]. However, *Entetococci* bacteria in freshwater irrigated soil were highly resistant to a greater number of antibiotics (erythromycin, tylosin, tetracycline, and ciprofloxacin) than long-term WW irrigated soil, which showed resistance to lincomycin and daptomycin [145]. Furthermore, no differences were found in the content of ARPs when WW and freshwater irrigated soils were compared, suggesting that ARPs rarely survive after they enter soil via WW. Although, it seems unlikely that development of antibiotic resistance to human pathogens in WW irrigated soil is related to the input of antibiotics and resistant organisms via WW, yet it is worthwhile to evaluate the exchangeable genetic material (e.g., plasmids), since such material can be assimilated by soil microorganisms, inducing antibiotic resistance. Many questions remains unanswered about the mechanisms leading to the transference of this type of genetic material [80].

6.3.1.3   Microbial Pollution in Crops

Crops are polluted by direct contact with WW during irrigation. Pollution of the edible parts of plants depends not only on the quality of water, but also on the quantity applied to soil, the irrigation method and the type of crop. For example, zucchini when spray-irrigated with WW accumulate higher levels of pathogens on their surface than other crops. Zucchini have a hairy and sticky cover and grows close to the ground, which favors the attachment of pathogens. Microbial contamination of crops can occur not only as a result of WW irrigation but also during washing, packing, transportation and marketing. These problems are frequently not addressed, giving the impression that irrigation is the only source of microbial pollution [93, 99]. It has been reported [156] that less microbial pollution is caused if crops were irrigated with subsurface drip irrigation compared with sprinkler irrigation, and gravity irrigation. Moreover, the subsurface irrigation does not pollute crops [153], even when using WW with 6–7 × 10^5 CFU/100 mL of fecal coliforms and 225 helminth ova/L. Microbial pollution of crops also depends on the crop type. Fruits from trees are rarely polluted when irrigation is not provided using sprinklers (this is not a common procedure used to apply WW, since sprinkler heads tend to become clogged). Fruits grow far from the watering sites when furrow and flood methods are used. The microbial contamination of crops in WW irrigation systems is closely related to the survival of microorganisms. Table 6.5 shows the survival times of some pathogens in agricultural soils and crops irrigated with WW.

TABLE 6.5   Survival of Selected Pathogens in Soil and Crops Irrigated with WW [66, 191]

Pathogen	Survival time, days	
	Soil	Crops
Ascaris lumbricodes eggs	180	30
Entamoeba histolytica	<20	<10
Enterovirus	<40	<20
Fecal coliforms	<70, but usually <20	<30, but usually <15
Salmonella spp.	80	25
Taenia saginata eggs	>180	<60, but usually <30
Trichuris trichiura eggs	>180	<60, but usually <30
Vibrio cholera	<20, but usually <10	<5, but usually <2

Both pathogenic and non-pathogenic microorganisms display differences in their survival in soil and crops. For instance, the non-pathogenic fecal coliform indicator *E. coli* can survive in soil for nearly a month, while the pathogenic strain of *E. coli* O157:H7 survives at most for 14 days in spinach leaves [158]. It is known in some detail that survival of pathogenic bacteria can increase by internalization within the plant tissues [99]. Previous studies indicate that *E. coli* can translocate from soil to leaves of lettuce through the root system [187]. In contrast, translocation of pathogenic bacteria to the edible parts of crops via the root system is quite unlikely [220]. It is more likely that pathogens enter to the edible parts of crops through wounds in vegetal tissues [148]. Wounded tissues have been demonstrated allow the entrance of *Salmonella* and *E. coli* to lettuce and tomato plants [110, 111]. Similarly, it is reported that *E. coli* can use the stomatal cavities in leaves to enter the internal structure of lettuce [88]. The pathway of this kind of entry is still unknown. Once inside the plant tissues, pathogen survival rates improve since they can use cellulose. Protozoa are larger in size than bacteria and thus they cannot access the internal parts of the plants. However, these pathogenic organisms can adhere to the surface of edible plants and remain there by the excretion of polymers, which facilitate adhesion. Table 6.6 shows some examples of the occurrence of protozoa in crops irrigated with treated/untreated WW.

6.3.1.4   Fate of Pathogenic Microorganisms in Soil

Upon their arrival to irrigated soils, microorganisms can either survive or be inactivated/killed by the physical and chemical processes naturally occurring in soil as well as by predation by indigenous soil organisms. Given the case that these microorganisms can survive in the soil, they may subsequently colonize soil particles, infiltrate the soil to the aquifer or migrate through across the landscape by runoff. Processes affecting the environmental fate of the pathogenic microorganisms in soil are shown in Fig. 6.4.

Past studies have demonstrated that some microorganisms can vertically and/or horizontally mobilize through the soil, travelling long distances from the initial point of contamination [42]. Bacterial migration in soil has been reported up to 830 meters, while for viruses such displacement is

TABLE 6.6   Occurrence of Some Pathogen Protozoa on the Surface of Crops Irrigated Using Treated/Untreated WW

Pathogen	Crop	Occurrence	Comments and citation
Giardia lamblia	Potatoes	5.1 cysts/kg	Crops irrigated using untreated WW in Marrakesh [17]
	Coriander	254 cysts/kg	
	Mint	96 cysts/kg	
	Carrots	155 cysts/kg	
	Radish	59.1 cysts/kg	
Ascaris lumbricoides	Potatoes	0.18 eggs/kg	
	Turnip	0.27 eggs/kg	
	Coriander	2.7 eggs/kg	
	Mint	4.63 eggs/kg	
	Carrots	0.7 eggs/kg	
	Radish	1.64 eggs/kg	
Enterobius vermicularis	Lettuce	10–40 cysts/kg	Crops irrigated using treated and untreated WW in Kahramanmaras, Turkey [63]
	Parsley	10–60 cysts/kg	
	Cress	10–20 cysts/kg	
	Spinach	1–3 cysts/kg	
Entamoeba hystolitica	Lettuce	10–50 cysts/kg	
	Parsley	10–50 cysts/kg	
Giardia lamblia	Lettuce	10–20 cysts/kg	
Ascaris lumbricoides	Lettuce	10–30 eggs/kg	
	Parsley	10–30 eggs/kg	
Trichuris trichiura	Spinach	3.3% of the analyzed samples.	Crops grown in soils irrigated with raw WW in West Bengal, India [94]
	Pudina	3.1% of the analyzed samples.	
	Coriander	5% of the analyzed samples.	
Hookworm	Lettuce	9.4% of the analyzed samples.	
	Parsley	3.3% of the analyzed samples.	
	Spinach	6.7% of the analyzed samples.	

TABLE 6.6   Continued

Pathogen	Crop	Occurrence	Comments and citation
	Pudina	9.4% of the analyzed samples.	
	Celery	3.6% of the analyzed samples.	
	Coriander	5% of the analyzed samples.	
Ascaris lumbricoides	Lettuce	43.8% of the analyzed samples.	
	Parsley	23.3% of the analyzed samples.	
	Spinach	36.7% of the analyzed samples.	
	Pudina	50% of the analyzed samples.	
	Celery	25% of the analyzed samples.	
	Coriander	35% of the analyzed samples.	
Helminth eggs	Leafy vegetables	100 eggs/kg	Vegetables irrigated with untreated WW in Faisalabad, Pakistan [61]

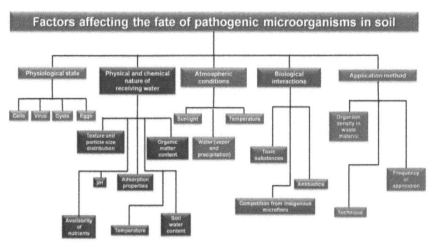

FIGURE 6.4   Factors affecting the environmental fate of pathogenic microorganisms in WW irrigated agricultural soils [modified from 1].

significantly lower, that is, up to 408 m [118, 190]. Survival of pathogens is related with their environmental fate since the longer the lifetime of the microorganisms the larger the distance they can travel. As indicated in Table 6.5, bacteria can survive for long periods compared to viruses, and thus bacteria can be transported farther. Climatic conditions also impact upon pathogen transportation. For instance, in frozen soils pathogens can survive longer and thus they can be transported farther than in tropical and desert soils [84]. Microorganisms can be more easily displaced through coarse textured soils than fine textured ones. The greater mobilization of coliforms in sand-gravel soil has been reported than in fine sand. In fact, in coarse sandy soils, the vertical movement of microorganisms can be as rapid as that observed for inorganic tracers [118]. In this regard, infiltration of streptomycin-resistant *E. coli* can be compared with that of the chloride tracer in undisturbed soil columns, even when different soil textures are compared [186]. Since the transportation of microorganisms is similar to that observed for tracers, the physical structure of soil is the determinant factor in reaching the aquifer. Therefore, a greater occurrence and interconnection of pores within the solid matrix may result in efficient infiltration of water and thus bacteria. Studies on the movement of pathogens in the field confirm the rapid movement of pathogenic bacteria observed in laboratory tests. These studies also found a high concentration of bacteria and viruses in groundwater [1]. In addition to the higher quantity and interconnection of pores, the increased transport of bacteria and viruses through the soil can be explained by the presence of preferential paths within the soil matrix. Such preferential paths are referred to cracks, fractures, worm holes and channels formed by plant roots or fauna in the soil. Larger microorganisms (*E. coli*) can mobilize deeper into soil than smaller coliphages [83]. It has been confirmed that bacterial cells smaller than 1 µm in diameter are more rapidly transported through soil than larger organisms [77].

The chemical properties of soil can also impact upon the vertical and horizontal transport of microorganisms. The mineral composition of soil can favor adhesion of microbial cells, eggs or cysts onto soil particles. Several types of bacterial cells have been shown to strongly adhere to the mineral domain of soil and aquifer material [178]; and once adhered, bacteria can replicate and form biofilms on the surface of soil particles. In WW irrigated soils, the accumulated organic matter as well as the continuous input of dissolved organic matter via WW may enhance the proliferation of bacteria. With regard to parasites, *Ascairs lumbricoides* eggs and *Giardia lamblia* cysts adhere to the mineral fraction of WW irrigated soils more rapidly and more strongly than to the organic domain. In the case of Ascaris eggs, adhesion occurs with the silica in sand particles [129].

In contrast to protozoan eggs, adhesion or adsorption of protozoan cysts may be related to soil organic matter rather than the mineral fraction of soil [100]. This has been attributed to the hydrophobic nature of the cysts walls. The detachment of bacteria from soil particles is effected by the composition of the irrigation water [76]. In that study, Pseudomonas sp. showed enhanced transport when distilled water was used for detachment in column experiments, compared with 0.01 M NaCl. Such results suggest that clean water can efficiently wash off the polysaccharides excreted by bacterial cells, which act as an adhesive between soil particles and bacteria. The opposite effect has been observed for Ascaris eggs. When soil is washed with NaOCl, eggs are effectively detached from soil particles; this is because sodium hypochlorite can destroy the albuminose layer that coats the surface of *helminth eggs* and which anchors with the soil particles [78]. The environmental relevance of studying the impact of this salt on the detachment of eggs from soil relies on the fact that NaOCl can be found in reclaimed water, as it is commonly used for disinfection of effluents [129].

Once microorganisms are retained by soil, either by adsorption/adhesion or straining, they can be inactivated or eliminated by desiccation. This phenomenon is particularly important in arid areas where high levels of solar radiation are reported. The environmental fate of microorganisms in soil also depends on the native microorganisms living in the solid matrix. Predators of WW-borne pathogenic bacteria in soil include Streptomycetes, Myxobacteria, Bdellovibrio and nematodes [1]. The presence of plants may affect the persistence and movement of microorganisms in soils. On the one hand, pathogen can found favorable conditions for survival in the rhizosphere due to the high content of nutrients in this zone; and on the other hand, native bacteria in rhizosphere can be natural predator of those pathogens, while roots may excrete antibiotics that inhibit or kill pathogenic microorganisms.

6.3.2   SOIL POLLUTION BY HEAVY METALS

Given that most agricultural WW irrigation is performed using municipal WW, which contains negligible amounts of heavy metals [13], the occurrence of these elements in WW irrigated soils is usually significantly lower than the maximum permissible concentrations established by international regulations. However, there are some cases where care should be taken when reusing WW in irrigation, for example, close to tanneries, metal processing or mining areas [112]. Different levels of risk are perceived for the different heavy metals. While some of them are nutrients for plants at trace concentrations, others have been shown to produce harmful effects on exposed organisms, or are absorbed by plants and accumulated through the food web. Table 6.7 presents the risks that are incurred by the presence of some heavy metals in soil.

Cadmium is the metal with the highest associated risk. It is toxic to humans and animals in doses much lower than those that visibly affect plants; furthermore crop uptake (which is notably high in acidic soils) can increase the dose consumed by organisms and in turn accumulation in animal tissue. Absorbed cadmium in animals is stored in kidney and liver, although meat and milk products have shown to be little affected by cadmium accumulation [162]. There is a relatively good knowledge to allow the setting of limits regarding the acceptable amount of heavy

TABLE 6.7   Heavy Metal Risk Characteristics During Irrigation [162]

Risks characteristics	Heavy metal
1. Low risk	Mn, Fe, Zn, Cu, Se, Sb
2. High risk	Cr, As, Pb, Hg, Ni, Al, Cd
3. Essential micronutrient to plants	Cu, Fe, Mn, Mo, Zn, Ni
4. Beneficial for some crops	Co, Na, Si
5. Can accumulate in crops to levels that are toxic for consumers	Cd, Cu, Mo
6. No human toxicological threshold established for WW intended for irrigation	Hg
7. Relatively high threshold for WW used in irrigation	Cu, Fe, Mn, Zn
8. Low absorption by plants	Co, Cu, Mn, Zn

metals contained in WW used to irrigate. In the study referred to in Ref. [37], numerical calculation of the limits for the maximum tolerable pollutant concentration in WW irrigated soils was carried out (health-based targets). This was based on the acceptable daily human intake (ADI) for selected heavy metals and the amount that can be "permitted" to accumulate in soil before harmful effects occur in consumers of crops (Table 6.8).

This analysis assumed: (a) only two exposure routes (WW → soil → plant → human; and, WW → crop → human); (b) a global diet in which the daily intake of grains/cereals, vegetables, root/tuber crops and fruit accounts for ~75% of daily adult food consumption; (c) a body mass for adults of 60 kg; (d) all of the food grain, vegetables, root/tuber crops and fruits are obtained from land irrigated with WW; and, (e) a total daily intake of pollutants by this consumption path of 50% of the ADI (the remaining 50% of the ADI was attributed to background exposure). Table 6.8 shows the inputs of heavy metals by WW to irrigated soils, assuming an application of treated WW of approximately 1.2 m/year, which is roughly the amount of water required to produce a crop cycle in an arid zone.

Health effects associated with the use of water heavily contaminated with industrial discharges for irrigation have been reported. In Japan, itai-itai disease, a bone and kidney disorder associated with chronic cadmium poisoning, occurred in areas where rice paddies were irrigated with water from the contaminated Jinzu River [207]. In some parts of China, the use of industrial WW for irrigation was associated with a 36% increase in

TABLE 6.8   Maximum Tolerable Concentration of Heavy Metals in WW Irrigated Soils [37].

Element	Maximum input by WW (kg/ha/year)	Maximum tolerable concentration (mg/kg)
Arsenic	0.6–12	9
Cadmium	0.06–0.24	7
Chromium	1.2–60	3200
Lead	1.2–60	150
Mercury	0.12–0.12	5
Nickel	0.24–12	850
Selenium	0.24–0.6	140
Silver	1.2	3

hepatomegaly (enlarged liver) and 100% increase in both cancer and congenital malformation rates [217].

An inventory of sources of some heavy metals (zinc, copper, nickel, lead, chromium and cadmium) in agricultural soils of England and Wales [155] has been reported. Results showed that the greatest contribution of heavy metals in those soils comes from the application of sludge from WW treatment plants, while irrigation appeared to be of little importance as a source of heavy metals in soils. According to this investigation, which followed the rates of deposition of heavy metals in the studied soils, the time required for metal concentrations to reach maximum values permitted by international regulations is 80 years for zinc and at least 1256 years for cadmium. In this respect, study referred to in Ref. [185] showed that concentrations of heavy metals in long-term untreated–WW irrigated soils in central Mexico were 10 times lower than the limits set by the Danish regulations; moreover, the authors estimated that another century of irrigation is necessary to exceed these values. In most cases, metals have little impact on aquifers. The most toxic metals to humans—cadmium, lead, and mercury—were absent in groundwater at five sites in the United States after 30–40 years of applying secondary and primary effluents at rates between 0.8 m/year and 8.6 m/year to different crops [132]. The researchers indicate that the pH values greater than 6.5 in soil and WW resulted in the precipitation of the entire amount of metals. Metals are normally bonded into the organic matter through the formation of organo-metallic complexes, which are not bioavailable to plants. The addition of lime and WW to soil assists the precipitation of metals, while the addition of chemical fertilizers has the opposite effect, since over the long term they tend to lower the soil pH and thus solubilize metals.

In contrast in agricultural soils, it has been reported concentration of heavy metals, such as cadmium and zinc, are close to reaching the maximum levels set out in international regulations. In these cases, the factors leading to an exacerbated soil contamination and thus increased risk of groundwater and crop pollution are: (a) sandy soil texture; (b) acidic to neutral soil pH; (c) low organic matter content; and/or, (d) the use of industrial WW for agricultural irrigation [140, 149]. In such cases, the cessation of irrigation with WW is recommended, together with allowing the recovery of soil through remediation techniques such as phytoremediation.

6.3.3 SOIL POLLUTION BY ORGANIC COMPOUNDS

The soil pollution by organic substances has been a matter of concern to scientists and organizations regulating the quality of soil/water resources/ food for several decades. An extensive research exists addressing the soil degradation by conventional organic pollutants (e.g., pesticides, polyaromatic hydrocarbons, organochlorides, paraffin, organic solvents, etc.). However in sites where treated/untreated WW is disposed of by irrigation, one can find organic substances different to those commonly studied and reported in literature treating oil spills, mining zones or soil polluted by industrial WW. Most of the dissolved and particulate organic matter contained in municipal WW is produced by the degradation of human and animal excreta, hence organic matter in WW is composed mainly by saccharides, lipids, amino acids and proteins. However, a tiny fraction of the organic material in WW originates from chemicals contained in everyday consumer products used and disposed of via sewage in urban and rural areas. According to Ref. [15], thousands of organic compounds are contained in municipal WW at trace levels and there is a lack of knowledge regarding the effects of such substances on organisms, either by themselves or in combination with other compounds or groups of organic compounds. This group of chemicals is referred as organic pollutants of emerging concern (OPEC) [16]. Though they should actually be listed as priority pollutants in cases where WW is used to irrigate crops, since these contaminants are in contact with soils, crops and water sources near the irrigated zone [125]. Over the last three decades, significant work in the field of analytical chemistry has been carried out in order to extract, isolate and quantify some of these pollutants in WW and soils. Frequently found OPECs in such complex matrices are pharmaceutically active compounds (PAC) and their metabolites, personal care products (e.g., disinfectants, fragrances, insect repellents, sunscreens, etc.), sweeteners, stimulants (e.g., caffeine and psychoactive drugs), detergents and their metabolites, plasticizers and industrial additives (e.g., additives in gasoline) [125]. Almost all of the studies addressing the removal of OPECs in WW treatment plants report that most of these substances are partially degraded/removed in primary and secondary treatment systems, and some pollutants are only partially removed even in tertiary treatment systems [27]. Because of this, OPECs occur in irrigated soils if either treated or untreated WW is used in irrigation. Effluents of WW treatment plants contain a small fraction of the parent substance as well as the by-products generated during treatment. However, some of the compounds may be retained and concentrated in sludge produced during WW treatment and reach the environment via the use of sludge (or biosolids) as soil amendments in agriculture. Due to continuous industrial development, the number of organic substances contained in WW is constantly increasing; in fact, most of these substances are not tested before they are released onto the market, and therefore their potential risks or the side effects they cause in non-target organisms in soils or water bodies is yet unknown.

6.3.3.2   Effects Caused by Domestic WW–Related Organic Pollutants in Irrigated Soils

As mentioned above, due to the ever-growing pool of organic compounds discharged to the soil via WW, there is a general lack of knowledge regarding the effects that such substances cause to exposed organisms. In general terms, municipal WW is the main vector of OPECs to reach the environment, so that these substances are ubiquitous at sites where WW streams occur. Pharmaceutically active compounds (PACs) are designed to cause a defined effect on target organisms. However, when trace amounts of these substances are transported by WW into environment, they can interact with non-target organisms. One effect that has captured the attention of the scientific community in recent years is the development of antibiotic resistance by pathogenic microorganisms due to the occurrence of antibiotics in WW, surface water bodies and soils receiving WW [126, 195]. However, a large number of studies on this subject report that proliferation of antibiotic-resistant pathogens is quite unlikely in WW irrigated soils [80, 145, 154]. Conversely, attempts have been made to relate the occurrence of sulfonamide and fluoroquinolone antibiotics with the emergence of antibiotic resistances in WW and long-term WW irrigated soils [50]. The authors reported a relationship between irrigation time and the frequency of detection of antibiotic resistance genes in soils. In the case of non-antibiotic PACs, the most studied compounds—because they are the most used worldwide—are

the analgesic and anti-inflammatory drugs [174]. Compounds such as ibuprofen, naproxen, diclofenac, paracetamol and ketoprofen have shown to
cause systemic damages in aquatic species; damages in liver, gills and kidney are commonly reported [68]. The non-steroidal anti-inflammatory drug
diclofenac has been demonstrated to cause visceral gout in vultures. In fact,
the presence of diclofenac in livestock was the cause of the mass death of
three species of vulture in India and Africa [193]. Other studies show that
chronic exposure to traces of anti-inflammatory drugs leads to a lessen-
ing in the development of human embryo cells [164]. The occurrence of
psychotropic agents at trace levels in water bodies polluted by WW discharges has shown to alter the behavior of some fish species, suppressing
their survival instincts against predators [146]. With regard to OPECs that
are not pharmaceutically active compounds, there is significant concern
that they may alter hormone homeostasis in organisms. These substances,
known as endocrine disruptors, can mimic or compete with natural hormones by binding with active sites on hormone receptors, causing reduced
or disproportionate hormonal responses in the affected organisms [206].
The most potent endocrine disruptors found so far in municipal WW are
the natural and artificial estrogenic hormones—the latter are used as birth
control agents—and the regulators of thyroidal function, followed by plasticizers (e.g., phthalates and bisphenols), surfactants and their metabolites
and some industrial additives [163]. Endocrine disruptors are suspected of
causing the feminization or masculinization of fish and reptile populations
as well as the occurrence of breast cancer, imbalances in thyroidal function, teratogenic effects (e.g., cryptorchid) in mammals, and even obesity
in mammals (obesogens) [24, 25, 91]. There is a lack of knowledge regarding the effects caused by OPECs in soil organisms. Studies in this field
have been developed very little compared to those for water bodies. Table
6.9 shows some examples of effects caused organism by the occurrence of
OPECs in soils.

The effects caused by this class of pollutants are not limited to soil
organisms and impacts can be observed in the soil matrix. For example,
surfactants can, on the one hand, decrease the capillarity and penetrability
of soil as well as increase the solid-liquid contact angle, the shape factor and the sorptivity of soil particles. On the other hand, the input of
these substances can increase the desorption of previously sorbed organic

TABLE 6.9   Summary of Negative Effects on Soil Organisms Caused by the Occurrence of Pollutants of Emerging Concern (OPECs) at Trace Levels

Compound	Effects on soil organisms	Citation
Estrone, 17β estradiol (hormones)	Negative impacts on the vegetative cycle of alfalfa (Medicago sativa).	[183]
Sex hormones	Shift in sex ratio of free life nematode communities in soil.	[101]
Triclosan (antibacterial agent)	Inhibition in plant growth (rice and cucumber). Effect concentrations 50 (EC50, i.e., 50% of exposed population was affected) were 57 and 108 mg/kg for rice and cucumber respectively. Inhibition of soil respiration and phosphatase activity at concentration levels higher than 10 mg/kg.	[138]
	Reduction in soil respiration 4 days after supplying the compound. The observed effects were dependent on the adsorption of the compound onto the soil.	[35]
Bisphenol A (plasticizer)	Shift in sex ratio to female individuals in isopod (soil)	[133]
Abamectin (anthelminthic)	Negative impacts on reproduction of Folsomia fimetaria and	[54]
Fenbendazole and cypermethrin (antiparasitic)	Negative impacts on degrading microorganisms of dung.	[188]
Sulfonamide and tetracycline antibiotics	Inhibition of the soil microbial activity by 10% (ED10) at concentrations of 0.003–7.35 µg/g of soil (dry mass). Shifts in fungi:bacteria ratio.	[196]
Sulfadiazine (antibiotic)	Decrease in denitrification rates when the input of antibiotic was 100 mg/kg of soil (dry mass).	[123]
	Significant decrease in the bacteria:fungi ratio	[96]
Chlortetracycline, tetracycline, tylosin, sulfamethoxazole, and sulfamethazine trimethoprim (antibiotics)	Decrease in crop growth (sweet oat, rice and cucumber). Inhibition of the microbial activity of soil (soil respiration and phosphatase enzyme activity).	[137]
Human and veterinary pharmaceutically active substances	Decrease in growth and development of Phaseolus vulgaris L., Glycine max, Medicago sativa, Zea mays, and several other crops.	[113]

molecules on the soil particles, which in turn increases the bioavailability and mobility of the desorbed compounds [2]. To evaluate the toxic effects caused by the occurrence of OPECs to soil organisms, two approaches are commonly used: acute and chronic toxicity studies. For the former, high concentrations of target pollutants are supplied to studied organisms under controlled conditions for a short period; chronic toxicity tests, on the other hand, are based on prolonged exposure of organisms to low (i.e., environmentally representative) doses of the studied pollutants. So far, most toxicity studies dealing with OPECs have been carried out using the acute toxicity approach. Even though these studies do not fully represent the field conditions, they provide valuable information on the subject of impacts caused by this kind of contaminants to soil organisms. Studies evaluating chronic toxic effects of pollutants are more representative of field conditions, i.e. toxic substances enter to soil in small doses over long periods. In this regard, conducting long-term toxicity studies are a priority to evaluate the chronic effects caused by OPECs in soil organisms. Several toxicity studies report that the effects of organic pollutants on soil organisms (i.e., reduction in soil respiration, enzymatic activity and nitrification/denitrification rates) are observed in the early days of exposition; then, after a short period (4 to 10 days), soil recovers to its basal conditions [35, 96, 123, 138]. The next step in toxicity studies for these emerging pollutants is to determine the dynamics of the toxic effects on soil organisms after tens or hundreds of growing cycles in which target contaminants are continuously supplied; i.e. under conditions similar to what occurs in long-term irrigated areas.

6.3.3.2   Occurrence of Domestic WW-Related Organic Pollutants in Irrigated Soils

In spite of the fact that WW is the main vehicle allowing OPECs to reach soil, very few studies reporting the presence of these pollutants in WW irrigated soils have been carried out.

This finds an explanation, on one hand, in the inherent difficulty of extracting and isolating organic compounds at trace levels from the soil matrix, and on other hand, in the fact that analyzing this type of pollutants is relatively expensive. Figure 6.5 shows the sites where monitoring

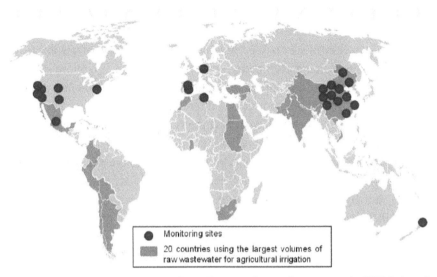

FIGURE 6.5   Monitoring studies for pollutants of emerging concern in WW irrigated soils throughout the world and comparison with the 20 countries using the largest volumes of raw WW for agricultural irrigation.

studies aimed at determining the occurrence of OPECs in WW irrigated soils have been performed. In this figure, the number of sites monitored is contrasted with the 20 countries with the highest use of untreated WW for agricultural irrigation. Most of the monitoring studies are concentrated in China, the country using the highest volume of untreated WW in agriculture [17], followed by the United States and Mexico, which ranks number two among all countries in terms of reuse of untreated WW for irrigation.

Efforts in monitoring emerging pollutants in developing countries, where the use of raw WW is widespread, are of value. This requires cooperation among research centers where analytical techniques are currently validated to perform soil analyses or by sharing "know how" and technology with developing countries in order to perform analysis on site. Determination of OPECs in soil requires an exhaustive extraction step, which in most cases has to be carried out at a moderately high temperature, particularly in the case of analysis of thermolabile compounds (e.g., sulfonamide antibiotics). Extraction methods such as pressurized fluids extraction, microwave assisted extraction and ultrasonic assisted extraction are preferred over traditional Soxhlet extraction techniques, since they guarantee greater contact between the solvent and the soil particles, resulting in higher recoveries of analytes. Analysis of OPECs is commonly accomplished using either liquid or gas chromatography techniques; although liquid chromatography is preferred as it is more suitable for the analysis of polar compounds, i.e. most PACs [20]. Monetary costs of these analyses are relatively high and analysis entails the use of potentially dangerous chemicals, which is in part the reason why monitoring studies for OPECs in soils are not carried out in poor countries. So far, the most reported emerging pollutants in WW irrigated soils are the pharmaceutically active substances (e.g., antibiotics, non-steroidal anti-inflammatory agents, anticonvulsants, anticoagulants and sex hormones), followed by plasticizers (e.g., phthalic acid esters and bisphenol A), metabolites of surfactants (e.g., nonylphenol, octylphenol) and antibacterial and antimycotic agents (e.g., triclosan and triclocarban). Table 6.10 shows the concentrations of some OPECs reported for WW irrigated areas.

Overall, higher concentrations of OPECs are found in the first 30 cm of the soil profile. Such behavior suggests that these compounds are retained by the organic matter accumulated over time of irrigation; which is consistent with the organic nature of these contaminants, although several of them display some polarity. Concentration levels reported for the monitored PACs range from below the detection limits of the analytical techniques to tens of μg/kg of soil (dry mass). The concentration levels of the pharmaceuticals ibuprofen, naproxen and carbamazepine in the range of 0.25 to 6.48 μg/kg of soil (dry weight) in Phaeozem and Leptosol soils have been reported in areas irrigated using untreated WW for eight decades [57]. Other studies [213] found average concentrations of 1.8 μg/kg for triclosan and 2.5 μg/kg for estrone. In contrast to PACs, concentrations reported for plasticizers and surfactants are of the order hundreds of μg/kg of soil (dry mass). For example, concentrations of 14–80 μg/kg are reported for nonylphenols, while concentrations of 140–2610 μg/kg were observed for some plasticizers [41]. Concentrations of up to 7110 μg/kg of the plasticizer di-2(ethylhexyl)phthalate, have been reported elsewhere [102]. High concentrations of plasticizers in soils are explained by the ubiquity of these compounds in environment. Phthalic acid esters are contained in almost all plastic products and can easily leach from the solid matrix (i.e., the plastic articles). Once phthalates are released from the solid matrix, they can get into environment not only via WW but by aerial deposition, using dust particles as carriers [95].

TABLE 6.10   Concentrations of Organic Pollutants of Emerging Concern (OPEC) in Treated/Untreated WW Irrigated Soils

Compound	Concentration (µg/kg)	Citation and comments
Carbamazepine	0.28–0.94	[57] Concentration range observed in the surface layer (0–10 cm depth) of a treated WW irrigated soil during an irrigation cycle (May to October). The lowest concentration was observed before irrigation started while the highest concentration was determined in soil at the end of the irrigation cycle. Irrigation at the site has been occurring for the last 30 years.
	5.14 and 6.48	[67] Concentrations found in the surface layer (0–10 cm depth) of Leptosol and Phaeozem soils, respectively that has been irrigated using untreated WW for 85 years.
	4.92, 2.9 and 1.92	[203] Concentrations found in forested, grass-covered and cultivated soil irrigated with treated WW for more than 25 years. Carbamazepine was found mainly in the first 30 cm of the soil profile.
Ibuprofen	<LOD–3; <LOD–3	[213] Concentration ranges observed in loamy sand and sandy loam turf soils (0–30 cm depth) irrigated with treated WW at an irrigation rate of 1.1–1.2 and 1.5–1.6–fold the evapotranspiration rate, respectively. WW irrigation has been occurring at the site for almost 20 years.
Naproxen	<LOD–12.5; <LOD–9.5	
Triclosan	<LOD–6; <LOD–2.8	
Bisphenol A	<LOD–1.25; <LOD–1	
Estrone	<LOD; <LOD–5.3	
Ibuprofen	<LOD and 0.25	[57] Concentrations found in the surface layer (0–10 cm depth) of Leptosol and Phaeozem soils that have been irrigated using untreated WW for 85 years.
Naproxen	0.73 and 0.55	
Nonylphenols	123 and 41	
Triclosan	18.6 and 4.4	
Bisphenol A	14.8 and <LOD	
Di-n-butylphthalate	552 and 244	
Butylbenzylphthalate	346 and 171	
Di-2-(etylhexyl) phthalate	2079 and 820	

TABLE 6.10   Continued.

Compound	Concentration (µg/kg)	Citation and comments
[169] Clofibric acid	<LOD–9	[214] Concentration range observed in soil from a golf course irrigated with reclaimed WW.
Triclocarban	<LOD –105	[41] Concentration ranges for pharmaceuticals and endocrine disrupting chemicals in agricultural soils of Hebei province, north China, which have been irrigated using treated WW for more than 50 years.
4-nonylphenol	14.2–60.3	
Salicylic acid	1.4–10.7	
Tetracycline	<LOQ –19.9	
Oxytetracycline	1.1 –16, maximum 212	
Trimethoprim	<LOQ –2.6	
Primidone	<LOQ –3.3	
Omeprazole	6.5–24.3	[161] Ranges of concentration found in two agricultural soils irrigated with treated WW in Spain Pollutants were found at higher concentrations in the surface layer of the soils.
Spironolactone	0.6	
Diazepam	4.65	[201] Concentration found in an agricultural soil irrigated with treated WW. Pollutants were accumulated in the surface layer of the studied soil (0–30 cm).
Carbamazepine	5.77	
Butylbenzylphthalate	59–1580	[218] Ranges of concentration of phthalate esters in agricultural soils irrigated with untreated WW in the peri-urban area of Guangzhou city.
Di-2-(etylhexyl) phthalate	107–29370	
Di-n-butylphthalate	9–2740	
Di-n-amylphthalate	1–80	
Caffeine	14	[86] Concentrations reported for volcanic soils (Vitric, Orthic, Allophanic soils) irrigated using treated WW for more than 15 years (at rates of 70 mm/ week) in Rotorua, New Zealand.
Amitriptyline	<5	
Carbamazepine	217	
Chlorpromazine 17α	<5	
ethynilestradiol	<5	
Diltiazem	<248	
Thioridazine	<259	

LOD = Limit of detection. LOQ = limit of quantification;

<LOD: concentrations below the LOD of the analytical method used for determination.

<LOQ: below the LOQ of the analytical technique.

Nonylphenols, the major by-products of the anaerobic biodegradation of surfactants [3], are commonly found in WW irrigated soils due to the significant presence of detergents in municipal WW in combination with the anaerobic conditions prevailing in sewerage systems. In contrast to PACs, plasticizers and surfactant metabolites are non-polar in nature and for this reason, higher adsorption can occur for these compounds in soil, causing not only their build up in the surface layer of soil but the potential decrease in their bioavailability to soil microorganisms. Most monitoring studies of OPECs in environmental solid matrices are focused on determining these contaminants in biosolids amended soils rather than in WW irrigated soils. This is necessary since: (a) biosolids in WW treatment plants concentrate organic pollutants during water depuration, hence a greater concentrations of contaminants are expected in biosolids amended soils than in treated/untreated WW irrigated soils; (b) the use of biosolids as agricultural soil amendment is a more socially acceptable practice than reusing WW, thus it tends to be more practiced (or at least more reported) than WW irrigation, and it therefore becomes necessary to determine the pollutant load reaching the soil in this manner; (c) since analysis of OPECs in soil is expensive, these types of monitoring studies are conducted mainly in developed countries, where the use of biosolids as soil amendment is practiced more intensively than the reuse of treated/ untreated WW in agricultural irrigation. At present there are no regulations that establish maximum permissible concentrations for organic pollutants of emerging concern in soils. The development of such regulations relies on the results obtained in both acute/chronic toxicity tests and in health risk assessments.

6.3.3.3   Environmental Fate of Organic Pollutants of Emerging Concern in WW Irrigated Soils

The environmental fate of *organic pollutants of emerging concern* (OPECs) in the soil is governed by the physical and chemical properties of both the compounds and the soil as well as by the climatic conditions of the site where reuse is taking place. The chemical properties of the organic pollutants significantly impacting the environmental fate of OPECs are polarity, hydrophobicity and volatility.

Table 6.11 shows some OPECs found in municipal WW, which serve as examples of the differences in the chemical properties affecting the environmental fate of OPECs in soil. Due to the organic nature of OPECs, soil organic matter, mainly its non-polar fraction, plays a determinant role in the retention of these pollutants in soil [4].

TABLE 6.11 Relevant Physical and Chemical Properties in Terms of the Environmental Fate of Emerging Pollutants in Soil [56]

Polarity (ionization state at commonly found soil pH values)		
Positive	Negative	Positive/ Negative (Zwitterions)
Erythromycin (antibiotic)	Naproxen (non-steroidal analgesic drug)	ofloxacin (antibiotic)
pK_a 8.91	pK_a 4.15	pK_a 6.27(COOH); pK_a 8.87 (NH$_2$$^+$)
Hydrophobicity		
Hydrophobic	Hydrophilic	
Di-2-(ethylhexy)phthalate (plasticizer)	Ciprofloxacin hydrochloride (antibiotic)	
pK_{ow} 7.5	pK_{ow} −0.82	
Solubility in water at 25°C: 4.1×10^{-2} g/L	Solubility in water at 25°C: 30 g/L	
Volatility		
Volatile	Non-volatile	
Galaxolide (fragrance used in detergents)	Bisphenol A (plasticizer precursor)	
Vapor pressure: 7.27×10^{-2} Pa	Vapor pressure: 9.33×10^{-6} Pa	

However, sorption onto soil organic matter does not occur equally for all contaminants, since polar molecules tend to remain soluble in water rather than be retained in the soil organic matter; conversely, non-polar molecules are instantaneously adsorbed by soil organic matter [43]. The polarity of organic compounds is determined by the presence of ionizable radicals within the molecules; carboxyl, phenol, amine and amide moieties may gain or lose protons, depending on soil pH values, acquiring a positive or negative charge, respectively. The compounds for which functional groups lose protons may be poorly retained by soil due to repulsion forces between the deprotonated radical and the negatively charged soil particles (i.e., organic matter and clay); this results in the facilitated leaching of organic pollutants into the aquifer [104]. However, when functional groups within organic molecules gain positive charge, they may be retained onto the soil particles by cation exchange—as occurs for some tetracycline antibiotics [175]. In both cases, the organic moiety within OPEC molecules may be held in the soil organic domain by hydrophobic affinity. In general, the pH of WW irrigated soils tends to be neutral to basic [127], which results in low retention of negatively charged compounds compared to neutral or positively charged organic compounds [104]. Non-steroidal anti-inflammatory drugs (NSAIDs) such as naproxen [55], which can produce negatively charged molecules after the ionization of the carboxyl functional group, are adsorbed to a lower extent than other compounds displaying higher hydrophobicity, such as carbamazepine or triclosan, in organic soils with high clay content.

Organic compounds lacking of ionizable functional groups or displaying non-ionizable functional groups express their hydrophobicity by spontaneously migrating from water to the soil organic domain [115]. In WW irrigation systems, dissolved and particulate organic matter contained in WW tends to accumulate in the surface soil horizons, significantly favoring the build up of these compounds in topsoil. A greater accumulation of hydrophobic compounds, such as carbamazepine and esters of phthalic acid, was found in surface horizons of the irrigated soils, whereas hydrophilic compounds, namely ibuprofen, naproxen and diclofenac, were found in subsurface horizons [85]. This behavior is explained, on the one hand, because hydrophilic compounds remain dissolved in water rather than being retained in soil and, on the other hand, because of hydrophilic compounds are more susceptible to desorption from soil either during further irrigation or heavy rain events, and thus tend to rapidly reach subsoil and the aquifer [55, 104].

The chemical structure also affects the environmental fate of OPECs in soil. Molecules displaying aromatic moieties, such as carbamazepine and naproxen, have been shown to be strongly retained by soil organic matter—both to the aliphatic and aromatic fractions of soil organic matter—compared with compounds that have no resonance structures [39, 40, 55]. This behavior is explained by the formation of bonds between aromatic rings within the solute molecules and the soil organic matter [39]. Nonylphenols and octylphenol compounds have surfactant properties as they possess an aliphatic chain and a phenol moiety at the edge of the molecule [3]. Due to this structure, these compounds can promote resolubilization of organic contaminants retained in soil, although the estimated risk of this occurrence is considerably low [72]. The presence of heteroatoms in organic molecules can impact upon their environmental fate in soil. For example, oxygen atoms within the ciprofloxacin molecule can form covalent bonds with aluminum and iron oxides in soil, resulting in irreversible adsorption of the compound onto the solid matrix [92].

With respect to volatile OPECs, artificial fragrances represent the best example of this feature; these compounds are contained both in personal care products and detergents. Typically, the more volatile compounds are also hydrophobic, so they can be spontaneously retained in topsoil and then volatilize when temperature increases [159]. Since irrigation using untreated WW, which contains large amounts of fragrances, is carried out in arid areas, it is expected that a significant fraction or all of the fragrance molecules are rapidly volatilized upon their input to soil via WW. Volatilization of OPECs in WW irrigated soils is still an unexplored issue; studies aimed at determining the fraction of organic contaminants that can be volatilized in soil enriched with organic matter via WW irrigation are still needed.

Natural attenuation processes leading to the removal and dissipation of OPECs in soil are shown in Fig. 6.6. Contaminants may either dissipate in soil by photodegradation, biodegradation or chemical degradation (hydrolysis, oxidation or reduction) mechanisms; they may be accumulated in soil by adsorption or removed from soil by volatilization. There is a significant lack of information in the literature with regard to the natural

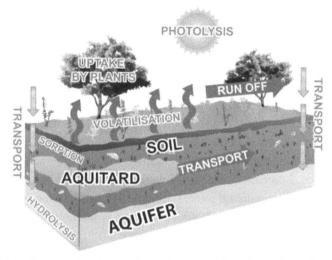

FIGURE 6.6   Processes involved in the environmental fate of emerging pollutants in soil.

photodegradation (i.e., photolysis of compounds by sunlight) of emerging pollutants in soil. The information available on the photodegradation of pesticides in agricultural soils is useful in elucidating, to some extent, the potential for photodegradation of OPECs in soil. Studies on natural photodegradation of the pesticide quinalphos showed that photodegradation takes place only in the first 2–5 mm of soil (photic layer); this photolysis takes place in two stages, each one at a different depth [89]. In the uppermost soil layer (the first 2 mm) direct photodegradation of the organic compounds (i.e., the transformation of compounds due to the direct incidence of photons) occurs; in this same layer, the production of free radicals (e.g., hydroxyl radicals and excited dissolved organic matter) occurs due to the breakdown of the soil organic matter. In the second stage, the free radicals migrate to the lower photic layer through facilitated transport by soil moisture; subsequently, photolytic transformation occurs below by the action of free radicals generated in the upper layer of soil (indirect photodegradation) [73]. As a result of the aforementioned aspects, soil moisture content and organic matter are determinant factors in the photodegradation of organic contaminants retained in the soil surface [73]. The physical structure of the soil can also significantly impact upon the photodegradation of organic pollutants, as it defines the depth to which solar radiation can penetrate the soil. In the study referenced in Ref. [73], sunlight photolysis of 4–nonylphenol in biosolids amended soils was studied. Photolysis resulted in 40% conversion of the compound within 30 days, with photodegradation observed in the first 5 mm of the soil. Since natural photodegradation occurs only in the soil surface layers, the organic compounds retained in the topsoil will be the most exposed to direct sunlight, although this does not necessarily imply increased rates of photodegradation. An example of this is the anticonvulsant agent carbamazepine, which is prominently retained in topsoil but has demonstrated poor photodegradation in water studies. Conversely, the anti-inflammatory drug diclofenac has been shown to present significant photoactivity [197], but it is less well retained in the topsoil. Due to this, photodegradation is unlikely to occur in soil by direct or indirect means. In spite of almost all of the studies evaluating the natural photodegradation of OPECs have been carried out in aqueous matrices [28], the results obtained in these experiments provide valuable information concerning the photoactivity of such compounds; which can be useful for studying the photolysis of organic pollutants in soil. For example, it is known that the NSAID ibuprofen and the anticonvulsants drugs carbamazepine and primidone are poorly photodegraded in water whereas the antibacterial agent triclosan, the antibiotic drug sulfamethoxazole and the NSAIDs diclofenac and naproxen are readily photodegraded [28]. These results can be the basis to establish experiments aimed at determining or modeling the photodegradation of organics in soil. In general terms, the natural sunlight reaching the troposphere (i.e., the surface of earth) does not possess enough energy to mineralize most photodegradable compounds [219]; therefore a wide variety of by-products occurs when organic contaminants in water and soil are photodegraded. It is known that, in some cases, more harmful compounds can be produced by photodegradation of some organic pollutants. For example, 2,8-dichlorodibenzo-p-dioxin is produced by the natural photodegradation of the antibacterial agent triclosan [130]. Differently to triclosan, its breakdown product has the potential to cause cancer in mammals. Another example is the antiepileptic drug carbamazepine, which photodegrades to acridone [36], a compound related to the occurrence of cancer in aquatic species.

Most photodegradation studies of OPECs (in water matrices) have been carried out in developed countries at latitudes higher than 30°N [28]. It is therefore necessary to investigate the intensity of photodegradation processes occurring at lower latitudes, in zones where higher incidence of sunlight occurs and treated/untreated WW irrigation is more intensively practiced.

Biodegradation of OPECs in soil has been studied in greater detail than photodegradation. Laboratory studies have found that biodegradation of emerging pollutants occurs optimally under aerobic conditions, while negligible transformations have been observed under anaerobic conditions [216]. This implies that biodegradation of this kind of contaminants is more likely to occur in well-structured soils, where tillage activities are frequently carried out, which allows better gas exchange through the soil matrix. The opposite behavior may be observed in anoxic/anaerobic soils, for instance in paddy fields. The antiepileptic drug carbamazepine is reported as one of the most refractory organic pollutants in soil, which has led researchers to consider this antiepileptic agent as a marker for anthropogenic contamination of surface and groundwater bodies [79]. In the study referenced in Ref. [134], mineralization of carbamazepine in soil was found to be less than 2%, after 120 days of incubation under aerobic conditions, while the reported in Ref. [204] show half-life times of 472 days in aerobic biosolids amended soils. Other compounds listed as recalcitrant in soil are the X–ray contrast media iopaminol, iomeprol and iohexol, whose biodegradation kinetic rate constants range from 0.29 to 0.46 μM/day [122].

Pharmaceuticals, such as the antiepileptic drug primidone and the psychoactive diazepam have shown recalcitrance in water [165]; further studies are necessary in order to elucidate whether such behavior may also occur in soil. Substances designed to exert an effect on microorganisms have been shown to be rapidly biodegraded in soil. Examples of these are antibacterial agents such as triclosan, triclocarban and antibiotic substances [121, 176]. Triclosan and triclocarban have been shown to be biodegraded in aerobic soils after 18 and 108 days, respectively [216], whereas the antibiotic compounds erythromycin, oleandomycin, tylosin, tiamulin and salinomycin displayed half-lives in aerobic soil of 20, 27, 8, 16 and 5 days, respectively [176]. Endocrine disrupting compounds, such as phthalate esters have been shown to efficiently biodegrade in agricultural soils, displaying half-lives of 7.8 to 8.3 days for di-butyl phthalate and 26–30 days for di–2–(ethylhexyl)phthalate [212].

Currently, few soil microorganisms have been identified as degraders of emerging pollutants. For example, the fungi Trametes versicolor has been demonstrated to degrade naproxen [142], while Rhodococcus rhodochorus bacteria [81] have been shown to degrade carbamazepine down to levels of 15% of its initial concentration in soil. In the case of phthalates (plasticizers) bacteria belonging to groups of Corynebacterium, Mycobacterium and Nocardia were demonstrated to degrade up to 90% of di-butyl phthalate within 48 hours in biodegradation experiments using isolated bacteria cultivated in saline solution [38]. Knowing the species of microorganisms that perform the biodegradation of OPECs in soil is useful in order to design engineered systems to treat WW and polluted soils based on the increased ability of degraders to degrade specific compounds by acclimatization and bioaugmentation. Such systems were tested in Ref. [107] using the fungus Trametes versicolor to degrade up to 94% of carbamazepine in WW after 6 days in an air pulsed bed bioreactor. Biodegradation of OPECs in soil is influenced by the sorption phenomenon, therefore soil characteristics such as the content of organic matter, soil texture and soil pH are crucial for this process to occur. Adsorption of the organic contaminants onto the surface of the soil particles may favor biodegradation when the sorbed compounds are still bioavailable; conversely, when strong adsorption occurs (chemisorption on soil organic matter, clay or soil micropores) it can result in decreased bioavailability of the compounds and thus in the confinement of the pollutants within the soil matrix. Other properties of soil involved in the biodegradation of OPECs are: (a) the climatic conditions of the site; (b) the physical structure of soil; (c) the soil moisture; and, (d) the adaptation of soil organisms to biodegrade the target pollutants. It is possible that microorganisms in long-term WW irrigated soils more efficiently biodegrade OPECs than those living in non-irrigated soils or soils irrigated for a shorter time. This is due to the ability of soil microorganisms to adapt to using emerging pollutants as a carbon source. In this sense, studies comparing the degradation efficiency of OPECs in long-term WW irrigated soils with that observed in non-irrigated soils or newly irrigated soils are needed in order to establish appropriate strategies to prevent contamination of groundwater. Very few efforts have been made to determine the nature and quantity of by-products generated in soil by the biodegradation of OPECs. As shown in Ref. [134], biodegradation of emerging pollutants can generate by-products that can be more harmful than the original substance [36, 130]; thus the presence of these by-products as well as their environmental fate should be priority for further research.

Those emerging pollutants that are not degraded by soil microorganisms may either accumulate in soil, be assimilated by plants (if they are bioavailable) or be degraded by other mechanisms (e.g., photodegradation or hydrolysis). In the case of carbamazepine, studies referred to in Ref. [85] explain that this compound is one of the most highly accumulated in WW irrigated soils. Moreover, carbamazepine can be assimilated by plants in WW irrigation systems at environmentally relevant concentrations (i.e., within the range 1–3 µg/L). The study referenced in Ref. [181] shows that cucumber (Cucumis sativus L.) can accumulate carbamazepine in different parts of the plant: 4.5 µg/kg in roots, 1.9 µg/kg in stem, 39.9 µg/kg in leaves and 2.1 µg/kg in fruit. According to the authors, phytotoxic effects were observed when carbamazepine was supplied to soil by irrigation at concentrations as high as 10,000 µg/L. Results of this study show that consumption of carbamazepine polluted cucumber results in doses of 1 ng of carbamazepine per gram of fruit. Other studies show that soybean (Glycine max L.) can take up carbamazepine, triclosan and triclocarban in roots, stems and leaves at concentrations of 1.3–3.4 µg/kg for carbamazepine and 2.4–13.7 µg/kg for the antibacterial agents triclosan and triclocarban. Concentrations of antibacterial agents in plants at a second harvesting were found to be higher than those obtained in the first one; this may be due to the accumulation of contaminants in the soil, as a bioavailable pool, between each irrigation events [210]. To date, the study of the assimilation of OPECs by plants in WW irrigated soils is still limited; moreover, priority should be given to develop health risk assessment studies related to the consumption of contaminated crops.

Adsorption (i.e., retention of solutes on the surface of the soil particles) of OPECs in the soil is a decisive process in their environmental fate, since through this process contaminants may either be retained or migrate into the aquifer. In cases where organic pollutants are retained in topsoil, photodegradation or volatilization phenomena can easily take place. The strength of the bonds that pollutants establish with soil particles determines the bioavailability of molecules to plants and soil microorganisms.

Adsorption of pollutants onto soil is measured by the distribution coefficient (Kd), which relates the amount of compound retained by soil to the mass remaining in the liquid phase [135].

Several models to determine the distribution coefficient of organic compounds have been developed; such models vary in complexity and the accuracy with which they represent the field conditions; yet simple adsorption models such as linear, Langmuir and Freundlich are the most used [135]. Due to their organic nature, OPECs tend to be rapidly and strongly adsorbed by soil organic matter; due to this effect, non-polar emerging pollutants, such as phthalates, have been shown to instantly adsorb onto organic soils [152]. On the other hand, OPECs displaying negative charge at the soil pH values, as occurs for NSAIDs, exhibit less adsorption by soil due to the repulsive forces between the negatively charged moiety within the molecule and the soil particles displaying negative charges (i.e., organic matter and clay) [55]. Accumulation of organic matter in WW irrigated soils increases the soil's ability to adsorb organic compounds. The proof of this can be found in the study referenced in Ref. [50], which reports greater adsorption of the antibiotics sulfamethoxazole and ciprofloxacin in long-term WW irrigated soils compared to non-irrigated ones from the same area. In addition to soil organic matter, OPECs may be retained by the inorganic domain of soil; for instance, ciprofloxacin showed strong and instantaneous adsorption by iron oxides and clay in agricultural soils, which was achieved by the formation of covalent bonds between metals in the soil and the oxygen atoms within ciprofloxacin molecules [92]. Furthermore, adsorption of carbamazepine by smectite type clays has been reported by Bi et al. [23]. According to studies referred to in Refs. [23, 39, 55], the adsorption of OPECs with multiple aromatic rings is more efficient in soils displaying a high content of humified organic matter, which displays higher aromaticity than labile organic matter. Polyaromatic compounds can establish π–π bonds between the aromatic rings within the pollutant molecules and aromatic compounds contained in soil organic matter. The formation of such bonds should be studied in future research in order to determine the optimum chemical characteristics of soil organic matter which enable better retention of contaminants, hence preventing their mobilization into the aquifer and/or making them available for uptake by plants. OPECs may be adsorbed by dissolved organic matter to soil via WW. Adsorption of organic pollutants to dissolved organic matter increases the solubility of the compounds and hence facilitates the lixiviation through soil. Studies referenced in Refs. [139, 192] report that compounds such as naproxen, carbamazepine and sex hormones can be adsorbed onto dissolved organic matter, notably to the hydrophobic and neutral hydrophilic fractions of dissolved organic matter. The speed of formation and strength of bonds between organic compounds and the dissolved organic matter varies depending on the quality of both WW and dissolved organic matter in soil [192]. The continuous occurrence of OPECs in WW irrigated soils can impact upon the adsorption of other organic pollutants; this is because at the time emerging pollutants enter to soil via WW, some of the active adsorption sites in soil are still occupied by previously adsorbed pollutants. In the study referred to in Ref. [55], the distribution coefficients of three OPECs, namely naproxen, carbamazepine and triclosan, were determined by an adsorption model which takes into account the previous presence of organic pollutants in the soil (the initial mass model [114]). The authors found modest differences between the values obtained in their study and those reported in the literature. However, it was observed that compounds previously adsorbed onto soil, i.e. naproxen and carbamazepine, were released from the solid matrix each time WW "washes" the soil in each irrigation event, resulting in a risk of contamination of the aquifer.

The transportation of OPECs through soil is closely related to their adsorption onto the solid matrix. Transport studies can be performed using different approaches, either packed soil columns or undisturbed soil columns tests. Transport of OPECs and pathogens is better described using the undisturbed soil column approach; through this approach, it is possible to evaluate the impact of both physical and chemical properties of soil on the transport of pollutants. In transport assays using undisturbed soil columns it is possible to assess the impact of preferential paths on the transport of solutes and particles, at the same time determining the effect of chemical properties of soil in the retention of solutes under dynamic flow conditions. The type of clays in soil significantly impacts on the transport of organic pollutants. The presence of expansive clays in soil results in the disappearance of preferential paths in the porous network of soil once clay becomes wet, which in turn provokes the decay in transport of contaminants contained in water. However, in such cases, soil conditions become anaerobic and thus organic pollutants are biodegraded with difficulty. The understanding of the environmental fate of OPECs in WW irrigated soil still has many gaps. It is therefore important to carry out studies on the laboratory scale and then in the field (plot level or landscape level) in order to determine the fate of these substances under real conditions. Results of these studies are of great importance, on the one hand, to allow more accurate and useful risk assessment studies and, on the other hand, to determine the characteristics of the sites suitable for irrigation with treated/untreated WW without posing a risk to the health of organisms and to the quality of crops and water sources. Lastly, regulations for OPECs in soil should be established in order to set maximum concentration limits for the accumulation of these compounds in terms not only of the effects caused to soil organisms, but also their potential to reach groundwater.

6.4 SCOPE FOR FURTHER RESEARCH

Reuse of WW in irrigation is a complex issue that requires the development of numerous studies in different disciplines; in this section some perspectives for further studies are presented.

1. Long-term studies aimed at determining the improvement of soil properties to produce food. Such studies should compare the rate of entry and conversion of carbon, nitrogen and phosphorus in irrigated soils in order to obtain a mass balance showing either the sustainability or the accumulation of organic matter in WW irrigated soils. Moreover, studies demonstrating the long-term increase in the soil's ability to treat WW used for irrigation should be carried out for each of the properties addressed in this chapter, as well as those considered appropriate in each system.
2. The determination of OPECs and pathogens in soils irrigated with WW. Such monitoring studies can be used to establish an inventory of contaminated sites that reflects the level of pollution in developed and developing countries. This can help in proposing ad hoc solutions for each site.

3. Determining feasibility and the mechanisms that can lead to horizontal propagation of antibiotic resistance genes in soil microorganisms (either innocuous microorganisms or opportunistic pathogens).

4. Chronic toxicity studies of OPECs in WW irrigated soils covering either several crop cycles, several generations of organisms or several years. Toxicity studies should address the effects of the presence of mixed contaminants at trace levels (environmentally relevant concentrations) on soil organisms. Such studies should be conducted including new emerging contaminants, for example, nanoparticles.

5. The study of the environmental fate of emerging contaminants using different model molecules in soil. Such environmental fate studies should be carried out at laboratory and field scale. In the case of environmental fate studies at laboratory scale, conditions used should be those that best emulate field conditions, for example, sunlight lamp intensities similar to those observed in the field for testing photodegradation or undisturbed soil columns in transport assays through soil.

6. The determination and quantification of the by-products appearing in soil upon dissipation of OPEC. Harmful compounds such as dioxins, chlorophenols and polyaromatics may be produced in soil from substances such as triclosan and carbamazepine. Discerning the occurrence and fate of these substances in soil should be addressed in future studies

7. Determination of the environmental fate of organic, inorganic and microbial contaminants in agricultural soil remaining after irrigation with WW has ceased. Worldwide, notably in developed countries, there are several sites where irrigation with WW has been stopped after a considerable time; in such cases, it is necessary to know the fate of the pool of pollutants that accumulated in soil during continuous input via WW. Phenomena such as the release of heavy metals confined in soil organic matter can occur when soil organisms start to mineralize organic carbon accumulated in the soil. In addition, the soil microorganisms can lose the capacity to treat pollutants in WW, leaving the soil vulnerable in cases where WW irrigation is restarted.

8. Studies elucidating the conditioning methods for agricultural soils newly irrigated with WW. Since in arid regions a considerable increase in the area under irrigation is being observed, it is necessary to use current knowledge to implement regulations establishing the optimal conditions for soils candidate to receive treated/ untreated WW. These are necessary to prevent soil degradation and contamination of water sources in the irrigated area.

9. Studies on the migration of contaminants in soil due to extreme events caused by climate change. Extreme rainfall events can cause an incremental increase in the mobilization of organic contaminants retained in the surface layers of soil into aquifers or to non-irrigated soils affected by runoff. However, increases in temperature can decrease the biodegradation of organic pollutants in the soil due to excessive drying of the solid matrix.

10. The development and implementation of WW treatment systems to remove organic, inorganic and biological pollution without reducing the content of organic matter in the water. These systems must be inexpensive for dissemination in developing countries. Advanced primary treatment systems may represent a plausible strategy in such cases.

11. The development and validation of environmentally-friendly analytical techniques for the determination of OPECs in soils.

12. Reuse of WW in micro irrigation is similar to using WW for other methods of irrigation In fact, there is no runoff and deep percolation is minimum. However, clogging of emitters and laterals are complex issues that require the development of numerous theoretical and field studies with WW in all core areas of drip irrigation. Use of WW in micro irrigation is a challenge to the educator and irrigator. Drip irrigation can save our planet from water scaricity, pollution by contaminants and health risks.

6.5 SUMMARY

The reuse of treated/untreated municipal WW for irrigation definitely has positive impacts on soil as a medium for the development of plants and animals. Additionally, this practice results in positive impacts on the welfare of farmers due to the monetary savings and profits that they obtain by the use of WW as a fertilizer and water source for crops. Similarly, the soil's ability to self-cleanse and treat the WW supplied at each irrigation event increases with the reuse of WW. The accumulation of organic matter in the soil surface results: in changes in soil pH to neutral and basic values; an improvement of soil physical structure; and an increase in the soil microbial activity. Together with this, soil organisms become acclimatized to the presence of contaminants and thus their resilience to the harmful effects caused by pollutants increase. These phenomena lead to an improvement in the ability of the soil to act as a filter and transforming medium for contaminants and thereby to an increase in its capacity to treat WW. Such an improvement in soil functions can be capitalized by the State and the conventional treatment regime can be changed to a cheaper one driven by natural attenuation mechanisms. This in turn improves the quality of life of people living in the area by increasing food production and the possibility of obtaining profit by sales of produce.

The responsible reuse of municipal WW for irrigation can help to mitigate three problems, which are a priority in developing countries: (a) water stress in arid areas where rain-fed agriculture makes development uncertain. In such areas fresh-water sources are used for agriculture rather than human consumption, and therefore the reuse of municipal WW not only results in savings of freshwater but also in the recharge of the aquifer in the irrigated area. Recharge is with good quality water produced by infiltration of WW through the soil; (b) the food crisis and the lack of jobs in rural and peri-urban areas in developing countries. Reuse of WW represents a way of producing food for consumption and sale; and (c) the treatment of municipal WW generated in urban and rural areas through a low cost natural treatment systems, which in turn generate profits for population.

In order to reuse WW responsibly and exploit its inherent benefits for soil and people living in the irrigated area, the occurrence of contaminants in WW –especially untreated WW-must be kept in mind. The presence of pathogenic microorganisms and the potential for antibiotic resistance dissipation via WW should be priority concerns in designing WW reuse schemes in agricultural areas, notably when using raw WW.

Attention should be paid to the fate of emerging contaminants in WW irrigation schemes including its transportation through irrigation canals, storage in dams and deposition in agricultural soils and transport to aquifers. Another priority is the elucidation of the chronic toxic effects caused by the continuous presence of traces of emerging contaminants in irrigated soils. Since the group of OPECs is quite broad, model compounds should be selected to determine the rate at which they are dissipated or retained/transported through soil, as well the risk of these compounds reaching the aquifer or being assimilated by plants. Despite the spread of antibiotic resistance in the environment, it has not been conclusively shown the role that irrigation with treated/untreated WW plays in this. To date, the concentrations of OPECs found in soil irrigated with WW are lower than the toxicity thresholds reported in literature. The precautionary principle states that WW must be minimally treated before irrigation in order to remove pathogenic microorganisms and trace of heavy metals, as well as to reduce as much as possible the concentration of emerging pollutants.

Other areas of opportunity to be developed in order to reduce the risk of soil degradation and effects on soil organisms are: (a) the development of environmentally friendly everyday-consumer products, containing organic compounds that have been proven to have no harmful effects on living organisms even at trace concentrations. Consumer products must follow strict risk assessments before release to the market; (b) an improvement in health systems in cities in order to reduce the incidence of infectious diseases that ultimately generate biological contamination of soil, especially in irrigation systems using raw WW; (c) the maintenance of WW irrigation schemes fed with municipal WW in order to avoid a high input of heavy metals and refractory organic compounds to soil and crops through irrigation; and (d) the ad hoc treatment of municipal WW to allow its reuse in agricultural activities. Low cost treatment systems aimed at removing microorganisms, suspended solids and trace heavy metals are recommended to treat WW without affecting its properties as a fertilizer and source of organic matter to improve physical, chemical and microbiological soil properties. Such an approach allows soil to fulfill its ecological functions as a generator of food and livelihoods and as a protective barrier to the aquifer.

KEYWORDS

- adsorbed contaminants
- adsorption
- aerobic conditions
- Africa
- aggregation
- agricultural irrigation
- agriculture
- alfalfa
- ammoniacal nitrogen
- amoebiasis
- anaerobic conditions
- anoxic
- antibiotic resistance gene, ARG
- antibiotic resistant pathogen, ARP
- antibiotics
- aquatic environment
- assessment
- bacterial migration
- barley
- beans
- beneficial impacts on soil
- bioavailability
- biodegradation
- biological degradation
- biological methylation
- biological pollutants
- biological treatment
- biosolids
- boron
- calcareous soils

- calcium phosphate
- cancer
- Carbon
- cation exchange
- chickpea
- chili
- China
- clean water
- coagulation
- Colombo
- colonization
- complexation
- contaminants of emerging concern
- contaminated water
- corn
- crop rotation
- daily human intake
- denitrification
- depuration
- desiccation
- developed countries
- developing countries
- dissipation
- drip irrigation
- dung
- ecosystem
- edible crops
- effluent water
- England
- environmental risk
- environmental strategy
- erodibility

- erosion
- *Escherichia coli*
- fertilizers
- flocculation
- food scarcity
- foodborne virus
- forest soils
- frequency of irrigation
- freshwater, FW
- Ghana
- grass
- gravity irrigation
- green tomatoes
- groundwater, GW
- health hazards
- heavy metal
- helminthes
- hormone
- hydrophobicity
- India
- infiltration
- inorganic pollutants
- irrigation
- irrigation water
- *IWMI*
- Japan
- lake Michigan
- lettuce
- lime
- livestock
- living filter
- macro-aggregates

- **water requirement**
- **water scarcity**
- **water science**
- **water stress**
- **water uptake**
- **waterborne diseases**
- **wheat**
- **WHO**
- **Zimbabwe**
- **zucchini**

REFERENCES

1. Abu–Ashour, J., Joy, D. M., Lee, H., Whiteley, H. R., Zelin, S. (1994). Transport of microorganisms through soil. *Water, Air, and Soil Pollution*, 75(1–2), 141–158.
2. Abu – Zreig, M., Rudra, R. P., Dickinson, W. T. (2003). Effects of application of surfactants on hydraulic properties of soils. Biosystems Engineering, 84(3), 363–372.
3. Ahel, M., Giger, W., Koch, M. (1994). Behaviour of alkylphenol polyethoxylate surfactants in the aquatic environment, I: Occurrence and transformation in sewage treatment. *Water Research*, 28(5), 1131–1142.
4. Ahmad, R., Kookana, R. S., Alston, A. M., Skjemstad J. O. (2001). The nature of soil organic matter affects sorption of pesticides, 1, Relationships with carbon chemistry as determined by 13C CPMAS NMR spectroscopy. *Environmental Science and Technology*, 35(5), 878–884.
5. Aleem, A., Isar, J., Malik, A. (2003). Impact of long-term application of industrial ww on the emergence of resistance traits in Azotobacter chroococcum isolated from rhizospheric soil. *Bioresource Technology*, 86(1), 7–13.
6. Altaf, M. M., Masood, F., Malik, A. (2008). Impact of long-term application of treated tannery effluents on the emergence of resistance traits in rhizobium sp. isolated from *trifolium alexandrinum*. *Turkish Journal of Biology*, 32(1), 1–8.
7. Amahmid, O., Asmama, S., Bouhoum, K. (1999). The Effect of ww reuse in irrigation on the contamination level of food crops by *giardia* cysts and *ascaris* eggs. *International Journal of Food Microbiology*, 49(1), 19–26.
8. Anderson, J. (2003). The Environmental benefits of water recycling and reuse. *Water Science and Technology: Water Supply*, 3(4), 1–10.
9. Androver, M., Farrus, E., Moya, G., Vadell, J. (2012). Chemical properties and biological activity in soils of Mallorca following twenty years of treated ww irrigation. *Journal of Environmental Management*, 95, 188–192.

10. Ansari, M. I., Malik, A. (2007). Biosoption of nickel and cadmium by metal resistant bacterial isolates from agricultural soil irrigated with industrial ww. *Bioresource Technology*, 98(16), 3149–3153.
11. Armstrong, A. S. B., Tanton, T. W. (1992). Gypsum application to aggregated saline-sodic clay topsoils. *Journal of Soil Science*, 43(2), 249–260.
12. Arora, M., Kiran, B., Rani, S., Rani, A., Kaur, B., Mittal, N. (2008). Heavy metal accumulation in vegetables irrigated with water from different sources. *Food Chemistry*, 111(4), 811–815.
13. Asano, T. (1998). *Waste Water Reclamation and Reuse: Water Quality Management Library*. Pennsylvania: Technomic Publishing Company.
14. Ayres, D., Westtcott, W. (1985). *Water Quality for Agriculture*. FAO Paper 29. Rome: Food and Agricultural Organization.
15. Baquero, F., Martínez, J. L., Cantón, R. (2008). Antibiotics and antibiotics resistance in water environments. *Current Opinion in Biotechnology*, 19(3), 260–265.
16. Barceló, D. (2003). Emerging pollutants in water analysis. *Trends in Analytical Chemistry*, 22(10), xiv–xvi.
17. Barker – Reid, F., Harapas, D., Engleitner, S., Kreidl, S., Holmes, R., Faggian, R. (2009). Persistence of *escherichia coli* on injured iceberg lettuce in the field, overhead irrigated with contaminated water. *Journal of Food Protection*, 72(3), 458–464.
18. Barwick, R. S., Mohammed, H. O., White, M. E., Bryant, R. B. (2003). Prevalence of *Giardia spp.* and *Cryptosporidium spp.* on dairy farms in Southeastern New York. State. *Preventive Veterinary Medicine*, 59(1–2), 1–11.
19. Beausse, J. (2004). Selected Drugs in Solid Matrices: A review of environmental determination, occurrence and properties of principal substances. *Trends in Analytical Chemistry*, 23(10–11, 753–761.
20. Beausse, J. (2003). Selected Drugs in Solid Matrices: A review of environmental determination, occurrence and properties of principal substances. *Trends in Analytical Chemistry*, 23(10), 753–761.
21. Berg, G., Eberl, L., Hartmann, A. (2005). The rhizosphere as a reservoir of opportunistic human pathogen bacteria. *Environmental Microbiology*, 7(11), 1673–1685.
22. Berg, G., Marten, P., Ballin, G. (1996). *Stenotrophomonas maltophilia* in the rhizosphere of oilseed rape: Occurrence, characterization and interaction with phytopathogenic fungi. *Microbiology Research*, 151(1), 19–27.
23. Bi, E., Schmidt, T. C., Haderlein, S. B. (2007). Environmental factors influencing sorption of heterocyclic aromatic compounds to soil. *Environmental Science and Technology*, 41(9), 3172–3178.
24. Bigsby, R., Chapin, R. E., Daston, G. P., Davis, B. J., Gorski, J., Gray, L. E., Howdeshell, K. L., Zoeller, R. T., vom Saal, F. S. (1999). Evaluating the effects of endocrine disruptors on endocrine function during development. *Environmental Health Perspectives*, 107(4), 613–618.
25. Birnbaum, L. S., Fenton, S. E. (2003). Cancer and developmental exposure to endocrine disruptors. Environmental health perspectives, 111(4), 389–394.
26. Boix–Fayos, C., Calvo–Cases, A., Imeson, A. C., Soriano–Soto, M. D., Tiemessen, I. R. (1998). Spatial and short term variations in runoff, soil aggregation and other soil properties along a Mediterranean climatological gradient. *Catena*, 33(2), 123–138.
27. Bolong, N., Ismail, A. F., Salim, M. R., Matsuura, T. (2009). A review of the effects of emerging contaminants in ww and options for their removal. *Desalination*, 239(1), 229–246.

28. Boreen, A. L., Arnold, W. A., McNeill, K. (2003). Photodegradation of pharmaceuticals in the aquatic environment: a review. *Aquatic Sciences*, 65(4), 320–341.
29. Bos, R., Carr, R., Keraita, B. (2010). Assessing and mitigating ww-related health risks in low-income countries: an introduction. In: *Drechsel Pay, Scott C. A., Raschid – Sally L., Redwood M., Bahri A. (eds.) WW Irrigation and Health: Assessing and Mitigating Risk in Low–Income Countries*. London/Ottawa/Colombo: Earthscan/IDRC/IWMI. Pages 29–51.
30. Bouwer, H. (1991). Groundwater recharge with sewage effluent. *Water Science and Technology*, 23(10–12), 2099–2108.
31. Bouwer, H. (1987). *Soil–Aquifer Treatment of Sewage*. Rome: Food and Agriculture Organization of the United Nations.
32. Bradford, A., Brook, R., Hunshal, C. (2003). WW irrigation: Hubli – Dharwad, India. International Symposium on Water, Poverty and Productive uses of Water at the Household Level, 21–23 January, Muldersdrift, South Africa.
33. Bronick, C. J., Lal, R., Soil structure and management. *Geoderma*, 124(1–2), 3–22.
34. Brunetti, G., Senesi, N., Plaza, C. (2007). Effects on amendments with treated and untreated olive oil mill ww on soil properties, soil humic substances and wheat yield. *Geoderma*, 138(1–2), 144–152.
35. Butler, E., Whelan, M. J., Sakrabani, R., van Egmond, R. (2012). Fate of Triclosan in field soils receiving sewage sludge. *Environmental Pollution*, 167, 101–109.
36. Calisto, V., Domingues, M. R. M., Erny, G. L., Esteves, V. I. (2011). Direct photodegradation of carbamazepine followed by Micellar electrokinetic chromatography and mass spectrometry. *Water Research*, 45(3), 1095–1104.
37. Chang, A. C., Page, A. L., Asano, T. (1995). *Developing human health-related chemical guidelines for reclaimed ww and sewage sludge applications in agriculture*. Geneva: World Health Organization.
38. Chao, W. L., Lin, C. M., Shiung, I. I., Kuo, Y. L. (2006). Degradation of Di-butylphthalate by soil bacteria. *Chemosphere*, 63(8), 1377–1383.
39. Chefetz, B., Mualem, T., Ben–Ari, J. (2008). Sorption and mobility of pharmaceutical compounds in soil irrigated with reclaimed ww. *Chemosphere*, 73(8), 1335–1343.
40. Chefetz, B., Xing, B. (2009). Relative role of aliphatic and aromatic moieties as sorption domains for organic compounds: a review. *Environmental Science and Technology*, 43(6), 1680–1688.
41. Chen, F., Ying, G. G., Kong, L. X., Wang, L., Zhao, J. L., Zhou, L. J., Zhang, L. J. (2011). Distribution and accumulation of endocrine-disrupting chemicals and pharmaceuticals in ww irrigated soils in Hebei, China. *Environmental Pollution*, 159(6), 1490–1498.
42. Chen, M. (1988). Pollution of ground water by nutrients and fecal coliforms from lakeshore septic tank systems. *Water, Air, and Soil Pollution*, 37(3–4), 407–417.
43. Chiou, C. T., Kile, D. E. (1998). Deviations from sorption linearity on soils of polar and non-polar organic compounds at low relative concentrations. *Environmental Science and Technology*, 32(3), 338–343.
44. Cifuentes, E., Gomez, M., Blumenthal, U., Tellez–Rojo, M. M., Romieu, I., Ruiz–Palacios, G., Ruiz–Velazco, S. (2000). Risk factors for *giardia intestinalis* infection in agricultural villages practicing ww irrigation in Mexico. *American Journal of Tropical Medicine and Hygiene*, 62(3), 388–392.
45. Colborn, T., vom Saal, F. S., Soto, A. M. (1993). Developmental effects of endocrine-disrupting chemicals in wildlife and humans. *Environmental Health Perspectives*, 101(5), 378–384.

46. Coleman, D. C., Crossley, J. D. A., Hendrix, P. F. (2004). *Fundamentals of soil ecology*. London: Academic Press.
47. Coppola, A., Santini, A., Botti, P., Vacca, S., Comegna, V., Severino, G. (2004). Methodological approach for evaluating the response of soil hydrological behavior to irrigation with treated municipal ww. *Journal of Hydrology*, 292(1–4), 114–134.
48. Cox, L., Celis, R., Hermosin, M. C., Becker, A., Cornejo, J. (1997). Porosity and herbicide leaching in soils amended with olive-mill ww. *Agriculture, Ecosystem and Environment*, 65(2), 151–161.
49. Czarnes, S., Hallett, P. D., Bengough, A. G., Young, I. M. (2000). Root and microbial derived mucilages affect soil structure and water transport. *European Journal of Soil Science*, 51(3), 435–443.
50. Dalkmann, P., Broszat, M., Siebe, C., Willaschek, E., Sakinc, T., Huebner, J., Amelung, W., Grohmann, E., Siemens, J. (2012). Accumulation of pharmaceuticals, enterococcus, and resistance genes in soils irrigated with ww for zero to 100 years in Central Mexico. *PLoS ONE*, 7(9), e45397.
51. Daughton, C. G. (2004). Non-regulated water contaminants: emerging research. *Environmental Impact Assessment Review*, 24(7–8), 711–732.
52. DeFeo, G., Mays, L. W., Angelakis, A. N. (2011). Water and ww management technologies in the ancient Greece and Roman civilizations. In: Wilder P. (ed.) *Treatise in Water Science*. Oxford: Elsevier. Pages 3–22.
53. Degens, B., Schipper, L., Claydon, J., Russell, J., Yeates, G. (2000). Irrigation of an allophanic soil with dairy factory effluent for 22 years: responses of nutrient storage and soil biota. *Australian Journal of Soil Research*, 38(1), 25–35.
54. Diao, X., Jensen, J., Hansen, A. D. (2007). Toxicity of the anthelmintic abamectin to four species of soil invertebrates. *Environmental Pollution*, 148(2), 514–519.
55. Durán – Álvarez, J. C., Prado–Pano, B., Jiménez–Cisneros, B. (2012). Sorption and desorption of Carbamazepine, Naproxen and Triclosan in a soil irrigated with raw ww: estimation of the sorption parameters by considering the initial mass of the compounds in the soil. *Chemosphere*, 88(1), 84–90.
56. Durán – Álvarez, Juan C., B. Jiménez–Cisneros, (2014). Beneficial and negative impacts on soil by the reuse of treated/untreated municipal ww for irrigation—a review of the current knowledge and future perspectives. Chapter 5, 137–197, In: *Environmental Risk Assessment of Soil Contamination by M. C. Hernandez (ed)*.<http://www.intechopen.com/books/environmental-risk-assessment-of-soil contamination/beneficial-and-negative-impacts-on-soil-by-the-reuse-of-treated-untreated-municipal-WW-for-a>
57. Durán–Álvarez, J. C., Becerril–Bravo, E., Castro, V. S., Jiménez, B., Gibson, R. (2009). The analysis of a group of acidic pharmaceuticals, carbamazepine, and potential endocrine disrupting compounds in ww irrigated soils by gas chromatography-mass spectrometry. *Talanta*, 78(3), 1159–1166.
58. *Earth Trends: Environmental Information*. http://www.wri.org/project/earthtrends/.
59. Edwards, A. P., Bremner, J. M. (1967). Micro-aggregates in soil. *European Journal of Soil Science*, 18(1), 64–73.
60. Elimelech, M. (2006). The Global challenge for adequate and safe water. *Aqua – Journal of Water Supply: Research and Technology*, 55(1), 3–10.
61. Ensink, J. H., Mahmood, T., Dalsgaard, A. (2007). WW-irrigated vegetables: market handling versus irrigation water quality. *Tropical Medicine and International Health*, 12(2), 2–7.

62. Ensink, J. H. J., Simmons R. W., van der Hoek, W. (2004). WW use in Pakistan: The cases of Haroonabad and Faisalabad. In: Scott C. A., Faruqui N. I., Raschid–Sally L. (eds.) *WW Use in Irrigated Agriculture: Confronting the Livelihood and Environmental Realities.* Wallingford: CAB International. Pages 91–99.

63. Erdogrul, Ö., Şener, H. (2005). The contamination of various fruit and vegetable with *enterobius vermicularis, ascaris* eggs, *entamoeba histolyca* cysts and *giardia* cysts. *Food Control*, 16(6), 557–560.

64. Fan, A. M., Steinberg, V. E. (1996). Health implications of nitrate and nitrite in drinking water: an update on methemoglobinemia occurrence and reproductive and developmental toxicity. *Regulatory toxicology and pharmacology*, 23(1), 35–43.

65. Fausto Cereti, C., Rossini, F., Federici, F., Quaratino, D., Vassilev, N., Fenice, M. (2004). Reuse of microbially treated olive mill ww as fertilizer for wheat (*Triticum Durum*). *Bioresource Technology*, 91(2), 135–140.

66. Feachem, R. G., Bradley, D. J., Garelick, H., Mara, D. D. (1983). *Sanitation and Disease – Health Aspects of Excreta and WW Management.* Chichester: The World Bank.

67. Fenet, H., Mathieu, O., Mahjoub, O., Li, Z., Hillaire–Buys, D., Casellas, C., Gomez, E. (2009). Carbamazepine, carbamazepine epoxide and di-hydroxycarbamazepine sorption to soil and occurrence in a ww reuse site in Tunisia. *Chemosphere*, 88(1), 49–54.

68. Fent, K., Weston, A. A., Caminada, D. (2006). Ecotoxicology of human pharmaceuticals. *Aquatic Toxicology*, 76(2), 122–159.

69. Filip, Z., Kanazawa, S., Berthelin, J. (1999). Characterization of effects of a long-term ww irrigation on soil quality by microbiological and biochemical parameters. *Journal of Plant Nutrition and Soil Science,* 162(4), 409–413.

70. Food and Agriculture Organization of the United Nations, FAO, (2012). Coping with water scarcity: an action framework for agriculture and food security. Rome: FAO.

71. Foppen, J. W., Schijven, J. F. (2006). Evaluation data from the literature on the transport and survival of *escherichia coli* and thermotolerant coliforms in aquifers under saturated conditions. *Water Research*, 40(3), 401–426.

72. Fox, K. K., Chapman, L., Solbe, J., Brennand, V. (1997). Effect of environmentally relevant concentrations of surfactants on the desorption or biodegradation of model contaminants in soil. *Tenside, Surfactants, Detergents*, 34(6), 436–441.

73. Frank, M. P., Graebing, P., Chib, J. S. (2002). Effect of soil moisture and sample depth on pesticide photolysis. *Journal of Agricultural and Food Chemistry*, 50(9), 2607–2614.

74. Friedel, J. K., Langer, T., Siebe, C., Stahr, K. (2000). Effects of long-term ww irrigation on soil organic matter, soil microbial biomass and its activities in Central Mexico. *Biology and Fertility of Soils*, 31(5), 414–421.

75. Gadd, G. M. (2004). Microbial influence on metal mobility and application for bioremediation. *Geoderma*, 122(2–4), 109–119.

76. Gannon, J., Tan, Y. H., Baveye, P., Alexander, M. (1991). Effect of sodium chloride on transport of bacteria in a saturated aquifer material. *Applied and Environmental Microbiology*, 57(9), 2497–2501.

77. Gannon, J. T., Manilal, V. B., Alexander, M. (1991). Relationship between cell surface properties and transport of bacteria through soil. *Applied and environmental microbiology*, 57(1), 190–193.

78. Gaspard, P. G., Wiart, J., Schwartzbrod, J.. Experimental study of the adhesion of *Ceufs d'Helminthes Ascaris* (Étude Expérimentale de l'Adhésion des Ceufs d'Helminthes Ascaris suum: Consequences pour l'Environnent). *J. of Soil Sciences,* 7, 367–376.

79. Gasser, G., Rona, M., Voloshenko, A., Shelkov, R., Tal, N., Pankratov, I., Elhanany, S., Lev, G. (2010). Quantitative evaluation of tracers for quantification of ww contamination of potable water sources. *Environmental Science and Technology*, 44(10), 3919–3925.

80. Gasser, G., Rona, M., Voloshenko, A., Shelkov, R., Tal, N., Pankratov, I., Elhanany, S., Lev, G., Cytryn, E. (2013). Impact of treated ww irrigation on antibiotic resistance in the soil microbiome. *Environmental Science and Pollution Research*, 20(6), 3529–3538.

81. Gauthier, H., Yargeau, V., Cooper, D. G. (2010). Biodegradation of pharmaceuticals by *rhodococcus rhodochrous* and *aspergillus niger* by cometabolism. *Science of the Total Environment,* 408(7), 1701–1706.

82. Gaye, M., Niang, S. (2002). Purification of ww resuse in urban agriculture. *Research Studies,* Dakar: ENDA–TM.

83. Gerba, C. P., Bitton, G. (1984). Microbial pollutants: their survival and transport pattern to groundwater. In: Bitton G., Gerba C. P. (eds.) *Groundwater Pollution Microbiology.* New York: John Wiley & Sons, Inc. pages 65–88.

84. Gerba, C. P., Melnick, J. L., Wallis, C. (1997). Fate of ww bacteria and viruses in soil. Journal of the irrigation and drainage division 1975;101(3) 157–174.

85. Gibson, R., Durán–Álvarez, J. C., Estrada, K. L., Chávez, A., Jiménez–Cisneros, B. (2010). Accumulation and leaching potential of some pharmaceuticals and potential endocrine disruptors in soils irrigated with ww in the Tula Valley, Mexico. *Chemosphere*, 81(11), 1437–1445.

86. Gielen, G. J. H. P. (2007). The fate and effects of sewage-derived pharmaceuticals in soil. PhD thesis. University of Canterbury.

87. Girovich, M. (1996). *Biosolids Treatment and Management: Processes for Beneficial Use.* New York: Marcel Dekker Inc.

88. Gomes, C., Da Silva, P., Moreira, R. G., Castell–Perez, E., Ellis, E., Pendleton, M. (2009). Understanding *E. coli* internalization in lettuce leaves for optimization of irradiation treatment. *International Journal of Food Microbiology*, 135(3), 238–247.

89. Goncalves, C., Dimou, A., Sakkas, V., Alpendurada, M. F., Albanis, T. A. (2006). Photolytic degradation of quinalphos in natural waters and on soil matrices under simulated solar irradiation. *Chemosphere,* 64(8), 1375–1382.

90. Goyal, Megh R. (Senior Editor-in-Chief), (2015). Book Series: Research Advances in Sustainable Micro Irrigation. Volumes 1 to 10. Oakville, ON, Ca: Apple academic Press Inc.,

91. Grün, F., Blumberg, B. (2009). Endocrine disrupters as obesogens. *Molecular and Cellular Endocrinology*, 304(1), 19–29.

92. Gu, C., Karthikeyan, and K. G. (2005). Sorption of the antimicrobial ciprofloxacin to aluminum and iron hydrous oxides. *Environmental Science and Technology*, 39(23), 9166–9173.

93. Guo, X., Chen, J., Brackett, R. E., Beuchat, L. R. (2001). Survival of *salmonellae* on and in tomato plants from the time of inoculation at flowering and early stages of fruit development through fruit ripening. *Applied and Environmental Microbiology*, 67(10), 4760–4764.

94. Gupta, N., Khan, D. K., Santra, S. C. (2009). Prevalence of intestinal *helminth eggs* on vegetables grown in ww-irrigated areas of Titagarh, West Bengal, India. *Food control,* 20(10), 942–945.

95. Halden, R. U. (2010). Plastics and health risks. *Annual Review of Public Health,* 31, 179–194.

96. Hammesfahr, U., Heuer, H., Manzke, B., Smalla, K., Thiele–Bruhn S. (2008). Impact of the antibiotic sulfadiazine and pig manure on the microbial community structure in agricultural soils. *Soil Biology and Biochemistry,* 40(7), 1583–1591.

97. Haynes, R. J., Beare, M. H. (1997). Influence of six crops species on aggregate stability and some labile organic matter fractions. *Soil Biology and Biochemistry,* 29(11–12), 1647–1653.

98. Haynes, R. J., Naidu, R. (1998). Influence of lime, fertilizer and manure applications on soil organic matter content and soil physical structure: a review. *Nutrient Cycling in Agroecosystems,* 51(2), 123–137.

99. Heaton, J. C., Jones, K. (2008). Microbial contamination of fruit and vegetables and the behaviour of enteropathogens in the phyllosphere: a review. *Journal of Applied Microbiology,* 104(3), 613–626.

100. Hsu, B. M., Huang, C. (2002). Influence of ionic strength and pH on hydrophobicity and zeta potential of *giardia* and *cryptosporidium. Physicochemical and Engineering Aspects,* 201(1), 201–206.

101. Hu, C., Hermann, G., Pen–Mouratov, S., Shore, L., Steinberger, Y. (2011). Mammalian steroid hormones can reduce abundance and affect the sex ratio in a soil nematode community. *Agriculture, Ecosystems and Environment,* 142(3), 275–279.

102. Hu, X. Y., Wen, B., Shan, X. Q. (2003). Survey of phthalate pollution in arable soils in China. *Journal of Environmental Monitoring,* 5(4), 649–653.

103. Hussain, I., Raschid–Sally, L., Hanjra, M. A., Marikar, F., van der Hoek, W. (2002). *WW use in agriculture: review of impacts and methodological issues in valuing impacts.* Colombo: International Water Management Institute, IWMI.

104. Hyland, K. C., Dickenson, E. R., Drewes, J. E., Higgins, C. P. (2012). Sorption of ionized and neutral emerging trace organic compounds onto activated sludge from different ww treatment configurations. *Water Research,* 46(6), 1958–1968.

105. Idelovitch, E., Michail, M. (1984). Soil-aquifer treatment: a new approach to an old method for ww reuse. *Journal Water Pollution Control Federation,* 56(8), 936–943.

106. Jastrow, J. D. (1996). Soil aggregate formation and the accrual of particulate and mineral associated organic matter. *Soil Biology and Biochemistry,* 28(4–5), 665–676.

107. Jelic, A., Cruz–Morató, C., Marco–Urrea, E., Sarrà, M., Perez, S., Vicent, T., Petrovic, M., Barcelo, D. (2012). Degradation of carbamazepine by trametes versicolor in an air pulsed fluidized bed bioreactor and identification of intermediates. *Water Research,* 46(4), 955–964.

108. Jiménez, B. (1993). WW reuse to increase soil productivity. *Water Science and Technology,* 32(12), 173–180.

109. Jiménez, B. (2007). WW use in agriculture: public health considerations. In: Trimble T. W., Trimble S. W. (eds.) *Encyclopedia of Water Science.* London: CRC Press. Pages 1303–1306.

110. Jimenez, B., Asano, T. (2008). Water reclamation and reuse around the world. In: Jimenez B., Asano T. (eds.) *Water Reuse – An international Survey of Current Practice, Issues and Needs.* London: IWA Publishing. Pages 3–27.

111. Jiménez, B., Chávez, A., Treatment of Mexico City WW for irrigation purposes. *Environmental Technology*, 18(7), 721–729.

112. Jiménez, B., Drechsel, P., Koné, D., Bahri, A., Raschid–Sally, L., Qadir, M. (2010). General ww, sludge and excreta use situation. In: Drechsel Pay, Scott C. A., Raschid-Sally L., Redwood M., Bahri A. (eds) *WW Irrigation and Health: Assessing and Mitigating Risk in Low–Income Countries*. London/Ottawa/Colombo: Earthscan/ IDRC/IWMI. Pages 3–29.

113. Jjemba, P. K. (2002). The Potential impact of veterinary and human therapeutic agents in manure and biosolids on plants grown on arable land: a review. *Agriculture, Ecosystems and Environment*, 93(1), 267–278.

114. Kaiser, K., Guggenberger, G., Zech, W. (1996). Sorption of DOM and DOM fractions to forest soils. *Geoderma*, 74(3), 281–303.

115. Karickhoff, S. W., Brown, D. S., Scott, T. A. (1979). Sorption of hydrophobic pollutants on natural sediments. *Water Research*, 13(3), 241–248.

116. Kay, B. D. (1998). Soil structure and organic carbon: a review. In: Lal R., Kimble J. M., Follett R. F., Stewart B. A. (eds.) *Soil Processes and the Carbon Cycle*. Boca Raton: CRC Press. Pages 169–197.

117. Keraita, B., Jiménez, B., Drechsel, P. (2008). Extent and implications of agricultural reuse of untreated, partially treated and diluted ww in developing countries. *CAB reviews: Perspectives in Agriculture, Veterinary Science, Nutrition and Natural Resources*, 3(58), 1–15.

118. Keswick, B. H., Gerba, C. P. (1980). Viruses in groundwater. *Environmental Science and Technology*, 14(11), 1290–1297.

119. Kim, D. Y., Burger, J. A. (1997). Nitrogen transformations and soil processes in a ww irrigated, mature Appalachian hardwood forest. *Forest Ecology and Management*, 90(1), 1–11.

120. Knapp, C. W., Dolfing, J., Ehlert, P. A. I., Graham, D. W. (2010). Evidence of increasing antibiotic resistance gene abundances in archived soils since (1940). *Environmental Science and Technology*, 44(2), 580–587.

121. Kookana, R. S., Ying, G. G., Waller, N. L. (2011). Triclosan: its occurrence, fate and effects in the Australian environment. *Water Science and Technology*, 63(4), 598–604.

122. Kormos, J. L., Schulz, M., Kohler, H. P. E., Ternes, T. A. (2010). Biotransformation of selected iodinated x-ray contrast media and characterization of microbial transformation pathways. *Environmental Science and Technology*, 44(13), 4998–5007.

123. Kotzerke, A., Sharma, S., Schauss, K., Heuer, H., Thiele–Bruhn, S., Smalla, K., Wilke, B. M., Schloter, M. (2008). Alterations in soil microbial activity and n-transformation processes due to sulfadiazine loads in pig-manure. *Environmental Pollution*, 153(2), 315–322.

124. Kretschmer, N., Ribbe, L., Gaese, H. (2002). WW reuse for agriculture: technology resource management and development. *Water Management*, 2(1), 35–61.

125. Kümmerer, K. (2009). *Pharmaceuticals in the Environment: Sources, Fate, Effects and Risks*. Berlin: Springer–Verlag.

126. Kümmerer, K. (2005). Significance of antibiotics in the environment. *Journal of Antimicrobial Chemotherapy*, 52(1), 5–7.

127. Kunhi krishnan, A., Bolan, N. S., Müller, K., Laurenson, S., Naidu, R., Kim, W. I., The influence of ww irrigation on the transformation and bioavailability of heavy

metals in soil. In: Sparks D. L. (ed.) *Advances in Agronomy*, Vol. 115. London: Academic Press. Pages 219–273.

128. Lal, R. (1991). Soil structure and sustainability. *Journal of Sustainable Agriculture*, 1(4), 67–92.

129. Landa – Cansigno, O., Durán–Álvarez, J. C., Jiménez, B. (2013). Retention of *escherichia coli, giardia lamblia cysts* and *ascaris lumbricoides* eggs in agricultural soils irrigated by untreated ww. *Journal of Environmental Management*, 128, 22–29.

130. Latch, D. E., Packer, J. L., Arnold, W. A., McNeill, K. (2003). Photochemical conversion of Triclosan to 2, 8-Dichlorodibenzo-p-dioxin in aqueous solution. *Journal of Photochemistry and Photobiology A: Chemistry*, 158(1), 63–66.

131. Lazarova, V., Levine, B., Sack, J., Cirelli, G., Jeffrey, P., Muntau, H., Salgot, M., Brissaud, F. (2001). Role of water reuse for enhancing integrated water management in Europe and Mediterranean countries. *Water Science and Technology*, 43(10), 25–33.

132. Leach, L., Enfield, C., Harlin, C. (1980). *Summary of long-term rapid infiltration system studies*. EPA Report EPA–600/2–80–165. Oklahoma: U. S. EPA.

133. Lemos, M. F., van Gestel, C. A., Soares, A. M. (2009). Endocrine disruption in a terrestrial iso-pod under exposure to Bisphenol-A and Vinclozolin. *Journal of Soils and Sediments*, 9(5), 492–500.

134. Li, J., Dodgen, L., Ye, Q., Gan, J. (2013). Degradation kinetics and metabolites of carbamazepine in soil. *Environmental Science and Technology*, 47(8), 3678–3684.

135. Limousin, G., Gaudet, J. P., Charlet, L., Szenknect, S., Barthes, V., Krimissa, M. (2007). Sorption isotherms: a review on physical bases, modeling and measurement. *Applied Geochemistry*, 22(2), 249–275.

136. Lin, C., Shacahr, Y., Banin, A., Lin, C. (2004). Heavy metal retention and partitioning in a large-scale soil-aquifer treatment (SAT) system used for ww reclamation. *Chemosphere*, 57(9), 1047–1058.

137. Liu, F., Ying, G. G., Tao, R., Zhao, J. L., Yang, J. F., Zhao L. F. (2009). Effects of six selected antibiotics on plant growth and soil microbial and enzymatic activities. *Environmental Pollution*, 157(5), 1636–1642.

138. Liu, F., Ying, G. G., Yang, L. H., Zhou, Q. X. (2009). Terrestrial ecotoxicological effects of the antimicrobial agent triclosan. *Ecotoxicology and Environmental Safety*, 72(1), 86– 92.

139. Maoz, A., Chefetz, B. (2010). Sorption of the pharmaceuticals carbamazepine and naproxen to dissolved organic matter: role of structural fractions. *Water Research*, 44(3), 981–989.

140. Mapanda, F., Mangwayana, E. N., Nyamangara, J., Giller, K. E. (2005). The Effect of long-term irrigation using ww on heavy metal contents of soils under vegetables in Harare, Zimbabwe. *Agriculture, Ecosystems & Environment*, 107(2), 151–165.

141. Mara, D. D., Sleigh, P. A., Blumenthal, U. J., Carr, R. M. (2007). Health risks in ww irrigation: comparing estimates from quantitative microbial risk analyses and epidemiological studies. *Journal of Water and Health*, 5(1), 39–50.

142. Marco – Urrea, E., Pérez–Trujillo, M., Blánquez, P., Vicent, T., Caminal, G. (2010). Biodegradation of the analgesic naproxen by *trametes versicolor* and identification of intermediates using HPLC–DAD–MS and NMR. *Bioresource Technology*, 101(7), 2159–2166.

143. Martens, D. A. (2000). Plant residue biochemistry regulates soil carbon cycling and carbon sequestration. *Soil Biology and Biochemistry*, 32(3), 361–369.
144. Martens, D. A., Frankenberger, W. T. (1992). Modification of Infiltration Rates in an Organic–Amended Irrigated. *Agronomy Journal*, 84(4), 707–717.
145. McLain, J. E. T., Williams, C. F. (2010). Development of antibiotic resistance in bacteria of soils irrigated with reclaimed WW. Proceedings of the 5th National Decennial Irrigation Conference by ASABE, 5–8 December, Phoenix, USA.
146. Mennigen, J. A., Stroud, P., Zamora, J. M., Moon, T. W., Trudeau, V. L. (2011). Pharmaceuticals as neuroendocrine disruptors: lessons learned from fish on prozac, Part B. *Journal of Toxicology and Environmental Health*, 14(5–7), 387–412.
147. Mikkelsen, R., Camberato. J. (1995). Potassium, sulfur, lime and micronutrient fertilizers. In: Rechcigl J. E. (ed.) *Soil Amendments and Environmental Quality*. Boca Raton: CRC Press. Pages 109–137.
148. Mitra, R., Cuesta–Alonso, E., Wayadande, A., Talley, J., Gilliland, S., Fletcher, J. (2009). Effect of route of introduction and host cultivar on the colonization, internalization, and movement of the human pathogen *escherichia coli* O157, H7 in spinach. *Journal of Food Protection*, 72(7), 1521–1530.
149. Muchuweti, M., Birkett, J. W., Chinyanga, E., Zvauya, R., Scrimshaw, M. D., Lester J. N. (2006). Heavy Metal content of vegetables irrigated with mixtures of ww and sewage sludge in Zimbabwe: implications for human health. *Agriculture, Ecosystems & Environment,* 112(1), 41–48.
150. Müller, K., Duwig, C., Prado, B., Siebe, C., Hidalgo, C., Etchevers, J. (2012). Impact of long-term ww irrigation on sorption and transport of *atrazine* in Mexican agricultural soils, Part B. *Journal of Environmental Science and Health*, 47(1), 30–41.
151. Müller, K., Magesan, G. N., Bolan, N. S. (2007). A Critical Review of the Influence of Effluent Irrigation on the Fate of Pesticides in Soil. *Agriculture, Ecosystems and Environment,* 120(2), 93–116.
152. Murillo–Torres, R., Durán–Álvarez, J. C., Prado, B., Jiménez–Cisneros, B. E. (2012). Sorption and mobility of two micropollutants in three agricultural soils: a comparative analysis of their behavior in batch and column experiments. *Geoderma*, 189–190, 462–468.
153. Najafi, P., Mousavi, S., Feizi, M. (2003). Effects of using treated municipal ww in irrigation of tomato. *Journal of Agricultural Science and Technology*, 15(1), 65–72.
154. Negreanu, Y., Pasternak, Z., Jurkevitch, E., Cytryn, E. (2012). Impact of treated ww irrigation on antibiotic resistance in agricultural soils. *Environmental Science and Technology,* 46(9), 4800–4808.
155. Nicholson, F. A., Smith, S. R., Alloway, B. J., Carlton–Smith, C., Chambers, B. J. (2003). An inventory of heavy metals inputs to agricultural soils in England and Wales. *Science of the Total Environment*, 311(1), 205–219.
156. Oron, G., DeMalach, Y., Hoffman, Z., Manor, Y. (1992). Effect of effluent quality and application method on agricultural productivity and environmental control. *Water Science and Technology,* 26(7–8), 1593–1601.
157. Pardo, A., Amato, M., Quaglietta Chiaranda, F. (2000). Relationships between soil structure, root distribution and water uptake of chickpea (*Cicer arietinum L.*), Plant growth and water distribution. *European Journal of Agronomy*, 13(1), 39–45.

158. Patel, J., Millner, P., Nou, X., Sharma, M. (2010). Persistence of enterohaemorrhagic and non-pathogenic *e. coli* on spinach leaves and in rhizosphere soil. *Journal of Applied Microbiology*, 108(5), 1789–1796.
159. Peck, A. M., Hornbuckle, K. C. (2004). Synthetic musk fragrances in lake Michigan. *Environmental Science and Technology*, 38(2), 367–372.
160. Pereira, L. S., Cordery, I., Iacovides, I. (2009). Coping with water scarcity: addressing the challenges. Paris: UNESCO.
161. Pérez–Carrera, E., Hansen, M., León, V. M., Björklund, E., Krogh, K. A., Halling–Sørensen, B., González–Mazo, E. (2010). Multiresidue method for the determination of 32 human and veterinary pharmaceuticals in soil and sediment by pressurized-liquid extraction and LC–MS/MS. *Analytical and Bioanalytical Chemistry*, 398(3), 1173–1184.
162. Pescod, M. (1992). *WW Treatment and Use in Agriculture*. FAO Irrigation and Drainage Paper 47. Rome: Food and Agriculture Organization (FAO).
163. Petrovic, M., Eljarrat, E., De Alda, M. L., Barceló, D. (2004). Endocrine disrupting compounds and other emerging contaminants in the environment: a survey on new monitoring strategies and occurrence data. *Analytical and Bioanalytical Chemistry*, 378(3), 549–562.
164. Pomati, F., Castiglioni, S., Zuccato, E., Fanelli, R., Vigetti, D., Rossetti, C., Calamari, D. (2006). Effects of a complex mixture of therapeutic drugs at environmental levels on human embryonic cells. *Environmental Science and Technology*, 40(7), 2442–2447.
165. Pomiès, M., Choubert, J. M., Wisniewski, C., Coquery, M. (2013). Modelling of micropollutant removal in biological ww treatments: a review. *Science of the Total Environment*, 443, 733–748.
166. Quanrud, D. M., Hafer, J., Karpiscak, M. M., Zhang, J., Lansey, K. E., Arnold, R. G. (2003). Fate of organics during soil-aquifer treatment: sustainability of removals in the field. *Water Research*, 37(14), 3401–3411.
167. Raimbault, B. A., Vyn, T. J. (1991). Crop rotation and tillage effects on corn growth and soil structural stability. *Agronomy Journal*, 83(6), 979–985.
168. Ramírez–Fuentes, E., Lucho–Constantino, C., Escamilla–Silva, E., Dendooven, L. (2002). Characteristics and carbon and nitrogen dynamics in soil irrigated with ww for different lengths of time. *Bioresource Technology*, 85(2), 179–187.
169. Read, D. S., Sheppard, S. K., Bruford, M. W., Glen, D. M., Symondson, W. O. C. (2006). Molecular detection and predation by soil-arthropods on nematodes. *Molecular Ecology*, 15(7), 1963–1972.
170. Rijsberman, F. R. (2006). Water scarcity: fact or fiction. *Agricultural Water Management*, 80(1–3), 5–22.
171. Rizzo, L., Manaia, C., Merlin, C., Schwartz, T., Dagot, C., Ploy, M. C., Michael, I., Fatta Kassinos, D. (2013). Urban ww treatment plants as hotspots for antibiotic resistant bacteria and genes spread into the environment: a review. science of the total environment, 447(1), 345–360.
172. Russell, J. M., Cooper, R. N., Lindsey, S. B. (1993). Soil denitrification rates at ww irrigation sites receiving primary-treated and anaerobically treated meat-processing effluent. *Bioresource Technology*, 43(1), 41–46.
173. Sala, L., Serra, M. (2004). Towards sustainability in water recycling. *Water Science and Technology*, 50(2), 1–8.

174. Santos, L. H., Araújo, A. N., Fachini, A., Pena, A., Delerue–Matos, C., Montenegro, M. C. B. S. M. (2010). Ecotoxicological aspects related to the presence of pharmaceuticals in the aquatic environment. *Journal of Hazardous Materials*, 175(1), 45–95.

175. Sassman, S. A., Lee, L. S. (2005). Sorption of three tetracyclines by several soils: assessing the role of pH and cation exchange. *Environmental Science and Technology*, 39(19), 7452–7459.

176. Schlüsener, M. P., Bester, K. (2006). Persistence of antibiotics such as macrolides, tiamulin and salinomycin in soil. *Environmental Pollution*, 143(3), 565–571.

177. Schmitt, H., Römbke, J. (2008). The Ecotoxicological effects of pharmaceuticals (antibiotics and antiparasiticides) in the terrestrial environment – a review. In: Kúmmerer K. (ed.) *Pharmaceuticals in the Environment: Sources, Fate, Effects and Risks*. Heidelberg: Springer. Pages 285–303.

178. Scholl, M. A., Mills, A. L., Herman, J. S., Hornberger, G. M. (1990). The Influence of mineralogy and solution chemistry on the attachment of bacteria to representative aquifer materials. *Journal of Contaminant Hydrology*, 6(4), 321–336.

179. Seymour, I. J., Appleton, H. (2001). Foodborne viruses and fresh produce. *Journal of Applied Microbiology*, 91(5), 759–773. Foodborne viruses

180. Shelef, G., Azov, Y. (1996). The Coming era of intensive ww reuse in the Mediterranean region. *Water Science and Technology*, 33(10–11), 115–125. 15

181. Shenker, M., Harush, D., Ben–Ari, J., Chefetz, B. (2011). Uptake of carbamazepine by cucumber plants-a case study related to irrigation with reclaimed ww. *Chemosphere*, 2(6), 905–910.

182. Shiklomanov, I. A. (1998). World water resources – a new appraisal and assessment for the 21st century. St Petersburg: UNESCO.

183. Shore, L. S., Kapulnik, Y., Ben-Dor, B., Fridman, Y., Wininger, S., Shemesh, M. (1992). Effects of eestrone and 17 β-estradiol on vegetative growth of medicago sativa. *Physiologia Plantarum*, 84(2), 217–222.

184. Shuval, H., Adin, A., Fattal, B., Rawutz, E., Yekutiel, P. (1986). WW irrigation in developing countries: health effects and technical solutions. Technical Paper No. 51. Washington: The World Bank.

185. Siebe, C., Cifuentes, E. (1995). Environmental impact of ww irrigation in central mexico: an overview. *International Journal of Environmental Health Research*, 5(2), 161–173.

186. Smith, M. S., Thomas, G. W., White, R. E., Ritonga, D. (1985). Transport of *escherichia coli* through intact and disturbed soil columns. *Journal of Environmental Quality*, 14(1), 87–91.

187. Solomon, E. B., Yaron, S., Matthews, K. R. (2002). Transmission of *escherichia coli* o157, h7 from contaminated manure and irrigation water to lettuce plant tissue. *Applied and Environmental Microbiology*, 68(1), 397–400.

188. Sommer, C., Bibby, B. M. (2002). The influence of veterinary medicines on the decomposition of dung organic matter in soil. *European Journal of Soil Biology*, 38(2), 155–159.

189. Steen, I. (1998). Phosphorus availability in the 21st century: management of a non-renewable resource. *Phosphorus and Potassium*, 217(1), 25–31.

190. Stewart, L. W., Reneau, R. B. (1981). Spatial and temporal variation of fecal coliform movement surrounding septic tank-soil absorption systems in two atlantic coastal plain soils. *Journal of Environmetal Quality*, 10(4), 528–531.

191. Strauss, M. Human waste (excreta and ww). *Swiss Federal Institute of Aquatic Science and Technology* (EAWAG). <http://www.eawag.ch/organisation/ abteilungen/sandec/publikationen/publications_wra/downloads_wra/human_waste_use_ETC_SIDA_UA.pdf>.

192. Stumpe, B., Marschner, B. (2007). Long-term sewage sludge application and ww irrigation on the mineralization and sorption of 17β–estradiol and testosterone in soils. *Science of the Total Environment*, 374(2), 282–291.

193. Swan, G. E., Cuthbert, R., Quevedo, M., Gree, n R. E., Pain, D. J., Bartels, P., Cunningham, A. A., Duncan, N., Meharg, A. A., Oaks, J. L., Parry–Jone, s J., Shultz, S., Taggart, M. A., Verdoorn, G., Wolter, K. (2006). Toxicity of diclofenac to gyps vultures. *Biology Letters*, 2(2), 279–282.

194. Tam, N. F. Y. (1998). Effects of ww discharge on microbial populations and enzyme activities in mangrove soils. *Environmental Pollution*, 102(2–3), 233–242.

195. Thiele-Bruhn, S. (2003). Pharmaceutical antibiotic compounds in soils – a review. *Journal of Plant Nutrition and Soil Science*, 166(2), 145–167.

196. Thiele – Bruhn, S., Beck, I. C. (2005). Effects of sulfonamide and tetracycline antibiotics on soil microbial activity and microbial biomass. *Chemosphere*, 59(4), 457–465.

197. Tixier, C., Singer, H. P., Oellers, S., Müller, S. R. (2003). Occurrence and fate of carbamazepine, clofibric acid, diclofenac, ibuprofen, ketoprofen, and naproxen in surface waters. *Environmental Science and Technology*, 37(6), 1061–1068.

198. Toze, S. (2006). Reuse of effluent water-benefits and risks. *Agricultural Water Management*, 80(1–3), 147–159.

199. United States Geological Survey (USGS), The USGS Water Science School, (2013). *The World's Water – Distribution of Earth's Water*. <http://ga.water.usgs.gov/edu/earth-wherewater.html>.

200. Van der Hoek, W., Ul–Hassan, M., Ensink, J. H. J., Feenstra, S., Raschid–Sally, L., Munir, S., Aslam, R., Ali, N., Hussain, R., Matsuno, Y. (2002). Urban ww: a valuable resource for agriculture. *Colombo: International Water Management Institute Research*.

201. Vazquez – Roig, P., Segarra, R., Blasco, C., Andreu, V., Picó, Y. (2010). Determination of pharmaceuticals in soils and sediments by pressurized liquid extraction and liquid chromatography tandem mass spectrometry. *Journal of Chromatography*, 1217(16), 2471–2483.

202. Von Oepen, B., Kördel, W., Klein W. (1991). Sorption of nonpolar and polar compounds to soils: processes, measurements and experience with the applicability of the modified OECD-guideline 106. *Chemosphere*, 22(3), 285–304.

203. Walker, C. W., Watson, J. E., Williams, C. (2012). Occurrence of carbamazepine in soils under different land uses receiving ww. *Journal of Environmental Quality*, 41(4), 1263–1267.

204. Walters, E., McClellan, K., Halden, R. U. (2010). Occurrence and loss over three years of 72 pharmaceuticals and personal care products from biosolids-soil mixtures in outdoor mesocosms. *Water Research*. 44(20), 6011–6020.

205. Wang. Z., Chang. A. C., Wu. L., Crowley. D. (2003). Assessing the soil quality of long-term reclaimed ww-irrigated cropland. *Geoderma*, 114(3–4), 261–278.

206. Welshons, W. V., Thayer, K. A., Judy, B. M., Taylor, J. A., Curran, E. M., Vom Saal, F. S. (2003). Large effects from small exposures. i. mechanisms for endocrine-disrupting chemicals with estrogenic activity. *Environmental Health Perspectives*, 111(8), 994–1006.

207. World Health Organization, (1992). WHO environmental health criteria 135, cadmium-environmental aspects. Geneva: World Health Organization.
208. World Health Organization, (2006). WHO guidelines for the safe use of ww, excreta and greywater. Geneva: WHO.
209. World Health Organization, (1989). WHO health guidelines for the use of ww and excreta in agriculture. Geneva: WHO.
210. Wu, C., Spongberg, A. L., Witter, J. D., Fang, M., Czajkowski, K. P. (2010). Uptake of pharmaceutical and personal care products by soybean plants from soils applied with biosolids and irrigated with contaminated water. *Environmental Science and Technology,* 44(16), 6157–6161.
211. Xia, K., Jeong, C. Y. (2004). Photodegradation of the endocrine-disrupting chemical 4–nonylphenol in biosolids applied to soil. *Journal of Environmental Quality*, 33(4), 1568–1574.
212. Xu, G., Li, F., Wang, Q. (2008). Occurrence and degradation characteristics of dibutyl phthaate (DBP) and di–(2–ethylhexyl) phthalate (dehp) in typical agricultural soils of China. *Science of the Total Environment*, 393(2), 333–340.
213. Xu, J., Chen, W., Wu, L., Green, R., Chang, A. C. (2009). Leachability of some emerging contaminants in reclaimed municipal ww-irrigated turf grass fields. *Environmental Toxicology and Chemistry,* 28(9), 1842–1850.
214. Xu, J., Wu, L., Chen, W., Chang A. C. (2008). Simultaneous determination of pharmaceuticals, endocrine disrupting compounds and hormone in soils by gas chromatography-mass spectrometry. *Journal of Chromatography*, 1202(2), 189–195.
215. Yamamoto, H., Nakamura, Y., Morigushi, S., Nakamura, Y., Honda, Yuta., Tamura, I., Hirata, Y., Hayashi, A., Sekizawa, J. (2009). Persistence and partitioning of eight selected pharmaceuticals in the aquatic environment: laboratory photolysis, biodegradation and sorption experiments. *Water Research*, 43(2), 351–362.
216. Ying, G. G., Yu, X. Y., Kookana, R. S. (2007). Biological degradation of triclocarban and triclosan in a soil under aerobic and anaerobic conditions and comparison with environmental fate modeling. *Environmental Pollution*, 150(3), 300–305.
217. Yuan, Y. (1993). Etiological study of high stomach cancer incidence among residents in ww irrigated areas. *Environmental Protection Science*, 19(1), 70–73.
218. Zeng, F., Cui, K., Xie, Z., Wu, L., Liu, M., Sun, G., Lin, Y., Lou, D., Zeng, Z. (2008). Phthalate esters (PAEs), Emerging organic contaminants in agricultural soils in periurban areas around Guangzhou, China. *Environmental Pollution*, 156(2), 425–434.
219. Zepp, R. G., Cline, D. M. (1977). Rates of direct photolysis in aquatic environment. *Environmental Science and Technology*, 11(4), 359–366.
220. Zhang, G., Ma, L., Beuchat, L. R., Erickson, M. C., Phelan, V. H., Doyle, M. P. (2009). Lack of internalization of *Escherichia coli* o157, h7 in lettuce (*lactuca sativa l.)* after leaf surface and soil inoculation. *Journal of Food Protection*, 72(10), 2028–2037.

CHAPTER 7

MICROORGANISM CONTAMINATION IN CROPS GROWN WITH WASTEWATER IRRIGATION: A REVIEW

PRERNA PRASAD and VINOD KUMAR TRIPATHI

CONTENTS

7.1   INTRODUCTION

Demand of more foods has been addressed in many forums due to increase in human populations. Production of more food raises its quality issues. The key for the successful provision of quality foods has depended on the availability of technologies to produce and preserve foods. More recently, the increase in the consumption of fresh-cut and leafy green (such as carrots, celery, and spinach) products that are usually consumed raw, have created a similar scenario where the industry and the government have to work on the appropriate minimum set of regulations to be put in place to control the occurrence of food borne diseases associated with these products. Several crucial changes have occurred in agricultural practices in the last 50 years. Food distribution has increased to cover large areas, and even different countries, it has become more difficult to keep track of where the food was produced and processed. In some cases, food is transported across different countries Therefore, a bacterial pathogen unique to some specific areas in the world may end up in a completely different area of the world. A good example of the latter is the 2008 outbreak of a virulent *Salmonella serotype Saintpaul* responsible for illnesses associated with the consumption of tomatoes. Suppliers of tomatoes normally rely on more than one grower to fill the orders. Also, tomatoes are not classified by origin but by ripeness, size, and grades during processing. Thus, tomatoes collected in Florida may be shipped to Mexico for packaging before they are sent back to the United States for final sale. In addition, the incorporation of sliced tomatoes in salad bars, deli counters, or supermarket salsas makes it extremely difficult to track where the tomatoes originated. The investigation into this particular outbreak of *Salmonella* Saintpaul resulted in suspicion that farms from Mexico and Florida were the ones involved in the production of the contaminated tomatoes. However, more than 1,700 samples collected from irrigation sources and packing, washing, and storage facilities were negative, and there was never a clear resolution of the actual source of the outbreak [16].

The international trade of food commodities and the ease with which people can move from different geographical areas have a long-term effect on food safety. The movement of foods increases the possibility of pathogens traveling in hiding from seemingly remote geographical. But humans also serve as carriers when they get infected in a country but develop the symptoms and suffer the disease in another country. An example is the case of salmonellosis in Sweden that still remain despite all the efforts to control the domestic cases of salmonellosis [13]. Most of these cases are associated with the contamination by travelers who return home with the infectious agents. As international food trade becomes more prevalent, countries that strive to control specific food borne agents may see their efforts curtailed and therefore will pressure international organizations to adopt more stringent international food safety regulations. Viruses are also opportunistic agents. The fact that we are still missing reliable techniques to isolate and identify some viruses makes it more difficult to study them than to study bacteria. The most recent examples of noroviruses affecting passengers on recreational cruises highlight the importance of food safety in new settings that were uncommon years ago.

This chapter reviews the status and advances related to microorganism contamination in crops grown with wastewater irrigation.

7.2 MICROORGANISMS IN WASTEWATER AND THEIR PATHWAY

Microorganisms are too small and not detected by the naked human eye. They are omnipresent, exist in different environments and are diverse in characteristics, behavior and resistance [11]. Waterborne microorganisms, which can be found in wastewater and surface waters, include bacteria, fungi, algae, protozoa and viruses [12]. Most microorganisms pose no harm to human health and hundreds of strains exist naturally in large amounts in the human intestines [21]. However, some microorganisms are pathogenic and can cause intestinal and other infectious diseases.

Pathogens found in wastewater can be classified into four main categories: bacteria, viruses, protozoa and helminthes [12]. In municipal wastewater, it is reasonable to assume that the pathogens mainly originate from human faces. As human faces contain large quantities of microorganisms [21], wastewater contains large amounts of various microorganisms as well. Pathogens entering the drinking water system can potentially cause large disease outbreaks. The main route for pathogen infection is ingestion of drinking water contaminated with human or animal faces (the faecal-oral route). However, microbiological drinking-water safety is not related only to faecal contamination as for example some pathogens may grow in water distribution systems [25]. But also, spreading by water is many times inferior to other pathways for transmission, including person-to-person contact, by food processing equipment, by inhalation of dust or aerosols and dermal or eye contact [21]. Some pathogens may be transmitted by multiple pathways. The route of interest in this study is though the faecal-oral route with transmission pathway from human faces, via wastewater and wastewater treatment plants to raw water sources and exposure through ingestion of drinking water.

Pathogens, transmitted by the faecal-oral route, are excreted in large amounts from faces by infected people and animals. As the pathogens are excreted from the body, there is an immediate reduction partly due to inactivation and partly due to the dilution with the flushing water [21]. The wastewater is normally collected in wastewater pipe systems to the wastewater treatment plants, where the inactivation of pathogens is continued. However, as wastewater treatment processes often have inadequate treatment with regard to pathogens, pathogens consequently are released to receiving waters through the discharge of the wastewater effluent. Typical concentrations of some common pathogens and indicator organisms vary significantly between different treatment facilities.

Once in the receiving water, an important characteristic for waterborne pathogenic microorganisms is that they will be further spread by the water flow [24]. The water transport means that the discharged pathogens can reach raw water intakes and enter the drinking water system. For a disease outbreak to occur, adequate quantities of pathogens need to enter the body. This is defined as the infection dose and varies between different species. Differences between humans, for example age, general health, disease record, vaccinations and gastric acid production also affect the infection dose. Virus and protozoa often have a low infection dose of 1–20 organisms whereas the infection dose for bacteria has a wide range between 102–109 organisms [21]. When an infection is established, pathogens multiply in the new host and large amounts are once again excreted with the faces.

Biofilms harbor pathogens and there is a great concern that engineered built water distribution systems along with climate change are supporting a robust amoeba population that harbors pathogens, such as, *Legionella* that remains one of the most significant waterborne diseases. In addition, the tragic death of a 4-year-old, in St. Bernard Parish – Louisiana during this year from the water dwelling parasite *Naegleria fowleri*, is a stark reminder of the critical importance of focusing on science (an understanding of the ecology of this parasite) and technology (proper municipal water disinfection) [19]. While these infections are rare, they are likened to lightning strikes, in which avoiding exposure is critical as the consequences are deadly. This free-living parasite lives in slimy films on surfaces in contact with water and sediments in human made and natural aquatic habitats. Yet our monitoring data and understanding of Naegleria's occurrence are poor. Table 7.1 highlights some of these biological, chemical, and physical contaminants of concern.

7.3   INDICATOR ORGANISM

Faecal indicator organisms are organisms, which normally exist in the human intestines [21] and therefore can be used to indicate faecal contamination of water. Faecal contamination of water in turn account for a greater risk of pathogenic microorganisms being present. Testing for presence of pathogens in water is relatively rare, as analyses are often complex, time-consuming, costly and face problems with sensitivity of detection due to normally low concentrations [25]. Instead faecal indicator organisms,

TABLE 7.1   Future Science Needs For Some Contaminants of Concern Associated with Impaired Water Quality and Impacts on Health [19]

Contaminants	Science Needs
Algae; Toxic Algae	Understand ability to accumulate and support pathogens (potentially chemicals); address processes associated with toxic blooms for species such as *Cylinderspermopsis*.
Amoeba and Biofilms	Monitor and understand processes for biofilm development and amoeba population dynamics associated and the harboring of pathogens. Focus on the ecology of *Naegleria*.
Bacterial Antibiotics Resistance	Develop a risk framework and examine occurrence of key genes, distribution, persistence, and transference from animal and human sources.
E. coli and Fecal Pathogens, Viruses, and Parasites	Characterize viral and parasitic pathogens in animal and pathogens, viruses, human wastes discharged to the environment using genetic tools. Determine the ability to use fecal indicators more effectively, including microbial source tracking tools. Study zoonotic potential. Develop persistence models.
Nanoparticles	Develop methods and transport and fate of trace nanoparticles including silver and titanium dioxide in natural waters.

usually bacteria, are normally analyzed for surveillance of water quality and for verification and operational monitoring of treatment processes. An ideal indicator organism should fulfill the following criteria [11]:

- They should be useful for all types of water.
- They should be present, whenever the pathogen of interest is present.
- They should survive longer in the environment than the most persistent pathogen.
- They should not reproduce in natural waters.
- The analysis method should be relatively easy to perform.
- The density of the indicator organism should have some direct relationship to the degree of faecal pollution.
- They should be naturally present in faces of warm-blooded animals including humans.

In reality no indicator organisms fulfill all criteria [11]. Therefore, testing for indicator organisms instead of pathogens is a somewhat uncertain method. Some pathogens are considerably more resistant than many indicator organisms so absence of indicators does not guarantee the absence of pathogens [25]. Analyses of more than one indicator organism can increase the certainty. Greater is the presence of indicator organisms, higher is the risk of presence of pathogens [21].

Commonly indicator bacteria's are coliform bacteria, enterococci and clostridia. The coliform group has been used as faecal indicator organism for a long time. A customary parameter is total amount of coliforms, which includes all types of coliform bacteria, both faecal and environmental. However, not all coliforms have faecal origin and use of the parameter as indication of faecal contamination is therefore limited. *E. coli*, which belongs to the *faecal coliform* bacteria, is considered as the most suitable indicator of faecal contamination. It is commonly used in monitoring programs for verification and surveillance of drinking-water systems. *E. coli* exist in faces in large amounts and are highly specific of faecal pollution. The disadvantage of *E. coli* is that they have less survival time than many pathogens and are less resistant to disinfection [25]. Coliform bacteria including *E. coli* are rod shaped in the size range of 0.5–2.0 μm [10]. Intestinal enterococci, which belong to the *faecal streptococci*, are more resistant to unfavorable conditions and disinfection than *E. coli* and tend to survive longer in aquatic environments [25]. Also, most species do not multiply in the aquatic environment. Therefore, this group has become more commonly used as indicator organism for faecal pollution. However, they exist in human faces in slightly less concentrations than *E. coli* and they could origin from sources other than human feces. Enterococci are cocci-shaped and are in the size range of 0.5–1.0 μm [10].

The most important indicator organism in the clostridia group is the *Clostridium Perfringens* [21]. *Clostridium Perfringens* are exclusively of faecal origin, whereas other members of the clostridium group are not. They exist in small amount in human faces but could also come from other sources. Closteridium produce spores, which are very resistant to disinfection and other unfavorable conditions and they have longer survival time in nature than other indicator organism as well as many pathogens. This makes them useful as indicator of old faecal contamination in raw water. They can also be used to assess the inactivation of protozoa and viruses in treatment processes. As the spores are very small, even smaller than protozoan cysts, *Clostridium Perfringens* can also be used as an indicator for verification of filtration process [25]. They are rod-shaped and in the size range of 0.6–1.3 × 2.4–19.0 μm [10]. Besides the common indicator bacteria, coliphages are also commonly used as indicator organism. Coliphages is a virus that infects *E. coli* bacteria [21]. They are suitable as indicator for human viruses in treatment processes as they are similar in size.

7.4 TRANSPORT MECHANISM

As soon as the pathogenic microorganisms enter a watercourse, several environmental and biological factors will influence the dilution and the decay of these organisms. The change in concentration of the pathogenic microorganisms along a distance of a watercourse depends both on different transport mechanisms in the flow and different factors affecting the inactivation [8]. Inactivation is defined as an event when the pathogens die-off or lose their ability to infect new hosts [21]. One important factor is the size properties of pathogenic microorganisms, which in aquatic environments influence the transport mechanisms in the way that most of the organisms have no other means of transport than by the water flow [5]. This transport can be either freely by advection or attached to particles in the water.

The concentration downstream an emission point can be defined in a simplified way as a function of the dilution factor and the transport time [20]. The dilution factor can be calculated by dividing the transport mechanism into transport by lateral diffusion and transport by longitudinal dispersion [8]. Lateral diffusion is caused by distribution and spreading by turbulence and molecular motions while longitudinal dispersion is caused by distribution and spreading due to the different water velocity at different cross-section due to continuity equation.

The inactivation of pathogens depends mainly on temperature, sunlight, pH and presence of predators [21]. According to Ferguson et al. [7], temperature is one of the most important factors that control the inactivation of pathogens and in general the die-off increases at higher temperatures. There are some variations between the species, however, the majority has a half-life of some days up to several months in colder water [17].

7.5   AGRICULTURAL PRACTICES FOR CROP CONTAMINATION

The agricultural practices can impact on the health of irrigators, consumers and people living in the area nearby field of irrigation. Farm workers are at high risk when furrow or flood irrigation methods are used because workers are not wearing any protective clothing [23]. Chemigation further increases the health risk. However treated wastewater can reduce pathogen risk by 2–3 log units. Sprinkler irrigation has the highest potential to spread contamination onto crop surfaces and affect nearby living communities. Bacteria and viruses can be transmitted through aerosols to nearby communities. To prevent the adverse impact on nearby communities, a buffer zone of 50–100 m from houses is necessary. Setting up an adequate buffer zone is equivalent 1 log unit of pathogen reduction. Spray drift, away from the site of application, can be reduced by using techniques such as low-throw sprinklers, micro-sprinklers, part-circle sprinklers, tree/shrub screens planted at field borders [15].

Micro irrigation techniques offer maximum health protection to field workers because the wastewater is applied directly to the plants. But it is expensive to implement. Recently, it has been adopted by some farmers in Cape Verde and India [6, 9, 18]. The benefits of these systems in terms of water usage and higher crop yields have been accepted at different forums.

Further research on viable approaches by making wide applicability may facilitate sustainability of this technology under low resource environment. Micro irrigation is estimated to provide an additional pathogen reduction of 2–4 log units, depending on whether or not the harvested part of the crop is in contact with the ground. Main problem associated with micro irrigation system is emitter clogging, if suspended solid content in wastewater is high. Emitter clogging also occurs as a result of soil-based algae migration to the emitters. Algae from waste stabilization ponds do not usually block emitters, although care is required to choose an emitter that does not block easily [3, 22, 23].

Cessation of irrigation with wastewater for one to two weeks prior to harvest can be effective in reducing crop contamination by providing time for pathogen die-off [1]. Enforcing period is likely to be difficult because many vegetables need watering nearly until harvest to increase the market value. However, it may be possible with some fodder crops that do not have to be harvested at the peak of freshness [2]. Alternatively, crops can be irrigated from non-contaminated water sources after cessation of wastewater use until harvest.

7.6 CASE STUDIES

7.6.1 CASE STUDY 1: EGGPLANT AND TOMATO

Cirelli et al. [4] have compared the performance of Black-Bell eggplant and Missouri Tomato in Italy. In the Southern regions of Italy, water shortage is a serious problem that has a major impact on agriculture dependent local economy. To add to this problem, there is excessive and uncontrolled groundwater withdrawal plus an inefficient water distribution network. These reasons lead to a requirement of an alternative source of irrigation water.

The solution to the problem was proposed in the form of usage of treated municipal wastewater (TWW) for micro irrigation. The experiment was conducted in an open field in San Michele di Ganzaria (Eastern Sicily, Italy) during 2008–2009. The experimental setup involved use of two different types of water. One was the treated municipal wastewater (TWW) and the control system involved the use of fresh water (FW).

The TWW was subjected to tertiary treatment using the constructed wetland (CW) technology due to its low operation and maintenance cost and efficiency in treating wastewater from small and medium communities. The physical, chemical and microbial characteristics of both TWW and FW were estimated. The parameters were TSS (105°C), BOD_5, COD, TP (Total Phosphorus), TN (Total Nitrogen), $CaCo_3$, EC, pH, DO, TC (Total Coliform), FC (*Faecal Coliform*), *Escherichia Coli* (*E. Coli*), FC (*Faecal Streptococci*), *Salmonella* and *Helminth eggs*. The parameters to study the effects on crop production of the quality of irrigation water on eggplant and tomato were: MY (Marketable Yield), MN (Marketable fruits), UMN (no. of unmarketable fruits), UMW (Unit mean weight), seed presence (using an index of 0–3), dry matter (by drying fruits in ventilated oven at 70°C until constant weight), and color. *Faecal Coliform* in the TWW in the two years of monitoring varied in the range of $3 \times 10 - 2 \times 10^5$ CFU/100ml. The *Faecal Streptococci* varied in the range of $3 - 3 \times 10^5$ CFU/100ml. *Salmonella* and *Helminth eggs* were not detected in the TWW samples (Table 2).

7.6.1.1 Presence of Microbial Parameters in Eggplant

Microbial contamination of fruits did not reach significant values for the eggplant irrigated by TWW in either year. For the microbial parameters investigated, the levels were less than 20 CFU/100 gms for 75% of cases in year 1 and less than 28 CFU/100 gms for 67% of cases in year 2. The only exception was the plot which was equipped with surface laterals (SL) where values of 10^2 CFU/100 gms of *Faecal Coliform* and *Faecal Streptococci* were found for the eggplant crop that were in contact with

TABLE 7.2   TWW and FW Quality [4]

Microbial parameters	TWW		FW	
in CFU/100ml	**Year 1**	**Year 2**	**Year 1**	**Year 2**
Faecal Coliform (FC)	6×10^2	5×10^4	18	3
Escherichia Coli (*E. Coli*)	4×10^2	4×10^4	16	1
Faecal Streptococci (FC)	2×10^2	5×10^4	13	1
Salmonella	absent	absent	absent	absent

the soil. For the plants that were not in contact with the soil or plastic mulch, values up to 10 CFU/100 gms were found.

7.6.1.2   Presence of Microbial Parameters in Tomato

Tomato fruit contamination was less than 20 CFU/100gms for 82% of the samples irrigated by TWW in year 2. The maximum contamination values were 10^2 CFU/100gms for *E. coli* and 10^3 CFU/100gms for *Faecal Coliform* and *Faecal Streptococci* for the fruits in contact with soil or plastic mulch. For the tomatoes not in contact with the soil or plastic mulch, zero contamination was found. The mulched surface enhanced the microbial biomass thus causing higher product contamination.

7.6.2   *CASE STUDY 2: LETTUCE*

Moyne et al. [14] have observed numerous outbreaks of *E. Coli* associated with the consumption of leafy green vegetables since 1996. This in turn led to the study of the bacterial population dynamics of the lettuce plant. The experiment was conducted in the Salinas Valley of California using drip and overhead sprinkler irrigation. With drip and furrow irrigation the lettuce foliage remains dry whereas with overhead sprinkler irrigation the water falls onto the edible portions of the plant and the soil surface.

There are two strains of *E. coli*, the toxigenic *E. coli* O157:H7 and the non-toxigenic *E. coli* O157:H7 ATCC 700728 also classified as BSL1. The non-toxigenic strain was used for this experiment. A rifampicin resistant mutant of *E. coli* O157:H7 ATCC 700728 was isolated by step wise exposure to increasing concentrations of rifampicin and selecting colonies that were resistant to rifampicin. Four field trials were conducted by Moyne et al. After seeding but prior to the first irrigation, a herbicide pronamide was applied @ 2.24kg/ha.

The drip irrigated plots were watered for 2 to 4 hours/irrigation, whereas the sprinkler irrigated plots were watered for 1.5 to 2.5 hours/irrigation. The total amount of water applied to the crops varied from 28 to 46 cm depending on the weather conditions. The sprinkler-irrigated crops were fertilized once with 80 kg of nitrogen/ha. On the other hand, nitrogen fertilizer was applied through the drip tape in 3 applications totaling approximately 80kg of N/ha for the drip-irrigated plots.

In this experiment, *E. coli* was inoculated on the lettuce or the soil between 9 to 11 AM each day for each of the field trials in either of the two ways: 1. Spray bottles were used to inoculate 4 week old lettuce plants, the bottles being calibrated to deliver a dose adjusted to 10^7 CFU/spray (and 10^5 CFU/spray for the summer of 2007), approximately delivering 1 ml in a single spray; and 2. Backpack sprayer containing inoculums adjusted to 10^7 CFU/ml was used to spray the soil surface of the lettuce bed just before the lettuce seed germination.

The detection limit by direct plating was 200 CFU/plant (2.3log CFU/plant) at day 0. This was increased to 400 CFU/plant (2.6 log CFU/plant) at days 2 and 7. By using a combination of plating and filtration, the detection limit was further reduced to 10 CFU/plant. When *E. coli* was not detected by plating, filtration or enrichment sample counts were treated statistically as zero. When cells were detected only by enrichment a value of 9 CFU/plant was assigned.

Higher levels of indigenous bacteria were present in the phyllosphere of non-inoculated plants during spring and fall of 2009 as compared to summer 2007 and spring 2008 at day 0. Plants irrigated by overhead sprinkler had significantly higher levels of indigenous bacteria than plants irrigated by drip. When plants were sampled immediately after inoculation (0 hours) the average *E. coli* O157:H7 population of 4 or 3 log CFU/plant was achieved with the maximum value being 10^7 CFU/ml and the minimum value being 10^5 CFU/ml. *E. coli* population declined rapidly after inoculation by 1–2 log CFU during the first hour and most plants had less than 10 CFU/plant at 2 days post-inoculation. At day 2 and 8, plants out of the 30 inoculated with the high level inoculums had *E. coli* ranging from 1 to 179 CFU and 2 plants out of the 30 inoculated with the low level inoculums had bacterial count ranging from 6 to 117 CFU.

7.7 CONCLUSIONS

The safe reuse of wastewater for agriculture is vital for the countries having shortage of fresh water. It has shown that the drip irrigation method, particularly with plastic soil covering, may be an effective measure. This method is not meant to replace proper microbiological standards. It is not recommended to use in highly contaminated water for practical agricultural production. Chemical and biological interventions can reduce surface pathogens and minimize cross-contamination. However, they are largely ineffective on internalized pathogens. In the event, internalization is a significant route of contamination in the field. The consequences of waterborne infections can be huge. The individuals will, except from the discomfort of being ill, also suffer from partial individual loss of income with great financial cost for the society. Finally, continuous research efforts to better understand the conditions necessary to control food borne pathogens and consistent consumer education efforts will provide solution from reemerging food borne pathogens outbreaks.

7.8 SUMMARY

Generally bacteria, viruses, protozoa and helminthes were found as pathogens in wastewater. These pathogens mainly originate from humans faces. The public health aspects of the use of wastewater in agriculture and the effects of the drip irrigation method on the contamination of vegetables have been reported. In first case study, the highest contamination values were probably due to a microbial accumulation of high bacterial loads in the treated wastewater effluent from constructed wetland. The maximum microbial contamination value found for fruits supplied with fresh water was 10^2 CFU/100gms for all the investigated contaminants. For every combination of tertiary treated wastewater and irrigation technologies (micro irrigation with surface lateral and subsurface lateral), a non-negligible level of contamination on fruits was found. These results confirm that constructed wetland effluents cannot be used for irrigation without additional disinfection or post treatment measures. Efforts made to maintain the microbial quality of the fruits but the *E. coli* content in treated wastewater was often over the limits set by competent authority. In case study 2 when 2 week old lettuce plants were inoculated and studied, a rapid decline of *E. coli* population from 5±0.5 log CFU/gm at 0 hours to 2.5±0.4 log CFU/gm at 2 hours was observed. At 21 dpi (days post inoculation), 1 plant out of 120 was tested positive for *E. coli*. At 28 dpi, no *E. coli* was detected.

KEYWORDS

- algae
- coliforms
- contamination
- crop growth
- crop practice
- dripper
- eggplant
- emitter
- freshwater, FW
- health
- heavy metal
- irrigation system
- lettuce
- micro irrigation
- microbial parameters
- nutrients
- parasite
- pathogen
- semi-arid
- tomato
- total solids
- total soluble solids, TSS
- transport
- turbidity
- virus
- wastewater, WW
- water quality
- World Health Organization, WHO

REFERENCES

1. Bastos, R. K. X., Mara, D. D. (1995). The bacteriological quality of salad crops drip and furrow irrigated with waste stabilization pond effluent: an evaluation of the WHO guidelines. *Water Science and Technology*, 31(12), 425–430.
2. Blumenthal, U. J. (2000). *Guidelines for wastewater use in agriculture and aquaculture: recommended revisions based on new research evidence.* Water and Environmental Health at London and Loughborough. London school of Hygiene and Tropical Medicine, WELL study task no. 68, Part 1.
3. Capra, A., Scicolone, B. (2004). Emitter and filter tests for wastewater reuse by drip irrigation. *Agricultural Water Management*, 68, 135–149.
4. Cirelli, G. L., Consoli, S., Licciardello, F., Aiello, R., Giuffrida, F., Leonardi, C. (2012). Treated municipal wastewater reuse in vegetable production. *Agricultural Water Management*, 104, 163–170.
5. Dechesne, M., Soyeux, E., Loret, J. F., Westrell, T., Stenström, T. A., Gornik, V., Koch, C., Exner, M., Stanger, M., Agutter, P., Lake, R., Roser, D., Ashbolt, N., Dullemont, Y., Hijnen, W., Medema, G. J. (2006). Pathogens in source water. Available at: <www.microrisk.com> Project results.
6. FAO, (2002). *Crops and drops: making the best use of water for agriculture.* Food and Agriculture Organization of United Nations, Rome.
7. Ferguson, C., Husman, A. M. de R, Altavilla, N., Deere, D., Ashbolt, N. (2003). Fate and transport of surface water pathogens in watersheds, critical reviews. *Environmental Science and Technology*, 33(3), 299–361.
8. Hartlid, C. (2009). *Microbiological risk analysis of drinking water supply in Lilla Edet: Potential causes of a waterborne disease outbreak* (Swedish). Master Thesis. Göteborgs Universitet, Göteborg.
9. Kay, M. (2001). *Smallholder irrigation technology: prospects for sub-Saharan Africa.* Food and Agriculture Organization of United Nations, Rome. International Program for Technology and Research in Irrigation and Drainage.
10. Levine, A. D., Harwood, V. J., Farrah, S. R., Scott, T. M., Rose, J. B. (2008). Pathogen and indicator organism reduction trough secondary effluent filtration: implications for reclaimed water production. *Water Environment Research*, 80(7), 596–608.
11. Maier, R. M., Pepper, I. L., Gerba, P. C. (2000). *Environmental Microbiology*. Academic Press, San Diego.
12. Metcalf and Eddy, (2004). *Wastewater Engineering – Treatment and Reuse*. 4th Edition, McGraw-Hill, New York.
13. Motarjemi, Y., Adams, A. (2006). *Emerging foodborne pathogens*. Boca Raton: CRC Press.
14. Moyne, A. J., Sudarshana, M. R., Blessington, T., Koike, S. T., Cahn, M. D., Harris, L. J. (2011). Fate of *Escherichia coli* O157, H7 in field inoculated lettuce. *Food Microbiology*, 28, 1417–1425.

15. NRMMC and EPHCA, (2005). *National guidelines for water recycling: managing health and environmental risks*. Canberra, Natural Resources Management Ministerial council and environment protection and heritage council of Australia.

16. Oyarzabal, O. A. (2012). *Emerging and Reemerging Foodborne Pathogens. Microbial Food Safety: An Introduction*. Food Science Text Series, Springer.

17. Pond, K., Rueedi, J., Pedley, S. (2004). Pathogens in drinking water sources. Available at: <www-microrisk.com>, Project results.

18. Postel, S. (2001). *Growing more food with less water. Scientific American*, February, 34–37 pages.

19. Rose, J. B. (2014). Water quality in the anthropocene: solving the problem of emerging, re-emerging, and recalcitrant contaminants. *Water Resources Impact*, 16(1), 10–12.

20. Sokolova, E., Åström, J., Pettersson, T., Bergstedt, O., Hermansson, M. (2012). Decay of bacteroidales genetic markers in relation to traditional fecal indicators for water quality modeling of drinking water sources. *Environment Science and Technology*, 46(2), 892–900.

21. Stenström, T. A. (1996). *Pathogenic microorganisms in wastewater systems – risk assessment of traditional and alternative drainage solutions* (Swedish). Environmental Protection Agency Report (4683). Stockholm.

22. Taylor, H. D., Bastos, R. K., Pearson, H. W., Mara, D. D. (1995). Drip irrigation with waste stabilisation pond effects: solving the problem of emitter fouling. *Water Science and Technology*, 31(12), 417–424.

23. Tripathi, V. K., Rajput, T. B. S., Patel, N. (2014). Performance of different filter combinations with surface and subsurface drip irrigation systems for utilizing municipal wastewater. *Irrigation Science*, 32(5), 379–391.

24. WHO, (2008). *Guidelines for drinking-water quality*. First addendum to Third Edition, Volume 1, Recommendations, World Health Organization, Geneva.

25. WHO, (2011). *Guidelines for Drinking-water Quality*. Fourth Edition, World Health Organization, Geneva.

CHAPTER 8

PRESENT STATUS OF WASTEWATER GENERATION, DISPOSAL AND REUSE CHALLENGES FOR AGRICULTURE

VINOD KUMAR TRIPATHI

CONTENTS

8.1   INTRODUCTION

In the present model of socio-economic development, concentration of investment and employment opportunities force the people to displace from rural to urban areas. Productive activities of industries and services cluster in cities. Eighty percentage of the world's gross domestic product (GDP) is generated by urban areas. As cities attract businesses and jobs, they bring together both the human and the entrepreneurial resources to generate new ideas, innovations and increasingly productive uses of technologies.

Consequently first time in the human history, the earth's population living in urban areas (3.42 billion) had surpassed the number living in rural areas (3.41 billion) and since then the world has become more urban than rural. The world urban population is increasing at the rate of 2.4% per year [37] and expected that it will increase by 84 per cent by 2050, from 3.4 billion in 2009 to 6.3 billion in 2050. By mid-century, the world urban population will likely be the same size as the world's total population in 2004. Virtually all of the expected growth in the world population will be concentrated in the urban areas of the less developed regions, whose population is projected to increase from 2.5 billion in 2009 to 5.2 billion in 2050. The 21 megacities in the world, each with at least 10 million inhabitants, accounted for 9.4 per cent of the world urban population. The number of megacities (population more than 10 million) is projected to increase to 29 in 2025, at which time they are expected to account for 10.3 per cent of the world urban population. Tokyo, the capital of Japan, is today the most populous urban agglomeration. Its population, estimated at 36.5 million in 2009, is higher than that of 196 countries or areas. Out of 10 highly populated cities of world, 3 cities (Delhi, 2nd; Kolkata, 4th; and Mumbai, 8th) are from India with the population of around 20 million.

Biologically clean and chemically under permissible limit potable water is must for healthy living. The Union Ministry of Works and Housing, Government of India recommends 125 to 200 liters per capita daily supply for Class I and Class II cities (above 50,000 population) [11]. Usually 80% of the water supply to the urban hydrologic system is spent for day-to-day household activities and better living standards. It generates greater volumes of wastewater [2, 19, 29]. It is a mixture of domestic effluent consisting of blackwater (excreta, urine and fecal sludge, i.e. toilet wastewater), greywater or washing water (personal, clothes, floors, dishes, etc.), water from commercial establishments and institutions, hospitals, industrial effluent (volatile or semi-volatile compounds, metal element), storm water and urban run-off. The composition of wastewater generated from different sources varies in their chemical and biological contaminants.

It has been observed that a small portion of wastewater is lost in evaporation, seepage in ground, leakage, etc. In many Asian and African cities, population growth has outpaced improvements in sanitation and wastewater infrastructure, making management of urban wastewater a tremendous challenge. In India, 38254 MLD wastewater is being generated from households and industries. Out of this only 31% is being treated [5]. A nationwide survey in Pakistan showed that an estimated 25% of all vegetables grown in the country were irrigated with untreated urban wastewater and that these vegetables, cultivated close to the urban markets, were considerably cheaper than the vegetables imported from different regions of Pakistan [8]. Likewise, 60% of the vegetables consumed in Dakar, Senegal were grown with a mixture of groundwater and untreated wastewater within the city limits [10]. In this context, the use of wastewater for periurban agriculture provided an opportunity and a resource for livelihood generation.

In this chapter, the current status of wastewater generation, its treatment and use in developing countries has been reviewed. The chapter presents a brief review of methods and approaches for risk reduction and management measures, public policies and institutional interventions that can improve wastewater management and minimize adverse impacts on irrigator and produce consumer health.

8.2   WHY WASTEWATER FOR IRRIGATION?

To protect the environment, the most prevalent practice for reuse of municipal wastewater is application to land. The use of wastewater in agriculture is started in the 19th century, when cities in Europe and North America introduced the water carriage system for domestic wastewater. Large sewage farms, as they were called, were established in the UK, USA, France and Germany, followed by India, Australia and Mexico [21]. Their purpose was to prevent contamination of rivers and to improve soil fertility. Most of these sewage farms were abandoned at the beginning of the 20th century for a number of reasons, notably the need for more land for expanding cities, increased awareness of the potential adverse human health impacts, the introduction of chemical fertilizers, and the development of wastewater treatment technologies. However, with a growing world population and unprecedented urbanization, especially in developing countries, the new driving force behind the use of wastewater in agriculture is that fresh water has become an increasingly scarce and contested resource. In the Near East Region of the world, some 16 countries out of 29 member states are classified as water-deficit, with less than 500 m^3 per capita of the annual renewable fresh water resources [9].

Availability of wastewater for irrigation in the water scarce peri-urban areas is an externality imposed by the fast growing urban areas in the developing countries like India. In many Asian and African cities, population growth has outpaced improvements in sanitation and wastewater infrastructure, making management of urban wastewater a tremendous challenge. Some specific examples include India where only 24% of wastewater from households and industry is treated, and in Pakistan only 2% is treated [16, 22]. In West African cities, usually less than 10% of the generated wastewater is collected in piped sewage systems and receives primary or secondary treatment [7]. In many developing countries, large centralized wastewater collection and treatment systems have proven difficult to sustain. Decentralized systems that are more flexible for long-term operation and financial sustainability and compatible with demands for local effluent use, have been promoted in many areas [30], although not without challenges. In Ghana, for example, only 7 of 44 smaller treatment plants are functional and probably none meets the designed effluent standards and a number of projects on wastewater and related risk mitigation significantly influenced farmers knowledge, while different media alerted policy-makers to take action [25].

Sometimes wastewater is the only water source available for urban or peri-urban agriculture. They even deliberately use undiluted wastewater as it provides nutrients or is more reliable or cheaper than other water sources [18, 31]. Despite farmers' good reasoning, this practice can severely harm human health and the environment [28] mainly due to not only the associated pathogens, but also heavy metals and other undesirable constituents depending on the source. Domestic sewage contains high percent of major (12.8 g N, 5.3 g P_2O_5, and 7.0 g K_2O per capita per day) and minor nutrients required for development of plants. When wastewater use is well managed, it helps to recycle nutrients and water and therefore diminishes the cost of fertilizers or simply makes them accessible to farmers. This in itself has environmental consequences (low energy is needed to produce fertilizers, less phosphorus needs to be mined, low-cost treatment method by taking the advantage of the soil's capacity to naturally remove contamination). Uncontrolled wastewater contributes to water resources degradation, reduces agricultural production and affects public health. On the opposite, the controlled use of wastewater, through treatment and planning, leads to water resources augmentation, particularly in the countries that suffer from water scarcity conditions, in addition to environmental protection. The use of wastewater in irrigation may also improve groundwater conditions, by recharging aquifers. Therefore, the use of wastewater in irrigation helps to reduce downstream health and environmental impacts that would otherwise result if the wastewater were discharged directly into surface water bodies.

8.3   HEALTH HAZARDS USING WASTEWATER IRRIGATION

The major health hazard with the use of wastewater is for the persons who are in contact with it and for those who consume the produce contaminated with wastewater. The products grown using wastewater finds its way to the consumers in the urban areas exposing them to the risks of eating crops contaminated with wastewater. Health impact due to wastewater differs according to the methods of use, which determine the degree of human exposure to it. The various types of wastewater use identified in literature include:

(1) Direct use of untreated wastewater: The application of wastewater to land directly from a sewerage system is called direct use of untreated wastewater. In this, the irrigation source is wastewater that is directly taken from the sewerage system or from storm water drains that carry large sewage flows. This type of use exists in countries like Pakistan and Kenya.

(2) Direct use of treated wastewater: On the other hand, the direct use of treated wastewater is the use of treated wastewater where control exists over the conveyance of the wastewater from the point of discharge to a treatment plant and to a controlled area where it is used for irrigation. Many countries in Middle East, which makes use of wastewater stabilization ponds to remove pathogens, widely adopt this method. This method is better known as the use of reclaimed water, meaning water received

at least secondary treatment, and is used after it flows out of a domestic wastewater treatment facility.

(3) Indirect use of wastewater: Apart from these two types of direct use, indirect use of wastewater is defined as the unplanned application to land of wastewater from a receiving water body. For example, municipal and industrial wastewater is discharged without treatment or monitoring into the watercourses draining an urban area from where farmers draw water for irrigation.

Asano [2] also makes a distinction between planned and unplanned use of wastewater. For example, imagine a situation where natural rivers passing through cities become so heavily polluted with wastewater and they become de facto sewers like the Musi River in Hyderabad, India. In such situations, diversion of water from a river downstream of a discharge of wastewater is an incidental or unplanned reuse. Indirect reuse normally constitutes unplanned reuse whereas direct reuse normally constitutes planned reuse.

Irrigation with wastewater is said to have both beneficial and harmful effects as it contains substantial amounts of beneficial nutrients and toxic heavy metals [4, 33]. Reliability and nutrient richness are considered as two important attributes of wastewater beneficial for agriculture. It is believed that nutrients present in wastewater results in higher crop yields and thereby considerably reduces the need to apply artificial fertilizers. Reliability of wastewater supply is yet another factor which makes it a valuable resource. The supply of water to the city ensures wastewater because the depleted fraction of domestic and residential water use is typically only 15–25 per cent with the remainder returning as wastewater [31]. On the other hand, wastewater use poses several threats both to environment and to human as well as livestock health. The contaminants present in the municipal and industrial wastewater are sequestered in the soils and thereby poses environmental problems. The presence of heavy metals is one of the major sources of concern. It has been reported that 45 per cent of wastewater irrigated areas in China are contaminated with heavy metal at the most serious level. Cadmium and lead are the elements most seriously contaminating soils. Not only in China, has this been a problem in several other countries like Germany, France and India as well [6, 15, 34]. The excessive accumulation of heavy metals in agricultural soils through wastewater irrigation, may not only result in soil contamination, but also lead to elevated heavy metal uptake by crops, and thus affect food quality and safety [24]. Humans are exposed to the risks through the consumption of food crops contaminated with heavy metals and are one of the important pathways for the entry of toxic substances into the human body. Some of the harmful impacts of intake of toxic metals become apparent only after several years of exposure [3, 14]. Chemical contaminants can be of concern especially in those countries where industrial development has started and industrial effluent enters domestic wastewater and natural streams. A survey along the Musi River in India, revealed the transfer of metal ions from wastewater to cow's milk through Para grass fodder irrigated with wastewater. Milk samples were contaminated with different metal ions like Cd, Cr, Ni, Pb and Fe ranging from 12 to 40 times the permissible levels [23]. Leafy vegetables accumulate greater amounts of certain metals like cadmium than do non-leafy species. Generally, metal concentrations in plant tissue increase with metal concentrations in irrigation water, and concentrations in roots usually are higher than concentrations in leaves. This challenge can only be addressed though water treatment. If data show increased metal levels in the food, wastewater irrigation is not encouraged. Some studies reports that the consumption of heavy metal contaminated food can deplete some essential nutrients in the body that are further responsible for decreasing immunological defenses, intrauterine growth retardation, impaired psycho-social faculties, disabilities associated with malnutrition and high prevalence of upper gastrointestinal cancer rates [17, 36].

Presence of microorganism in wastewater poses a serious risk to the health of those who are directly or indirectly exposed to wastewater of which the greatest concern are pathogenic micro- and macro-organisms. Pathogenic viruses, bacteria, protozoa and helminthes present in the wastewater pose health problems. The pathogens and associated health risks can be ranked by taking into account factors such as persistence in the environment, infective dose, immunity, and transmission routes that contribute to the transmission of pathogens by raw wastewater irrigation, in developing countries. For example, Shuval et al. [32] have ranked the pathogens and their associated health risks in the following manner. They are: (i) helminthes, the intestinal nematodes constitute a risk to agricultural workers and to consumers of wastewater irrigated produce; (ii) bacteria and protozoa for the transmission of dysentery, cholera, typhoid and other bacterial and amoebic diseases to consumers of wastewater irrigated produce; and (iii) viruses for the transmission of viral infections to agricultural workers or to those living close to wastewater irrigated fields. In addition to these, the organic or non-organic toxicants such as heavy metals and pesticides contained in the water also pose health risks to workers and farmers.

8.4 BEST MANAGEMENT PRACTICES FOR REUSE OF WASTEWATER

The risks of using untreated or only partially treated wastewater in agriculture can be reduced through wastewater treatment and non-treatment options or a combination of both [39]. They include: (i) water quality improvements, (ii) human exposure control, (iii) irrigation management, (iv) crop restriction and diversification, and (v) harvest and post-harvest interventions.

8.4.1 WATER QUALITY IMPROVEMENTS

Wastewater treatment processes are divided into primary, secondary and advanced processes. Primary treatment includes basic treatment such as screening of coarse solids and grit removal. Secondary treatment includes low-rate processes such as stabilization ponds with high land and low capital and energy inputs, and high-rate processes such as activated sludge with low land and high capital and energy inputs [27]. Tertiary stages further improve quality by nitrification-denitrification (to reduce the nitrogen level) and soil and aquifer treatment [13]. Initial improvements in water quality can be achieved in many developing countries by at least primary treatment of wastewater, particularly where wastewater is used for irrigation. Secondary treatment can be implemented at reasonable cost in some areas, using methods such as waste-stabilization ponds, constructed wetlands, infiltration-percolation, and up-flow anaerobic sludge blanket reactors [20]. Important concept is to aim at treated wastewater quality standards, which can be achieved in the local context. The recent WHO guidelines provide complementary options for wastewater treatment and control of human exposure [39]. Storing reclaimed water in reservoirs improves microbiological quality and provides peak-equalization capacity, which increases the reliability of supply and improves the rate of reuse. Long retention times in the King Talal Reservoir in the Amman-Zarqa Basin of Jordan reduced *faecal coliform* levels in water downstream of the dam, although it was not initially intended for that purpose [12].

8.4.2   HUMAN EXPOSURE CONTROL

Farm workers are at high risk to microorganism infections. Such risk can be eliminated, by use of less contaminating irrigation methods and by the use of appropriate protective clothing (shoes or boots for fieldworkers and gloves for crop handlers). These health protection measures have not been quantified in terms of pathogen exposure reduction but are expected to have an important positive effect. Fieldworkers should be provided with access to sanitation facilities and adequate water for drinking and hygienic purpose in order to avoid the consumption of, and any contact with wastewater. A study conducted in Peru indicated that wastewater-irrigated crops with acceptable levels of bacteria at the farm were frequently recontaminated in the market [26]. Effective hygiene promotion programs are almost always needed. These should target fieldworkers, produce handlers, vendors and consumers. It may be possible to link hygiene promotion to agricultural extension activities or other health programs [39].

8.4.3   IRRIGATION MANAGEMENT

An important consideration for sustainability of wastewater application in agriculture is selection of irrigation method. Commonly, flood, furrow, drip or watering cans are adopted for irrigation. Choice of irrigation methods does not depend on water quality only but also on affordability, tenure security, labor availability and other production factors. Flood irrigation is the lowest cost method, if the topography is favorable or farmers can afford a pump. However, water use efficiency is low, thus successful where water is not a limiting factor. Furrow irrigation provides a higher level of health protection in comparison to flood method, but requires favorable topography and land leveling. Irrigation with sprinklers and watering cans are not recommended as this spreads the water on the crop surface, although cans are usually the cheapest investment option and favored for fragile vegetable beds. Sprinklers require in addition a pump and hose, have medium to high cost, and medium water use efficiency. Irrigating at night and not irrigating during windy conditions are important considerations when using sprinklers. Drip irrigation, especially with sub-surface drippers, can effectively protect farmers and consumers by minimizing crop and human exposure, but drip irrigation components (pump, filters, fertilizer injecting devices, pipes, laterals, emitters etc.) according to planting density and pre-treatment of wastewater is needed for operation of irrigation method [23]. Drip system running with wastewater requires repair, maintenance and flushing of system to avoid chocking of filters and clogging of emitters. It also increases expenditure on cultivation [35].

8.4.4   CROP RESTRICTION AND DIVERSIFICATION

Water management and demand of crop produce is the key for crop diversification. Crops have different sensitivities to water supply. For example, groundnuts and safflower have low sensitivities to water supply, while paddy and bananas have high sensitivities to water supply. It is difficult to enforce crop restrictions because demand for vegetables is high in cities and farmers need to maintain their livelihoods [7]. Crops may be adopted, which are not for human consumption (for example cotton, sisal). Selected crops normally processed by heat or drying before human consumption (grains, oilseeds, sugar beet). Vegetables and fruit grown exclusively for canning that effectively destroy pathogens. Fodder crops and other animal feed crops that are sun-dried and harvested before consumption by animals may be selected. Landscape irrigation in fenced areas without public access (nurseries, forests, green belts) will be a better option for crop diversification.

8.4.5   HARVEST AND POST-HARVEST INTERVENTIONS

Harvested produce should be free from pathogens contamination. Pathogens die-off is possible after providing post-harvest interventions.

Cooking vegetables is the most effective way of achieving complete reduction of pathogens [39], but washing is important with vegetables like lettuce, which are served uncooked. In West Africa it was shown that washing methods vary widely between Anglophone and Francophone countries with significant differences depending on the disinfectant, contact time and water temperature used. In general, a reduction of *E. coli* levels by 2–3 log10 units can be achieved. The effective removal of helminthes eggs requires good agitation and rubbing of the leaves [1]. Another measure is the cessation of irrigation, prior to harvest to allow natural pathogen die-off. Field trials in Tunisia with forage crops and sorghum showed that the bacterial contamination after irrigation with secondary treated wastewater varied with crop species, season, the number of days after cessation of irrigation, and weather. For both sorghum and alfalfa, 7–10 days between the last irrigation and cutting were needed to achieve natural decontamination [38]. In field tests conducted in Ghana on lettuce, cessation resulted in a significant loss of fresh weight. Although 4–5 days without irrigation would significantly increase food safety, a yield loss of about 25% was not acceptable to farmers. A compromise with higher adoption potential would be a cessation of 2 days, with a yield reduction of 10% [7].

8.5   ASPECTS OF PLANNING AND POLICIES

The safe management of wastewater use in agriculture is facilitated by appropriate policies, legislation, institutional framework and regulations at the international, national and local levels. In many countries, where wastewater use in agriculture takes place, these frameworks are lacking. Policy leads to the creation of relevant legislations. Institutional framework determines which agencies have the lead responsibility for creating regulations and who has the authority to implement and enforce the regulations. Policy priorities for each country are necessarily different to reflect local conditions [39]. Following issues should be considered for the national policy on use of wastewater in agriculture:

- Setting of appropriate standards and regulation by health impact assessment of wastewater use in agriculture.
- Water scarcity considering wastewater availability now and in the future.

- Acceptability of wastewater use in agriculture by considering locations where wastewater is generated.
- Type and extent of wastewater use practice.
- Downstream impacts if wastewater is not used for agriculture.
- Trade implications of exporting crops produced with wastewater.
- Livelihood of the people's dependent upon wastewater use in agriculture.

Wastewater resource requires good management, which can be implemented by giving direct and indirect subsidy. Where the use of treated wastewater is promoted, incentives for its use are helpful for allowing water users to choose among different water sources. Lower water prices and subsidies for purchasing new equipment can speed the pace at which farmers begin using wastewater. Incentives can be combined with monitoring to ensure compliance with incentive programs and safe use of wastewater. Countries should have realistic criteria of water quality for wastewater reuse. Policies and legislations are to protect and regulate long term wastewater reuse. Meaningful criteria need to be established in accordance with local, technical, economic, social, and cultural contexts. In addition, improving water quality requires new approaches to wastewater management in cities-subdividing cities into manageable units, like the example of Bangkok, which has made step-wise successes possible [29]. In Pakistan large number of court cases initiated by local water utilities or sanitation agencies has been brought against local farmers, challenging their rights to use wastewater resources. The outcome of these court cases was that farmers were forced either to pay for wastewater or to abandon its use. In Faisalabad, a group of wastewater farmers successfully appealed against one of these court orders once they proved that they have no access to another suitable water source [8].

Information on present status and extent of wastewater use for irrigation can enhance the efforts of public agencies and researchers to address actual opportunities and threats. There is also a need for more holistic risk assessments. Public agencies can improve the coordination of policy targets and methods to ensure that public goals regarding wastewater management are achieved. For example, coordination among the ministries of agriculture, water resources, public health, and economic development is needed to ensure that the goals and programs of one agency are not in conflict with the goals and programs of another. The total cost of achieving public goals will be minimized with effective inter-ministry coordination [29].

8.6 CONCLUSIONS

Untreated or partially treated wastewater is widely used for irrigation in water scarce regions of several countries including India. Nutrients contained in the wastewater are considered as beneficial to agriculture but the contaminants present in the wastewater poses health risks directly to agricultural workers and indirectly to consumers of the wastewater grown produce. This paper briefly reviews the health risks of using wastewater for irrigation and elicits the health problems of those who are directly exposed to wastewater based livelihood activities. Wastewater can work as an asset by reducing health risk and its efficient management. It is possible with the support of research, strong political will and investments in infrastructure. Implementation of economic incentives, training, and coordination among the ministries of agriculture, water resources, public health, and economic development can play vital role for expansion of agriculture using wastewater for irrigation. The study recommends adequate treatment of wastewater and public health education for adopting precautionary and preventive measures for those directly exposed to the wastewater.

8.7 SUMMARY

Increasing demand for freshwater in domestic, industrial and irrigation sector putting pressure to water managers for adopting easily available alternative water resources. Burgeoning urban population and industries are generating huge amount of wastewater round the year. It is available in outskirt of cities to fulfill irrigation demand and save finite fresh water resources in arid and semi-arid regions. The argument for consumption of treated, untreated and partially treated wastewater finds risk managers pondering the question of what types of water quality standards might be set in order to provide the proper level of safety associated with the use of wastewater. The promising alternative for disposal of wastewater is its utilization for irrigation after treatment. Health effects of wastewater irrigation can be both direct as well as indirect and even affect unsuspecting people. Wastewater irrigated vegetables and fodder may serve as the transmission route for heavy metals in the human food chain.

While farmers can suffer from harmful health effects from the contact with wastewater, consumers are at the risk from eating vegetables and cereals irrigated with wastewater. The long-term health effects of wastewater use are not yet well documented. Although there are studies on soil and crop contamination with heavy metals and their associated health risks to our knowledge there are not many studies, especially in the Indian context which has tried to estimate the health costs of using wastewater. Effective hygiene education and promotion are required for better livelihoods of farmers. Controlled use of wastewater, through treatment and planning, leads to water resources augmentation, particularly in the countries that suffer from water scarcity conditions, in addition to environmental protection. This paper is an attempt to fill up the gap in the literature and focuses on the present status, planning and policies of wastewater use for agriculture.

KEYWORDS

- best management practices
- crop
- crop restrictions
- environment
- guidelines
- harvest
- health
- heavy metals
- human exposure control
- hygiene
- institutional aspect
- irrigation

- **irrigation management**
- **livelihood**
- **nutrients**
- **planning**
- **policies**
- **population**
- **post -harvest**
- **quality**
- **treatment**
- **urban**
- **wastewater**
- **wastewater management**
- **wastewater reuse**
- **water quality**

REFERENCES

1. Amoah, P., Drechsel, P., Abaidoo, R., Klutse, A. (2007). Effectiveness of common and improved sanitary washing methods in selected cities of West Africa for the reduction of coliform bacteria and *helminth eggs* on vegetables. *Trop. Med. Int. Health,* 12(s2), 40–50.

2. Asano, T., Burton, F., Leverenz, H., Tsuchihashi, R., Tchobanoglous, G. (2007). *Water Reuse: Issues, Technologies, and Applications.* (McGraw-Hill), Metcalf & Eddy Inc.

3. Bahemuka, T. E., Mubofu, E. B. (1999). Heavy metals in edible green vegetables grown along the sites of the Sinza and Msimbazi Rivers in Dares Salaam, Tanzania. *Food Chemistry,* 66, 63–66.

4. Chen, Y., Wang, C., Wang, Z. (2005). Residues and source identification of persistent organic pollutants in farmland soils irrigated by effluents from biological treatment plants. *Environment International,* 31, 778–783.

5. CPCB, (2009). Status of water supply, wastewater generation and treatment in class-I cities and class-II towns of India. *Control of urban pollution, series: CUPS/70/2009–10.* Central Pollution Control Board, Ministry of Environment and Forest, Government of India.

6. Dere, C, Lamy, I., van Oort, F., Baize, D., Cornu, S. (2006). Trace metal inputs reconstitution and migration assessment in a sandy Luvisol after 100 years of massive irrigation with raw wastewaters. CR *Geoscience,* 338(8), 565–573.

7. Drechsel, P., Graefe, S., Sonou, M., Cofie, O. (2006). Informal irrigation in urban West Africa: an overview. In: *Research Report 102*, IWMI, Colombo, Sri Lanka.
8. Ensink, J. H. J., Mahmood, T., van der Hoek, W., Raschid-Sally, L., Amerasinghe, F. P. (2004). A nationwide assessment of wastewater use in Pakistan: an obscure activity or a vitally important one? *Water Policy*, 6, 197–206.
9. FAO, (1997). *Irrigation in the Near East in figures*. Water Reports No. 9. Rome, Italy.
10. Faruqui, N., Niang, S., Redwood, M. (2004). Untreated wastewater reuse in market gardens: a case study of Dakar, Senegal. In: *Wastewater Use in Irrigated Agriculture: Confronting the Livelihood and Environmental Realities* (eds. CA Scott, NI Faruqui and L Raschid-Sally). CABI Publishing, Wallingford, pages 113–125.
11. Ghosh, G. K. (2002). *Water of India (Quality and Quantity)*. A. P. H. Publishing Corporation, 5, Ansari Road, Darya Ganj, New Delhi, pages 161.
12. Grabow, G., McCornick, P. G. (2007). Planning for water allocation and water quality using a spreadsheet-based model. *J. Water Resources Planning and Management*, American Society of Civil Engineers. Manuscript number: WR/2005/022968.
13. Haruvy, N. (1997). Agricultural reuse of wastewater: nation-wide cost-benefit analysis. *Agriculture Ecosystem and Environment*, 66, 113–119.
14. Ikeda, M., Zhang, Z. W., Shimbo, S., Watanabe, T., Nakatsuka, H., Moon, C. S., Matsuda-Inoguchi, N., Higashikawa, K. (2000). Urban population exposure to lead and cadmium in east and south-east Asia. *Science of the Total Environment*, 249, 373–384.
15. Ingwersen, J., Streck, T. (2006). Modeling the environmental fate of cadmium in a large wastewater irrigation area. *Journal of Environmental Quality*, 35(5), 1702–1714.
16. IWMI (International Water Management Institute), (2003). Confronting the realities of wastewater use in agriculture. Water Policy Briefing 9. IWMI, Colombo, Sri Lanka.
17. Iyengar, V., Nair, P. (2000). Global outlook on nutrition and the environment: meeting the challenges of the next millennium. *Science of the Total Environment*, 249, 331–346.
18. Keraita, B. N., Drechsel, P. (2004). Agricultural use of untreated urban wastewater in Ghana. In: Scott, C. A., Faruqui, N. I., Raschid-Sally, L. (Eds.), *Wastewater Use in Irrigated Agriculture*. CABI Publishing, Wallingford, UK, pages 101–112.
19. Lazarova, V., Bahri, A. (2005). *Water Reuse for Irrigation: Agriculture, Landscapes, and Turf Grass*. CRC Press, Boca Raton, USA.
20. Mara, D. (2003). *Domestic Wastewater Treatment in Developing Countries*. Earthscan, UK.
21. Mara, D., Cairncross, S. (1989). *Guidelines for the Safe Use of Wastewater and Excreta in Agriculture and Aquaculture*. World Health Organization (WHO), Geneva.
22. Minhas, P. S., Samra, J. S. (2003). *Quality Assessment of Water Resources in Indo-Gangetic Basin Part in India*. Central Soil Salinity Research Institute, Karnal, India, 68 pages.
23. Minhas, P. S., Samra, J. S. (2004). *Wastewater Use in Peri-urban Agriculture: Impacts and Opportunities*. Central Soil Salinity Research Institute, Karnal, India, 75 pages.
24. Muchuweti, M., Birkett, J. W., Chinyanga, E., Zvauya, R., Scrimshaw, M. D., Lester, J. N. (2006). Heavy metal content of vegetables irrigated with mixtures of wastewater and sewage sludge in Zimbabwe: Implications for human health. *Agriculture, Ecosystems and Environment*, 112, 41–48.

25. Obuobie, E., Keraita, B., Danso, G., Amoah, P., Cofie, O. O., Raschid-Sally, L., Drechsel, P. (2006). Irrigated urban vegetable production in Ghana: characteristics, benefits and risks. IWMI-RUAF-IDRC-CPWF, Accra, Ghana: IWMI, 150 pages. Available at <www.cityfarmer.org/GhanaIrrigateVegis.html>.

26. Peasey, A. (2000). A review of policy and standards for wastewater reuse in agriculture: a Latin American perspective. London, Water and Environmental Health at London and Loughborough (WELL Study no. 68, Part II at <http://www.lboro.ac.uk/well/>).

27. Pettygrove, G. S., Asano. T. (Eds.), (1985). *Irrigation with reclaimed municipal wastewater-a guidance manual.* Lewis Publishers Inc., Chelsea, Michigan.

28. Qadir, M., Sharma, B. R., Bruggeman, A., Choukr-Allah, R., Karajeh, F. (2007). Non-conventional water resources and opportunities for water augmentation to achieve food security in water scarce countries. *Agricultural Water Management,* 87, 2–22.

29. Qadir, M., Wichelns, D., Raschid-Sally, L., McCornick, P. G., Drechsel, P., Bahri, A., Minhas, P. S. (2010). The challenges of wastewater irrigation in developing countries. *Agricultural Water Management,* 97(4), 561–568.

30. Raschid-Sally, L., Parkinson, J. (2004). Wastewater reuse for agriculture and aquaculture-current and future perspectives for low-income countries. *Waterlines J.,* 23(1), 2–4.

31. Scott, C. A., Faruqui, N. I., Raschid-Sally, L. (2004). Wastewater use in irrigated agriculture: management challenges in developing countries. In: Scott, C. A., Faruqui, N. I., Raschid-Sally, L. (Eds.), *Wastewater Use in Irrigated Agriculture.* CABI Publishing, UK.

32. Shuval, H. I., Adir, A., Fattal, B., Rawitz, E., Yekutel, P. (1986). *Wastewater Irrigation in Developing Countries: Health Effects and Technical Solutions.* The World Bank Technical Paper No. 51, Washington DC, USA.

33. Singh, K. P., Mohon, D., Sinha, S., Dalwani, R. (2004). Impact assessment of treated/untreated wastewater toxicants discharge by sewage treatment plants on health, agricultural, and environmental quality in wastewater disposal area. *Chemos,* 55, 227–255.

34. Singh, S., Kumar, M. (2006). Heavy metal load of soil, water and vegetables in peri-urban Delhi. *Environmental Monitoring and Assessment,* 120, 79–91.

35. Tripathi, V. K., Rajput, T. B. S., Patel, N. (2014). Performance of different filter combinations with surface and subsurface drip irrigation systems for utilizing municipal wastewater. *Irrigation Science,* at <DOI 10.1007/s00271–014–0436–2>.

36. Turkdogan, M. K., Fevzi, K., Kazim, K., Ilyas, T., Ismail, U. (2003). Heavy metals in soil, vegetables and fruits in the endemic upper gastrointestinal cancer region of Turkey. *Environmental Toxicology and Pharmacology,* 13, 175–179.

37. UN, (2010). *World Urbanization Prospects: The 2009 Revision.* United Nations Department of Economic and Social Affairs/Population Division, pages 47.

38. UNDP, (1987). Reclaimed wastewater reuse in agriculture. Technical Report (in French), vol. 1, 3rd Part, 69 pages. Project RAB/80/011, Water Resources in the North African Countries, Rural Engineering Research Center (Tunisia)-United Nations Development Program (UNDP).

39. WHO, (2006). Guidelines for the safe use of wastewater, excreta and greywater. In: Volume 2, *Wastewater Use in Agriculture,* World Health Organization (WHO), Geneva, Switzerland.

CHAPTER 9

IMPACT OF SEMI-ARID CLIMATE ON WASTEWATER QUALITY PARAMETERS

VINOD KUMAR TRIPATHI

CONTENTS

9.1   INTRODUCTION

Worldwide unsystematic and untreated disposal of wastewater is the biggest challenge to environmental scientist and engineers. This problem started long back but intensified during the last few decades, and now the situation has become alarming in India [7]. During the past three decades, the effects of municipal wastewater and effluents, the point source pollution, on the water quality of canals, streams and rivers have received some attention but very little attention has been reported about the sewage water quality of the receiving water bodies. The input of large quantities of nutrients mainly nitrates and phosphates in to river waters caused eutrophication and its related effects [9]. These nutrients particularly phosphorus is often the limiting nutrient in such systems [13]. Sewage effluent containing sulphate may produce hydrogen sulphide that in the form of sulphur is poisonous; so the untreated sewage should be treated in the sewage treatment plant (STP) before letting into the receiving water bodies.

Rapid trend of global urbanization causes wastewater generation in the tune of 38254 Million Liter Per Day (MLD) from Class I cities and Class II towns of India [4]. Wastewater is a mixture of domestic effluent consisting of blackwater (excreta, urine and fecal sludge, i.e. toilet wastewater), greywater (kitchen and bathing wastewater), water from commercial establishments and institutions, hospitals, industrial effluent where present, storm water and urban run-off. Contribution of different sources, water utilized in those sources affects the quality of wastewater. Its quality in drain is dependent upon rainfall, soil properties, agricultural activities, exploitation of water resources, and sort of industries. The quantitative contribution of different sources also affects the quality of mixed wastewater. The indiscriminate release of municipal sewage has severely deteriorated the aquatic environment leading nutrient enrichment of the receiving water body [2], affecting environment health worldwide. Indian rivers and canals especially in their urban portions are grossly polluted due to release of municipal sewage and industrial wastewater [4, 7].

Worldwide wastewater is emerging as an alternative water resource for irrigation. The main driving forces for the growth of wastewater irrigation are global scarcity of freshwater resources with the problem of wastewater disposal. Estimates on the extent of this practice range widely, but figures point at about 20 million ha of land irrigated in this way, most of it in Asia, Latin America, and sub-Saharan Africa [11]. In the world more than 4–6 million hectares (M ha) are irrigated with wastewater or polluted water [10, 21]. A separate estimate indicates 20 million ha globally, an area that is nearly equivalent to 7 per cent of the total irrigated land in the world [22]. The wastewater for irrigation poses risk to the environment and endangering the health of the farmers and produce consumers.

Studies on water quality in Indian rivers receiving municipal sewage and effluents have been carried out by many earlier investigators, for example, River Ganga [19] and Kathajodi River in Cuttack city in Orissa [5]. Girija et al. [7] also reported the water quality assessment of an urban stream polluted with untreated effluents at Guwahati in Assam (India). The present study assessed the seasonal effects of point source pollution, untreated and treated sewage on the water quality parameters. It is imperative to characterize wastewater resources on the basis of physiochemical properties with respect to time for effective pollution control and better water resource management. This results in huge and complex data matrix comprised of a large number of physiochemical parameters, which are often difficult to interpret and draw meaningful conclusions [6].

In this chapter, physiochemical water quality parameters of wastewater were determined. The data set is used to present spatial and temporal variations of the municipal wastewater quality. Identification of wastewater quality parameters responsible for temporal variations due to effect of semi-arid climate was done through multivariate cluster analysis. Correlation study between the identified parameters was also done.

9.2 MATERIALS AND METHODS

9.2.1 STUDY AREA

Study was conducted at Indian Agricultural Research Institute (IARI), Pusa, New Delhi, India, which is located within 28°37′22″ N and 28°39′ N latitude and 77°8′45″ E and 77°10′24″ E longitude covering an area of about 475 ha at an average elevation of 230 m above mean sea level. On east side Arawali range indicates the highest point while on western side natural drain forms the last western boundary of the farm. Soil of the research farm is alluvial in origin and sandy loam in texture (Typic mixed hypothermic ustocrept). Water table in the farm area is about 5–7 m [14]. The climate of IARI is subtropical, semiarid with hot dry summer and cold winter and the area falls in Agro-eco-region-IV (Semiarid with alluvium derived soil). The mean annual temperature is 24°C with months of May and June having 30 years normal maximum of 40°C, and January with the normal minimum temperature of 14°C and the minimum temperature dips as low as 1°C. The mean annual rainfall is 710 mm of which as much as 75% is received during the monsoon season (July to September). The wastewater (WW) samples were collected from natural drain passing through IARI near the western boundary. The contribution of wastewater in this drain is urban runoff, domestic effluent, institutions and small-scale industrial effluent.

9.2.2 WATER SAMPLING AND PRESERVATION

Municipal wastewater samples were collected from the drains in each month during the study period (January to December, 2009). Collection of wastewater samples from the drain was done across the drain at the depth of 15 cm below the surface and at three points then mixed. The preservation and transportation was performed according to the standard methods [1].

9.2.3 ANALYSIS OF WASTEWATER SAMPLES

Wastewater samples were analyzed for the level of pH, electrical conductivity (EC), total solids (dissolved and undissolved), turbidity, total soluble salts (TSS), ammonical nitrogen (NH_4-N), nitrate nitrogen (NO_3-N), phosphate (P), potassium (K), sodium (Na^+), calcium (Ca^{2+}), magnesium (Mg^{2+}), carbonate (CO_3^-), bicarbonate (HCO_3^-), copper (Cu), zinc (Zn), manganese (Mn), iron (Fe), lead (Pb), chromium (Cr), arsenic (As), molybdenum (Mo), sulphate (SO_4^-), and chloride (Cl^-) (Table 9.1). All the water quality parameters were expressed in milligram per liter (mg L^{-1}), except EC (dS m^{-1}). To prepare all the reagents analytical grade chemicals and double glass distilled deionized water were used during analysis. The glassware was washed with dilute nitric acid followed by distilled water. Using few water quality parameters temporal variation in three indices namely Sodium adsorption ratio (SAR), Residual sodium carbonate (RSC), and Mg/Ca ratio were estimated. SAR and RSC were estimated using equations 1, and 2.

$$SAR = \frac{Na^+}{\left[\dfrac{Ca^{2+} + Mg^{2+}}{2} \right]^{1/2}} \tag{1}$$

TABLE 9.1 Water Quality Parameters and Analytical Methods Used For Analysis

Parameters	Unit	Analytical methods	Instruments
Ca, Mg	mgL^{-1}	Titrimetric	Titration assembly
Cl	mgL^{-1}	Titrimetric	Titration assembly
CO$_3$, HCO$_3$	mgL^{-1}	Titrimetric	Titration assembly
Cu, Zn, Mn, Fe, Pb, Cr, As, Mo	mgL^{-1}	Instrumental	Atomic Absorption Spectrophotometer
EC	dS m^{-1}	Instrumental	pH meter
K	mgL^{-1}	Instrumental	Flame Photometer
Na	mgL^{-1}	Instrumental	Flame Photometer
NH$_4$-N, NO$_3$-N	mgL^{-1}	Instrumental and Titrimetric	Kjeldahl distillation apparatus
P	mgL^{-1}	Instrumental	UV and VIS Spectrophotometer
pH	-	Instrumental	EC meter
SO$_4$	mgL^{-1}	Instrumental	UV and VIS Spectrophotometer
Total Solids	mgL^{-1}	Gravimetric	Temperature controlled oven
TSS	mgL^{-1}	Estimation	EC meter
Turbidity	NTU	Instrumental	Sonar Turbidity meter

$$RSC\left(mel^{-1}\right) = \left(CO_3^{2-} + HCO_3^{-}\right) - (Ca^{2+} + Mg^{2+})$$ (2)

9.2.4 STATISTICAL ANALYSIS

Descriptive statistical parameters were analyzed for different municipal wastewater (MWW) quality parameters. To get the range of any quality parameter minimum and maximum values were analyzed. On the basis of total values for a particular parameter arithmetic mean was also determined. To make the knowledge of dispersion from mean of any quality data standard deviation and standard error was calculated through MS Excel spreadsheet. Correlation among water quality parameters were calculated and presented in tabular form. This analysis was helpful to determine the relation between two parameters.

9.2.5   CLUSTER ANALYSIS

Clustering is a multivariate technique of grouping rows together that share similar values. It can use any number of variables. The variables must be numeric for which numerical differences makes sense. The identification of these clusters goes a long way towards characterizing the distribution of values. Hierarchical agglomerative clustering (a combining process) was applied on entire parameters of water quality MWW. The method starts with each point (row) as its own cluster. At each step the clustering process calculates the distance between each cluster (monthly, row) and combines the two clusters (month) that are closest together. This combining continues until all the points (month) are in one final cluster. The number of clusters can be chosen that seems right and cuts the clustering tree at a given point. The combining record is portrayed as a tree, called a dendrogram, with the single points as leaves, the final single cluster of all points as the trunk, and the intermediate cluster combinations as branches.

The distance between two clusters was calculated using Ward's minimum variance method (which has got merits over other clustering methods), in which the ANOVA sum of squares between the two clusters added up over all the variables. At each generation, the within-cluster sum of squares is minimized over all partitions obtainable by merging two clusters from the previous generation. The sums of squares are easier to interpret when they are divided by the total sum of squares to give the proportions of variance (squared semipartial correlations).

9.3   RESULTS AND DISCUSSION

9.3.1   PHYSIOCHEMICAL PROPERTIES OF MUNICIPAL WASTEWATER

Descriptive statistics (minimum, maximum, mean, standard deviation, and standard error) of MWW quality for 24 parameters is presented in Table 9.2. Maximum electrical conductivity (EC) 1.90 dS m^{-1} was observed with small standard error (SE) as shown in Table 9.2. The pH is the indicator of acidic and alkaline condition of water status. pH was in the range of 6.6 to 7.32 with the mean of 6.87. It shows the acidic

TABLE 9.2 Descriptive Statistics of Municipal Wastewater Quality Parameters

Quality parameters	Unit	Minimum	Maximum	Mean	Standard deviation	Standard error
As	mg/L	0.03	0.23	0.13	0.05	0.02
Ca	mg/L	68.00	136.00	94.67	20.84	6.02
Cl	mg/L	173.90	339.50	262.44	56.45	16.30
CO_3	mg/L	12.00	78.00	43.50	22.76	6.57
Cr	mg/L	0.00	0.46	0.06	0.13	0.04
Cu	mg/L	0.03	0.19	0.09	0.05	0.01
EC	dS/m	1.63	1.90	1.74	0.08	0.02
Fe	mg/L	0.25	2.00	0.98	0.62	0.18
HCO_3	mg/L	440.00	610.00	516.13	53.76	15.52
K	mg/L	15.80	60.70	30.00	13.44	3.88
Mg	mg/L	25.20	38.40	32.13	4.52	1.31
Mn	mg/L	0.03	0.16	0.08	0.04	0.01
Mo	mg/L	0.00	0.01	0.01	0.00	0.00
Na	mg/L	108.80	226.11	169.94	33.30	9.61
NH_4-N	mg/L	19.72	37.52	28.66	5.42	1.57
NO_3-N	mg/L	3.19	8.31	5.45	1.54	0.45
P	mg/L	1.18	11.36	6.91	3.21	0.93
Pb	mg/L	0.02	0.59	0.18	0.19	0.06
pH		6.60	7.32	6.87	0.21	0.06
SO_4-S	mg/L	19.93	62.14	40.47	11.20	3.23
Total Solids	mg/L	733	1297	989	211	60.90
TSS	mg/L	1043	1217	1116	52	15.11
Turbidity	NTU	33	68	55.33	11.46	3.31
Zn	mg/L	0.01	0.11	0.05	0.02	0.01

nature of MWW. pH range 6.5–8.5 is recommended by BIS [3] and CPCB [4] for irrigation purpose. Maximum standard deviation 211 mg L^{-1} and standard error was observed for total solids. Maximum and minimum variation was also higher with total solids. It may be due to presence of more contaminants like heavy metals and soil particles from the drains. Turbidity of MWW was high and in the range of 33 to 68 NTU. The higher turbidity of MWW was due to presence of surface runoff and foreign particles from anthropogenic pollution and stream bed material.

Sodium content in MWW was in the range of 109 to 226 mgL^{-1}. Ammonical nitrogen (NH_4-N) in MWW was observed in the range of 19.7 to 37.5 mgL^{-1} but nitrate-nitrogen (NO_3-N) was in the range of 3.19 to 8.31 mgL^{-1}. Higher content of ammonical nitrogen in MWW was due to conversion of nitrate nitrogen into ammonical nitrogen under reduction process. Lower dissolved oxygen in MWW is also evident for this fact. This represents influence of agricultural runoff from the soil as nitrogenous fertilizers are extensively used in this region. Mean Phosphorus (P) content in MWW was 6.91 mgL^{-1}. It may be due to presence of industrial effluent, sewage water and urban runoff. The similar result was observed by Goldman and Horne [8]. Potassium (K) content was higher in comparison to phosphorus (P) content. Calcium (Ca) content was higher than Mg content but bicarbonate content was higher than carbonate content. It shows the presence of calcium bicarbonate and magnesium carbonate in MWW. It was due to conversion of carbonate into bicarbonate under slightly acidic pH condition. The mean value of pH was 6.6 for MWW.

Highest copper content of 0.19 mg L^{-1} was observed in MWW with mean value 0.08 mgL^{-1}. It is evident from the fact that urban industries are contributing their poor quality effluent in MWW. Highest mean value 0.04 of Zinc (Zn) was observed for MWW due to higher metal content in urban runoff and industrial wastewater stream joining the MWW drain. Manganese (Mn) content with mean value 0.078 mgL^{-1} was detected in MWW. Iron (Fe) content was highest among heavy metals (Cu, Zn, Mn, Mo, Cr, As, Pb, Fe) with mean value 1 mg L^{-1}. Maximum concentration of chromium (Cr) was 0.46 mg L^{-1}. Highest concentration for Arsenic (As) and Molybdenum (Mo) with mean value 0.007 and 0.125 mgL^{-1} respectively was observed. Lead (Pb) was observed with mean value 0.18 mg L^{-1} with maximum SE among heavy metals (Cu, Zn, Mn, Mo, Cr, As, Pb, Fe) except Fe. Sulphate (SO_4) content was in the range of 19.93 to 62.14 mg L^{-1} with mean value of 40.47 mg L^{-1}. Chloride (Cl) content was in the range of 173.9 to 339.5 mg L^{-1} with mean value of 262.44 mg L^{-1}. SD and SE both were observed higher with chloride in comparison to sulphate.

9.3.2 TEMPORAL VARIATION IN WATER QUALITY

Seasonal variation in municipal wastewater (MWW) quality is presented in Fig 9.1. Electrical conductivity (EC) was in summer season with little variation. It may be the reason that supply for domestic and industrial water consumption was from the Yamuna river having lower EC in other than summer months. There is another possibility for lower water supply and higher evaporation from drainage channels causes increase in concentration

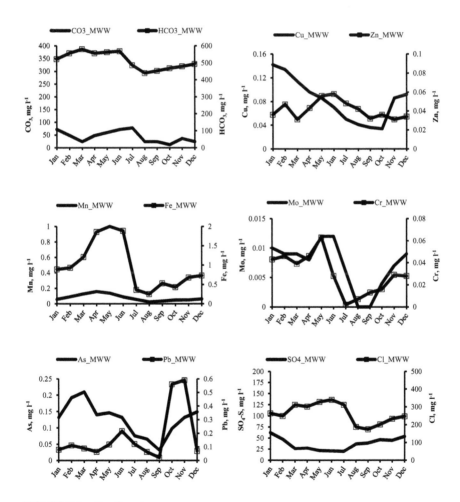

FIGURE 9.1 Continued

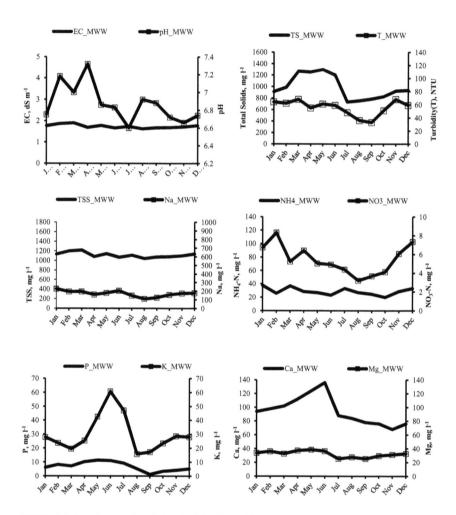

FIGURE 9.1   Temporal variation in MWW quality

of salts. pH of wastewater also followed the same trend i.e. higher hydrogen ion concentration during rainy season, least during summer season and moderate during winter season. In summer season maximum exploitation of GW causes such increment. pH of MWW was lower than GW for entire year due to addition of acidic nature industrial effluent and surface runoff. Total solid (TS) of MWW was observed higher in Nov, Dec, and Jan to May. It may be due to higher concentration of pollutant. In summer season highest concentration of TS, 1300 mg L^{-1} was observed. Turbidity (T) in

MWW was in the range of 33–68 NTU. It may be due to uneven distribution of rainfall in the catchment area of MWW drainage channel and higher amount of rainwater added to the MWW. Total soluble salt (TSS) concentration is directly related to electrical conductivity of water. Hence similar seasonal variation was observed for EC and TSS.

Sodium (Na) content was higher during summer season with wastewater due to higher concentration and excess exploitation of GW. It was added to MWW after utilization in domestic and industrial sector. Ammonical nitrogen was higher during winter season (November to February) but after that the concentration was reduced. It may be due to reduction in volatilization loss and lower demand of fertilizers by microorganism because there growth rate decreases due to effect of drop in temperature during winter season. Maximum nitrate nitrogen content 8.3 mg L^{-1} was observed in the month of February of winter season but ammonical nitrogen was always higher in comparison to nitrate nitrogen for all the seasons of a complete year. The nitrate ion is usually derived from anthropogenic sources like untreated domestic sewage and its effluents, agriculture watershed, and storm water containing nitrogenous compounds [12]. Ravindra et al. [15] also reported NO_3 in winter and summer seasons were 5 and 9 mg/L, respectively in the Yamuna river in Haryana before entering to Delhi. During the study P and K concentrations were higher during the month of May and June in summer season with P = 11.4 mg.L^{-1} and K = 62.6 mg.L^{-1}. It may be due to slow rate of flow and stagnant water in WW drains.

Higher concentration of Calcium and Magnesium was observed during summer season and starting of the rainy season. The main reason for higher concentration in July and August can be explained that runoff from agricultural fields and watersheds were affecting the quality of water. Carbonate and bicarbonate followed the identical trend for seasonal variation with higher concentration during winter and summer and lower during rainy season. It may be due to dilution of WW after addition of good quality fresh water in the form of rainfall. They are also highly correlated. Presence of metals (Cu, Zn, Mn, Fe, Mo, Cr, As and Pb) with maximum value was observed during summer, moderate during winter and minimum during rainy season. The main reason for such behavior in summer was due to hot climate, which causes increase in evaporation rate resulting higher concentration. In rainy season, dilution factor due to rainwater and moderate effect during winter. Sometimes non-uniform trend was also observed in both types of WW (Fig. 9.1). It may be due to uneven distribution of rainfall and variation in quality of industrial effluents contributing to stream. Sulphate (SO_4) concentration was lower during summer and pre-monsoon seasons. It may be due to decomposition of organic matter by anaerobic bacteria might have taken place. Rest of the period sulphate content was higher. In general chloride (Cl) content in water is not harmful. Maximum Cl content was observed in summer season and minimum during rainy season.

9.3.3 TEMPORAL VARIATION IN WATER QUALITY INDICES

The Fig. 9.2 indicates water quality indices, such as: Sodium adsorption ratio (SAR), Residual sodium carbonate (RSC), and Mg/Ca ratio and its limit for safe irrigation. SAR was less than 10 for the entire year of study. Hence on the basis of index SAR the MWW can be used directly for irrigation [16] without any intervention. The RSC is important for carbonate and bicarbonate rich irrigation waters. It indicates their tendency to precipitate Ca^{2+} as $CaCO_3$. In the present study lower carbonate content and higher bicarbonate was observed in MWW. Richards [16] has categorized RSC (mel^{-1}) under safe (RSC<1.25), moderate (RSC 1.25–2.5), and unsafe (RSC>2.5) with the view of irrigation. On the basis of RSC, MWW were suitable for irrigation, except during the month of July. In July MWW were unsafe due to start of rainfall causes maximum pollution came into MWW drain as runoff. Another point of view irrigation requirement will be minimum or nil during July as rainy month. Behavior of calcium and magnesium is not same in the soil system. Magnesium deteriorates soil structure particularly when waters have dominated sodium so that highly saline. High level of Mg usually promotes higher developments of exchangeable Na in the irrigated soils. Based on the ratio of Mg to Ca, waters were categorized safe if ratio is less than 1.5, moderate from 1.5 to 3.0, unsafe if more than 3.0 [20]. Entire year Mg/Ca ratio was less than 1.5 shows its suitability for irrigation.

9.3.4 TEMPORAL SIMILARITY AND PERIOD GROUPING

Cluster analysis applied to detect temporal similarity grouping of twelve months on the basis of water quality parameters for MWW. Dendogram of

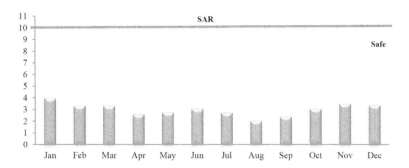

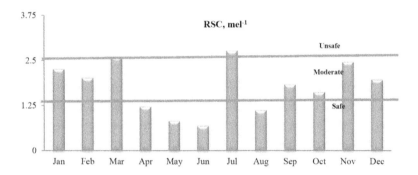

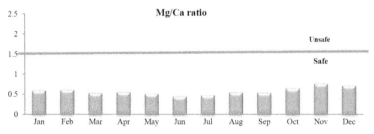

FIGURE 9.2   Water quality indices (SAR, RSC and Mg/Ca ratio) and their limit for suitability to irrigation.

cluster analysis is shown in Fig. 9.3. The clustering procedure generated three groups have similar characteristic features. Similarity in water quality has been shown through Cluster 1 (July, August, September, October, November, and December) shows similar water quality during rainy season to mid-winter. After middle of the winter (January and February) and

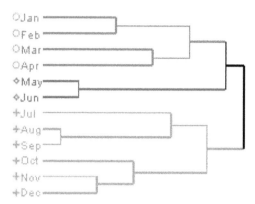

FIGURE 9.3   Dendogram showing temporal similarity for MWW through 3 clustering.

March, April shows Cluster 2. It means more pollutant (untreated sewage and industrial effluents) in MWW made effect in water quality from January. Cluster 3 (May and June) is same due to unchanged weather condition. Spatial grouping on the basis of water quality by cluster analysis was performed by Singh et al. [18] and Simeonov et al. [17].

9.3.5   CORRELATION AMONG WATER QUALITY PARAMETERS

Knowledge of association of water quality parameters is important for efficient utilization of water resources. Significantly ($P<0.05$) correlated water quality parameters were presented in form of diagonal matrix as Table 9.3. Correlation coefficient gives inherent association among different water quality parameters. EC was highly correlated to TSS ($P<0.01$) and arsenic ($P<0.01$). pH is not related to any water quality parameter except Mn. Total solids in MWW were highly ($P<0.01$) correlated to Ca, Mg, HCO_3, and chloride because these elements were found at the same time in the WW. Turbidity is not correlated to any quality parameters except sodium and heavy metals (As and Mo). P is highly ($P<0.01$) correlated to Ca, CO_3, HCO_3, Mn, Fe, Mo, and chloride. K is related ($P<0.01$) to CO_3 and chloride only. Na is highly related ($P<0.01$) to heavy metal Mo and As only. Carbonate, bicarbonate, Mn, and Fe were correlated to Calcium. The reason may be conversion of CO_3 into HCO_3 under acidic environment.

TABLE 9.3 Correlation Matrix of Municipal Wastewater (MWW) for Different Water Quality Parameters

	A	B	C	D	E	F	G	H	I	J	K	L	M	N	O	P	Q	R	S	T	U	V	W	X
A	1	ns	ns	*	**	ns	*	ns	*	*	ns	ns	Ns	*	*	*	ns	ns	ns	ns	**	ns	ns	ns
B		1	ns	ns	ns	ns	ns	ns	ns	ns	ns	ns	Ns	ns	ns	ns	*	ns	ns	ns	ns	ns	ns	ns
C			0	*	ns	ns	ns	*	ns	ns	**	**	Ns	**	*	ns	**	**	ns	ns	*	**	ns	**
D				1	*	ns	ns	ns	*	**	ns	*	Ns	*	*	ns	ns	ns	ns	ns	**	**	ns	ns
E					1	ns	*	ns	ns	*	ns	ns	Ns	*	*	*	ns	ns	ns	ns	**	ns	ns	ns
F						1	*	ns	ns	ns	ns	ns	Ns	ns	ns	ns	ns	ns	ns	*	*	*	ns	ns
G							1	ns	*	*	ns	*	Ns	ns	*	ns	ns	ns	ns	*	*	*	ns	**
H								1	*	ns	**	*	**	**	ns	ns	**	**	ns	ns	ns	**	ns	**
I									1	ns	*	*	**	ns	ns	ns	ns	ns	ns	ns	ns	*	ns	**
J										1	ns	**	Ns	**	ns	ns	**	**	ns	ns	*	**	ns	ns
K											1	**	*	**	**	ns	**	**	ns	ns	*	*	ns	*
L												1	Ns	**	ns	ns	ns	ns	ns	ns	ns	ns	ns	*
M													1	ns	ns	ns	ns	ns	ns	ns	ns	ns	ns	ns
N														1	*	ns	ns	ns	ns	ns	**	**	ns	**
O															1	ns	*	*	ns	ns	*	**	ns	ns
P																1	ns	ns	ns	ns	ns	ns	ns	*
Q																	1	**	ns	ns	*	*	ns	**
R																		1	ns	ns	ns	**	ns	**

TABLE 9.3   Continued

	A	B	C	D	E	F	G	H	I	J	K	L	M	N	O	P	Q	R	S	T	U	V	W	X
S																			1	ns	ns	ns	ns	ns
T																				1	ns	ns	ns	ns
U																					1	**	ns	ns
V																						1	ns	**
W																							1	ns
X																								1

* Correlation is significant at the 0.05 level (2-tailed).

** Correlation is significant at the 0.01 level (2-tailed).

ns Correlation is not significant at less than 0.05 level

A=EC, B=pH, C=Total solids, D=Turbidity, E=TSS, F=NH$_4$-N, G=NO$_3$-N, H=P, I=K, J=Na, K=Ca, L=Mg, M=CO$_3$, N=HCO$_3$, O=Cu, P=Zn, Q=Mn, R=Fe, S=Pb, T=Cr, U=As, V=Mo, W=SO$_4$, and X=Cl.

Mg is related to HCO_3, and heavy metals (Cu, Mn, Fe, and Mo). Carbonate is correlated ($P<0.05$) to chloride only. It shows the presence of Calcium chloride in wastewater. Metals were highly or moderately correlated. Like Cu was correlated to Mn, Fe, As, and Mo. Similarly Mn was correlated to Fe, As, Mo, and Cl ($P<0.01$). Ar and Mo are highly correlated but Mo is highly correlated to chloride only.

9.4 CONCLUSIONS

Utilization of treated municipal wastewater for irrigation is a viable alternative to protect environment, save freshwater and saving of nutrients for plant production. It is indispensable to study the impact of semi-arid climate on wastewater with respect to its quality parameters so that it can be used extensively for irrigation purpose in sustainable manner. The municipal wastewater passing through perennial drains was analyzed for different physiochemical properties (EC, pH, turbidity, total solids, NH_4-N, NO_3-N, P, K, Na, Ca, Mg, CO_3, HCO_3, SO_4, Cl and heavy metals (Cu, Zn, Mn, Fe, Pb, Cr, As, Mo). The spatio-temporal quality of wastewater shows acidic nature. Availability of major nutrients and micro elements in wastewater provides the opportunity for its use in irrigation to fulfill crop water and nutrient demand. It also provides the opportunity for alternative disposal of wastewater. Most of the water quality parameters concentration was higher during summer season, moderate during winter season and least during rainy season. Correlation study between quality parameters shows the presence of chloride with calcium and bicarcarbonate with calcium and magnesium. Presence of calcium, magnesium, bicarbonate and chloride play important role for the quantum of total solids in municipal wastewater. Heavy metals arsenic and molybdenum were highly correlated. Wastewater is under safe limit throughout year in terms of irrigation water quality indices SAR and Mg/Ca ratio. Cluster analysis is useful to provide information about seasonal distribution of water quality data and showing the cluster of months with similar quality. Information about different clusters, correlation among different water quality parameters and their seasonal variations provides information for integrated water resources planning to water resource planners and policymakers.

9.5 SUMMARY

Rapid urbanization causes increase in urban population. Over half of the world's population lives in cities. By 2050, seven out of every 10 people will be city dwellers. India is a part of this global trend. Nearly 28 percent of India's population lives in cities and this is expected to increase to 41 percent by the year 2020. Urban population and industries will generate huge amount of municipal wastewater. The promising alternative for disposal of wastewater is its utilization for irrigation after treatment. To utilize wastewater it is vital to generate the information about of different quality parameters and their variations due to impact of seasonal weather conditions. Physiochemical water quality parameters (EC, pH, turbidity, total solids, NH_4-N, NO_3-N, P, K, Na, Ca, Mg, CO_3, HCO_3, SO_4, Cl and heavy metals (Cu, Zn, Mn, Fe, Pb, Cr, As, Mo) of wastewater were determined for the period of one year. The data set is used to present spatial and temporal variations of the municipal wastewater quality. Identification of wastewater quality parameters responsible for temporal variations due to effect of semi-arid climate was done through multivariate cluster analysis. Correlation study between the identified parameters was also conducted. Wastewater was slightly acidic in nature with mean value of pH 6.87. Highest concentration was observed for total solids. Concentration of ammonical nitrogen was higher than nitrate nitrogen similarly bicarbonate concentration was higher in comparison to carbonate concentration. In the category of heavy metals highest concentration with mean value 0.98 mg L^{-1} was observed with iron and least with molybdenum with mean value 0.01 mg L^{-1}. Most of the water quality parameters concentration was higher during summer season, moderate during winter season and least during rainy season. Correlation study between quality parameters shows the presence of chloride with calcium and bicarbonate with calcium and magnesium. Presence of calcium, magnesium, bicarbonate and chloride play important role for the quantum of total solids in municipal wastewater. Heavy metals arsenic and molybdenum were highly correlated. Wastewater quality was under safe limit throughout year in terms of irrigation water quality indices SAR and Mg/Ca ratio. But it was under safe to moderate limit in terms of RSC index. Cluster analysis divides the months of a complete year in three clusters. First cluster have six months (July, August, September, October, November and December), second cluster have four months (January, February, March and April) and third cluster have two months (May and June).

KEYWORDS

- cluster
- cluster analysis
- correlation
- dendogram
- descriptive statistics
- heavy metals
- magnesium calcium ratio
- nutrients
- residual sodium carbonate
- semi-arid
- sodium adsorption ratio
- standard deviation
- standard error
- seasonal variation
- total solids
- turbidity
- wastewater
- water quality
- water quality indices

REFERENCES

1. APHA, (2005). Standard methods for the examination of water and wastewater (21st ed.). Washington, DC: American Public Health Association.

2. Akpan, A. (2004). The water quality of some tropical freshwater bodies in Uyo (Nigeria) receiving municipal sewage effluents, slaughter house washing and agriculture land drainage. *The Environmentalist*, 24, 49–55.

3. BIS, (1986). Guideline for irrigation waters. Bureau of Indian Standards, IS: 11624.

4. CPCB, (2005). Assessment of pollution – case study: Highlights. Ministry of Environment and Forest, Govt. of India, Delhi.

5. Das, J., Acharya, B. C. (2003). Hydrology and assessment of lotic water quality in Cuttack city, India. *Water, Air, and Soil Pollution*, 150, 163–175.

6. Dixon, W., Chiswell, B. (1996). Review of aquatic monitoring program design. *Water Research*, 30(9), 1935–1948.

7. Girija, T. R., Mahanta, C., Chandramouli, V. (2007). Water quality assessment of an untreated effluent imparted urban stream: The Bharalu Tributary of the Brahmaputra River, India. *Environmental Monitoring and Assessment*, 130, 221–236.

8. Goldman, R. C., Horne, J. A. (1983). *Limnology*. Book, New York: McGraw-Hill Company.

9. House, W. A., Denison, F. H. (1997). Nutrient dynamics in a lowland stream impacted by sewage effluent: Great of use, England. *The Science of the Total Environment*, 205, 25–49.

10. Jimenez, B., Asano, T. (2008). Water reclamation and reuse around the world. In: B. Jimenez and T. Asano (eds.) *Water Reuse: An International Survey of Current Practice, Issues and Needs*. IWA Publishing, London, pages 648.

11. Keraita, B., Jiménez, B., Drechsel, P. (2008). Extent and implications of agricultural reuse of untreated, partly treated and diluted wastewater in developing countries. *Perspectives in agriculture, veterinary science, nutrition and natural resources*, 3(58), 15.

12. Luiza, A., Alex, V., Reynaldo, L., Plinio, B., De Camargo, P. B. (1999). Effects of sewage on the chemical composition of Piracicaba River, Brazil. *Water, Air, and Soil Pollution*, 110, 67–79.

13. Neal, C., Jarvie, H. P., Williams, R. J., Neal, M., Wickham, H., Hill, L. (2002). Phosphorus-calcium carbonate saturation relationships in a lowland chalk river impacted by sewage inputs and phosphorus remediation: An assessment of phosphorus self-cleansing mechanisms in natural waters. *The Science of the Total Environment*, 282–283 and 295–310.

14. Natrajan, E., Yadav, B. R., Tomar, S. P. S., Chandrasekharan, H. (2004). Bio-Chemical characterization of ground water of IARI farm. *J. of Agricultural Physics*, 4(1–2), 40–43.

15. Ravindra, K., Ameena, M., Monika, R., Kaushik, A. (2005). Seasonal variations in physoico-chemical characteristics of River Yamuna in Haryana and its ecological best design use. *Journal of Environmental Monitoring*, 5, 419–426.

16. Richards, L. A. (1954). *Diagnosis and improvement of saline and alkaline soils*. USDA Handbook No. 60, pages 160.

17. Simeonova, P., Simeonov, V., Andreev, G. (2003). Water quality study of the Struma River Basin, Bulgaria (1989–1998). *Central European Journal of Chemistry*, 1, 136–212.

18. Singh, K. P., Malik, A., Mohan, D., Sinha, S. (2004). Multivariate statistical techniques for the evaluation of spatial and temporal variations in water quality of Gomti River (India), A case study. *Water Research*, 38, 3980–3992.
19. Sinha, A. K., Pande, D. P., Srivastava, K. N., Kumar, A., Tripathi, A. (1991). Impact of mass bathing on the water quality at the Ganga River at Houdeshwarnath (Pratapgarth) India: A case study. *The Science of the Total Environment,* 101(3), 275–280.
20. Tandon, H. L. S. (2005). *Methods of analysis of soils, plants, waters, fertilizers and organic manures.* Fertilizer Development and Consultation Organization, New Delhi, India. pages 204.
21. UNHSP, (2008). Global atlas of excreta, wastewater sludge, and biosolids management: moving forward the sustainable and welcome uses of a global resource. In: R. LeBlanc, P. Matthews and P. Roland (eds.), *UN-Habitat,* Nairobi, pages 632.
22. WHO, (2006). Guidelines for the safe use of wastewater, excreta and greywater. In: *Volume 2, Wastewater Use in Agriculture.* World Health Organization, Geneva, Switzerland.

PART III

WASTEWATER MANAGEMENT PRACTICES

CHAPTER 10

QUALITY OF MUNICIPAL WASTEWATER FOR MICRO IRRIGATION

VINOD KUMAR TRIPATHI, T. B. S. RAJPUT, NEELAM PATEL, and LATA

CONTENTS

10.1 INTRODUCTION

Worldwide, availability of good quality water for irrigation sector is expected to decline as the requirement of fresh water for all other sectors (domestic, industry, power, inland navigation, ecology) increases [29].

Reprinted from Vinod Kumar Tripathi, T.B.S. Rajput, Neelam Patel, and Lata, 2015. Quality of municipal wastewater for micro irrigation. Chapter 7, In: *Sustainable Micro Irrigation Management for Trees and Vines, volume 3* by Megh R. Goyal (ed.). Oakville, ON, Canada: Apple Academic Press Inc.

Therefore, it is greatly essential to reduce the fresh water consumption in irrigation sector by adopting efficient methods of irrigation and making reuse of waste water (WW) generated as byproduct from other sectors for irrigation. In India, WW generation from Class I and Class II cities are 38,254 million liters per day [10]. This huge quantity of WW gives opportunity for its reuse in agricultural sector to mitigate water demand for irrigation in water scares areas. Numerous groups have described future issues that must be addressed to ensure water quantity, quality, security and controlling emerging contaminants and health risk with protection of environment. There is an urgent need to focus on integrated management of WW use to ensure sustainability of water quality and quantity for future generation [26].

It was estimated that about 20 million hectares of land worldwide was irrigated with untreated WW [21, 27]. The use of untreated WW (or polluted water) poses risks to human health since it may contain excreta related pathogens (viruses, bacteria, protozoan and multi cellular parasites), skin irritants and toxic chemicals like heavy metals and pesticide residues. When WW is used in agriculture, pathogens and certain chemicals are the primary hazards to human health. The risk for human health is mainly with consumption of WW grown produce. Outbreaks of food borne illness throughout the world are increasingly linked to consumption of contaminated fruits and vegetables [8, 17, 18]. Bacterial human pathogens such as *Escherichia coli* O157:H7, *Salmonella* and *Listeria monocytogenes* have been demonstrated to be involved in such outbreaks of food borne illness [6, 7]. But WW reuse for agricultural purposes is now considered an important resource for either regions with high demand or low supply or areas vulnerable to macronutrients in several European countries [24].

Oron et al. [25] concluded that poliovirus can penetrate into the plant through the root system. Water as a medium plays a tremendous role in differential distribution of pathogens in soil and plant tissues. Vaz da Costa Vargas et al. [32] observed that when poor quality WW (trickling filter effluent with 10^6 thermotolerant coliforms per 100 mL) was used to spray-irrigate lettuces, the initial concentrations of indicator bacteria exceeded 10^5 coliforms per 100 g fresh weight. Once the irrigation ceased, no *Salmonella* could be detected after five days, and after

7–12 days, thermotolerant coliform levels were similar to or just above the level seen in lettuces irrigated with fresh water. The crop quality was better than that of lettuces irrigated with surface waters on sale in the local markets (10^6 thermotolerant coliforms per 100 g), presumably because of recontamination in the market through the use of contaminated water for spray of vegetables. The lettuces irrigated in uncovered plots had high level of bacterial contamination, unless a period of cessation of irrigation occurred 7–12 days before harvest. Islam et al. [20] observed no detectable populations at harvest for onions (day 140) but detectable populations at harvest for carrots (day 126). Pre-harvest contamination of carrots and onions with *E. coli* O157:H7 for several months can occur through contaminated manure (compost) and irrigation water. Hence, the type of crop, its texture and type of leaves/fruits can influence the retention of coliforms and their differential distribution.

Studies on drip and furrow irrigation of radishes and lettuces by Bastos and Mara (1995) with waste stabilization pond effluent (1.7×10^3 to 5.0×10^3 coliforms per 100 mL) indicated that crop quality was better under dry weather condition with 10^{3}–10^4 *E. coli* per 100 g for radishes and lettuces and no *Salmonella* was present. In Israel, Armon et al. [4] undertook an study where sprinkler irrigation of vegetables and salad crops with poor quality effluent from WW storage reservoirs (up to 10^7 thermo tolerant coliforms per 100 mL) resulted in high levels of fecal indicator bacteria on crop surface (up to 10^5 thermo tolerant coliforms per 100 mL). However, when vegetables were irrigated with better quality effluent (0–200 thermo tolerant coliforms per 100 mL) from a different storage reservoir, thermo tolerant coliform levels on crops were generally less than 10^3 per 100 g and often lower. Many research workers have focused their attention on survival of pathogens in irrigation water, soil and vegetable produced in different countries and climatic conditions [9, 11, 15, 22, 25, 30].

In India, municipal WW for irrigation is being used mainly for growing vegetables in periurban areas [33], which may pose serious risk of coliform outbreak [12]. Therefore, this research study investigates the possible accumulation of coliforms bacteria in soil under placement of drip laterals at surface and subsurface; and assesses the quality of eggplant fruits in terms of coliforms.

10.2 MATERIALS AND METHODS

10.2.1 EXPERIMENTAL SITE

The experiment was conducted at research farm of Water Technology Centre, Indian Agricultural Research Institute (IARI), Pusa, New Delhi, India which is located within 28°37'22" N and 28°39' N latitude and 77°8'45" E and 77°10'24" E longitude. The mean annual rainfall is 710 mm of which 75% is received during the monsoon season (July to September). WW used for irrigation of crops were collected from the drain of IARI, which is fed by domestic effluents from individual houses, group houses, hostels and runoff from agricultural field particularly in rainy season and sewage water [31]. The groundwater (GW) was collected from the tube well, which provides water from more than 30 m below the ground level. Location of WW collection point is shown in Fig. 10.1.

10.2.2 NURSERY RAISING AND CROP PRACTICES

Seedlings of eggplant (cv: *Supriya*) were raised in the month of September, in the plastic tray with the mixture of coco peat, vermiculite

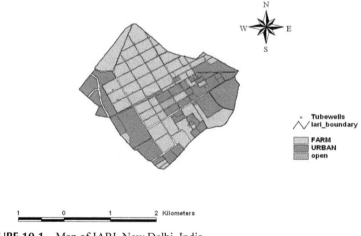

FIGURE 10.1 Map of IARI, New Delhi, India.

and perlite in the ratio of 3:2:1. Experiments were conducted during 2009–2010. WW was not used in the nursery. No Farm Yard Manure (FYM) was added to avoid precontamination with coliforms in the nursery. The 25 days old seedlings were transplanted in the field. Herbicides and pesticides were not applied. Water requirement of eggplant was estimated by calculating reference Evapotranspiration (ET_0) using the Penman-Monteith method and the crop coefficient (K_c) suggested by Allen et al. [2].

10.2.3 DESCRIPTION OF IRRIGATION SYSTEM

Drip irrigation system was installed for WW and GW (ground water) application separately. In-line lateral (J-Turbo Line) with 40 cm dripper spacing was laid on ground for surface and were buried at 15 cm and 30 cm depths from ground surface for subsurface drip. System included sand media filter (F1, flow rate 30 $m^3.h^{-1}$, 50 mm size, silica sand 1.0 to 2.0 mm, thickness 80 cm) with back flush mechanisms, Disc filter (F2, flow rate 30 $m^3.h^{-1}$, 20 mm size, 130 micron, disc surface 1.198 cm^2 screen surface 815 cm^2 AZUD helix system, model 2NR), and venturi injection system for chemigation. Water was passed through filter F1 and F2 alone as well as combination of both the filters (F1 and F2) to improve the quality of WW. The velocity of water was kept minimum to improve the efficiency of filters. Main lines (50 mm diameter, PVC pipe) were connected to submains (35 mm diameter, PVC pipe) for each of the plots through a gate valve.

10.2.4 SAMPLING OF WATER AND SOIL

WW samples from the drain were collected across the drain at the depth 15 cm below the surface and at three points, and then mixed. The preservation and transportation was performed according to the standard methods [3]. Soil samples were collected at time intervals of 25 and 50 days after transplanting and immediately after harvesting from each plot. Approximately 100 g of soil was aseptically collected in a sterile plastic bag from randomly selected plant at the depth of 0, 15, 30 and 45 depth for the surface and subsurface (15 cm) placed drip lateral. However, in case of subsurface drip (30 cm depth of lateral from surface), samples were collected up to 60 cm with the interval of 15 cm from ground surface. Fruit samples of eggplant were also collected randomly from each plot and transported in sterilized bags to the laboratory for analysis. All the soil samples were stored in refrigerator and analyzed within 48 h of collection.

10.2.5   DETERMINATION OF TOTAL COLIFORMS IN SOIL AND FRUIT SAMPLES

Total coliforms were analyzed by Most Probable Number (MPN) method and presented as per gram weight of dry soil/ fruit. The MPN values were determined by MPN table [1]. Soil/ water/ fruit samples were diluted tenfold. Graduated amounts of samples (10, 1 and 0.1 mL) were placed in 5–5 tubes of BCP (Bromo cresol purple) lactose broth with durhams tube. Five tubes for each dilution was incubated at 37°C for 24 h and individual tubes were checked for acid (yellow color) and gas production (Fig. 10.2). If no gas was present in any of the tubes, the incubation was continued for an additional 24 h.

FIGURE 10.2   Total coliform detection.

10.3   RESULTS AND DISCUSSIONS

10.3.1   CHEMICAL AND BIOLOGICAL PROPERTIES OF WATER

The WW and GW were analyzed for the physico-chemical and biological properties (Table 10.1). WW was highly turbid compared to GW but presence of total solids was higher in GW due to higher soluble salts. Higher EC, sodium and chloride content was observed in GW. Macronutrients N, P and K were found to be higher in WW. Available Mg was almost same in both the water samples. Carbonate content in WW and GW were 119 and 58 mg L^{-1} respectively. Population of total coliforms an indicator of fecal contamination was found to be in the range of 1.8×10^1 to 2.6×10^4 MPN mL^{-1}. No coliforms were detected in GW samples. Worldwide, research in 23 laboratories with 1000 strains of coliforms from various types of water has proven that only 61% of the total numbers examined were non-fecal in origin [14].

TABLE 10.1   Physicochemical and Biological Properties of Water Used For Irrigation

Properties	Unit	Waste water	Ground water
		Mean ±SD	Mean ± SD
EC	dS m^{-1}	1.48±0.23	2.17±0.25
pH		7.33±0.35	7.4±0.43
Total Solids	mgL^{-1}	849.8±148.5	967.4±212.6
Turbidity	NTU	46.25±10.23	1.50±0.52
NO$_{3-}$N	mgL^{-1}	4.57±1.91	5.22±0.44
P	mgL^{-1}	2.68±1.45	0.35±0.15
K	mgL^{-1}	26.83±14.78	10.3±2.98
Na	mgL^{-1}	139.91±37.4	287.8±62.4
Mg	mgL^{-1}	33.61±5.5	35.28±5.81
CO$_3$	mgL^{-1}	119.5±41.69	58.0±8.23
DO	mgL^{-1}	7.37±0.84	7.65±0.92
BOD$_5$	mgL^{-1}	95.17±19.71	0.725±0.339
COD	mgL^{-1}	139.25±30.7	16.67±4.03
Total coliforms	MPN mL^{-1}	2040± 1085	nd

nd = not detected.

10.3.2 EMITTER PERFORMANCE

Primary treatment of collected WW was done by sedimentation and filtration. WW was allowed to settle for 24 h and upper portion of settled water was used for filtration before application to plants. The coefficient of variation of emitter discharge (CV_q) for different filters and for their combination is presented in Table 10.2. Maximum CV_q's of 3.49% and 7.28% were observed with WW in surface and subsurface (30 cm) treatments, respectively. The performance of emitters under combination of both filters with WW and GW (ground water) was excellent with less than 5% of variation. The effect of chemical deposition in the emitters did not cause much variation in the emitter discharge.

10.3.3 COLIFORM POPULATION IN SOIL SAMPLES

Coliforms population was evaluated for eggplant crop (Fig. 10.3) at three stages of crop growth (initial, middle and maturity). The variation in population of coliforms was also quantified in soil samples collected from different depth of soil (Fig. 10.3). The variation was not detected in the soil irrigated with GW. However, the presence of coliforms was detected in the plots irrigated with WW. In soil, total coliform count increased up to middle stage of crop (50 days after transplanting) and

TABLE 10.2 Coefficient of Variation of Emitter Discharge Under Different Filtrations

Filter	Time	Lateral placement		
		Surface	15 cm	30 cm
Sand media	Beginning	1.78	1.78	1.78
for WW	End	2.36	6.08	7.28
Disk for WW	Beginning	1.29	1.29	1.29
	End	3.03	6.77	4.01
Sand media and	Beginning	1.26	1.26	1.26
disk for WW	End	3.49	4.57	3.60
Sand media and	Beginning	1.26	1.26	1.26
disk for GW	End	2.40	3.03	2.99

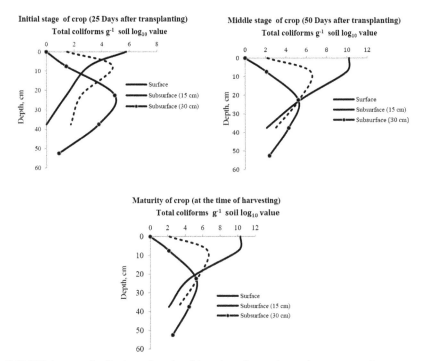

FIGURE 10.3   Distribution of total coliform in soil samples at three stages of crop.

then stabilized up to its maturity. This may be due to availability of limited nutrients in soil to maintain the threshold population of coliforms. Maximum population of coliforms (36.10×10^5, 2.01×10^{10}–2.03×10^{10} g^{-1} soil for initial, middle and maturity stages, respectively) was observed at soil surface.

Presence of fecal contamination in soil with the subsurface placement of drip lateral at 15 and 30 cm depth was also estimated at different crop stages by analyzing the soil samples collected from surface to 60 cm depth at an interval of 15 cm (Fig. 10.3). Subsurface drip irrigated soil showed the vertical distribution of coliforms bacteria in bell-shaped curve with maximum population adjacent to the lateral while lower population was observed below the lateral and higher population above the lateral. This may be due to bulk density and tilth of the soil. Interestingly no coliforms were observed at the surface of soil when placement of lateral was at 30 cm below the ground level. The results in this chapter are in agreement with research studies by Ijzerman et al. [19] and WHO [34]. Hassan et al. [16] also observed that the movement of water in unsaturated soil up to the depth of 30 cm was effective in the removal of fecal coliforms.

10.3.4 COLIFORM POPULATION IN EGGPLANT FRUIT

The Table 10.3 indicates data on the presence of total coliforms in eggplants fruits after washing and crushing. Maximum concentration of total coliforms was observed in crushed eggplant fruit. In general, the fruit after crushing showed higher concentration of coliforms compared to that seen in other treatments. This could be attributed to the contamination from soil. Fruit washing with sterilized water indicated higher contamination of coliforms in the treatments having lateral pipe on ground surface. It may be

TABLE 10.3 Coliform Population in Eggplant Fruit

S.No.	Treatment	Population of total coliforms (MPN g^{-1} $\log_{10}$ value)	
		Washing	Crushing
1.	WW through media filter and surface placement of drip laterals	0.12	1.21
2.	WW through media filter and subsurface placement of drip laterals at 15 cm depth	0.18	1.30
3.	WW through media filter and subsurface placement of drip laterals at 30 cm depth	nd	1.24
4.	WW through disk filter and surface placement of drip laterals	0.14	1.08
5.	WW through disk filter and subsurface placement of drip laterals at 15 cm depth	0.19	1.19
6.	WW through disk filter and subsurface placement of drip laterals at 30 cm depth	nd	1.23
7.	WW through media and disk filters and surface placement of drip laterals	0.19	1.18
8.	WW through media and disk filters and subsurface placement of drip laterals at 15 cm depth	0.14	1.26
9.	WW through media and disk filters and subsurface placement of drip laterals at 30 cm depth	nd	1.24

TABLE 10.3 Continued

S.No.	Treatment	Population of total coliforms (MPN g^{-1} $\log_{10}$ value)	
		Washing	Crushing
10.	GW through media and disk filters and surface placement of drip laterals	nd	nd
11.	GW through media and disk filters and subsurface placement of drip laterals at 15 cm depth	nd	nd
12.	GW through media and disk filters and subsurface placement of drip laterals at 30 cm depth	nd	nd
13.	After boiling	nd	nd

nd = not detected.

due to the spread of pathogens from surface through aerosol. No coliforms were observed in fruit washing, with subsurface placement of drip lateral at 30 cm depth. It may be due to the smooth surface of eggplant, that lets only minimum attachment with the skin of the fruit. In crushed fruit, maximum concentration of total coliforms were observed with subsurface placement of drip lateral at 15 cm depth. Fardous and Jamjoum [13] found high number of coliforms on the leaves of a corn plant irrigated with treated WW. In fruit crushing with subsurface placement of drip lateral at 15 cm depth have show maximum concentration of total coliforms. No coliforms were detected after boiling the fruit crush and wash sample. Availability of moisture may have prolonged the survival of bacteria or even allowed their regrowth. Kirkham [23] reported that pathogens may survive on the surface of a plant irrigated with WW because a warm, dark, and moist place could harbor bacteria. High levels of organic matter in treated effluent can also enhance the regrowth of bacteria [28]. WHO [34] recommended a bacteriological standard of 1000 fecal coliforms per 100 g of food/vegetable.

10.4 CONCLUSIONS

This research study examined the presence of coliform bacteria in soil and plant system of eggplant with the impact of municipal WW irrigation under surface and subsurface drip irrigation system. Maximum reduction in coliforms population was observed with sedimentation and combination of both filters. Total coliform count in soil at different depth from soil surface depends upon the placement of drip lateral. Maximum population of coliforms was observed at soil surface under surface irrigated drip system. Subsurface system with placement of drip lateral at 30 cm shows no coliform bacteria at soil surface. Higher concentration of total coliforms was observed in crushed eggplant fruit in comparison to washed fruit. No coliforms were observed in fruit wash with subsurface placement of drip lateral at 30 cm depth. Uncooked fruit grown under municipal WW is not advisable for consumption. It must be stressed that agricultural manipulation of the irrigation method described above can be used as an auxiliary means of public health protection.

Long-term study is suggested to evaluate the effects of wastewater on the survival of other beneficial soil microorganisms. Its impact on soil fertility should be assessed. There is also a need to develop suitable filtering mechanism so that transmission of harmful coliforms can be prevented to enter in drip irrigation system laterals and subsequently soil as well as plant system.

10.5 SUMMARY

Application of waste water (WW) with efficient irrigation methods is a viable option to protect the environment and mitigate the irrigation demand in arid and semi arid regions. A research study on vegetable crop eggplant (*Solanum melongena* cv. *Supriya*) with surface and subsurface drip system was conducted at PFDC field of Water Technology Centre, Indian Agricultural Research Institute, New Delhi, India. Irrigation was done with WW and it was fed to the drip system after 24 h of settlement of foreign material as primary treatment. Municipal WW was applied through sand media type filter, disk type filter and combined sand media and disk type filters, under surface and subsurface drip irrigation system.

Total coliforms population in WW was in the range of 7 $\log_{10}$ value. This population was reduced to 5 $\log_{10}$ after primary treatment (i.e., sedimentation and passing through the filters). Maximum population of coliforms 2.03×10^{10} MPN g^{-1} were observed in surface soil at maturity stage of crop growth. Presence of harmful pathogens on soil surface were not detected by placement of irrigation lateral at the depth of 30 cm under subsurface irrigation but in case of surface and subsurface (15 cm) population was 10 $\log_{10}$ value and 2 $\log_{10}$ value respectively on soil surface. In crushed fruit of eggplant, maximum concentration of total coliforms were observed with subsurface placement of drip lateral at 15 cm depth. But no coliforms were observed in fruit washing, with subsurface placement of drip lateral at 30 cm depth.

ACKNOWLEDGEMENTS

The authors are thankful for the financial support by the National Committee on the Plasticulture Applications in Horticulture (NCPAH), Department of Agriculture and Cooperation, Ministry of Agriculture, Government of India.

KEYWORDS

- coliforms
- crop growth
- drip laterals
- dripper
- eggplant
- emitter
- ground water, GW
- media and disk filters
- micro irrigation
- most probable number, MPN
- National committee on the plasticulture applications in horticulture India, NCPAH
- waste water, WW
- World Health Organization, who

REFERENCES

1. Alexander, M. (1982). Most probable number method for microbial population. Methods of soil analysis, Part 2. *Chemical and Microbiological properties-Agronomy monograph no. 9* (2nd edition), 815–820.

2. Allen, R. G., Pereira, L. S., Raes, D., Smith, M. (1998). *Crop evapotranspiration. Guidelines for computing crop water requirements.* FAO Irrigation and Drainage Paper 56, FAO, Rome, Italy.

3. APHA, (2005). *Standard methods for the examination of water and wastewater* (21st ed.). American Public Health Association, Washington, DC.

4. Armon, R., Dosoretz, C. G., Azov, Y., Shelef, G. (1994). Residual contamination of crops irrigated with effluent of different qualities: a field study. *Water Science and Technology*, 30(9), 239–248.

5. Bastos, R. K., X. Mara, D. D. (1995). The bacteriological quality of salad crops drip and furrow irrigated with waste stabilization pond effluent: an evaluation of the WHO guidelines. *Water Science and Technology*, 31(12), 425–430.

6. Beuchat, L. R. (1996). Pathogenic microorganisms associated with fresh produce. *Journal of food Protection*, 59, 204–216.

7. Brandl, M. T. (2006). Fitness of human enteric pathogens on plants and implications for food safety. *Annual Review of Phytopathology, 44, 367–392.*

8. Campbell, V. J., Mohle-Boetani, J., Reporter, R., Abbott, S., Farrar, J., Brandl, M. T., Mandrell, R. E., Werner, S. B. (2001). An outbreak of *Salmonella* serotype Thompson associated with fresh cilantro. *J. Infect. Dis.*, 183, 984–987.

9. Cools, D., Merckx, R., Vlassak, C., Verhaegen, J. (2001). Survival of *E. coli* and Enterococcus spp. derived from pig slurry in soils of different texture. *Appl. Soil Ecol.*, 17, 53–62.

10. CPCB, (2009). Status of water supply, wastewater generation and treatment in class-I cities and class –II towns of India. *Control of urban pollution, series: CUPS/ 70 / 2009–10.* Central Pollution Control Board (CPCB), Ministry of Environment and Forest, Government of India.

11. Deshmukh, S. K., Singh, A. K., Dutta, S. P., Annapurna, K. (2010). Impact of long-term wastewater application on microbiological properties of vadose zone. In: *Environmental Monitoring and Assessment.* Springer Netherlands, published online first, DOI 10.1007/s10661–010–1554–9.

12. Doyle, M. P., Erickson, M. C. (2008). The problems with fresh produce: an overview. *J. of Appl. Microbiol.*, 105, 317–330.

13. Fardous, A., Jamjoum, K. (1996). Corn production and environment effects associated with the use of treated wastewater in irrigation of Khirbet al-Samra Region. Ann. Rep., NCARTT, Amman, Jordan.

14. Gavini, F., Leclerc, H., Mossel, D. A. A. (1985). Enterobacteriaceae of the coliform group in drinking water: Identification and worldwide distribution. *Syst. Appl. Microbiol.*, 6, 312–318.

15. Gerba, C. P., Wallis, C., Melnick, J. L. (1975). Fate of wastewater bacteria and viruses in soil. *J. Irrig. Drain. E-ASCE IR3*, 157–174.

16. Hassan, G., Reneau Jr., R. B., Hagedorn, C. C. (2008). On-Site Waste Treatment and Disposal by Sequencing Batch Reactor – Drip Irrigation: Effluent Distribution and Solute Transport. *Communication in Soil Sci., Plant Analy.,* 39(1), 141–157.

17. Hedberg, C. W., Angulo, F. J., White, K. E., Langkop, C. W., Schell, W. L., Stobierski, M. G., Schuchat, A., Besser, J. M., Dietrich, S., Helsel, L., GriYn, P. M., McFarland, J. W., snd Osterholm, M. T. (1999). Outbreaks of salmonellosis associated with eating uncooked tomatoes: implications for public health. *Epidemiol. Infect.,* 122, 385–393.

18. Hilborn, E. D., Mermin, J. H., Mshar, P. A., Hadler, J. L., Voetsch, A., Wojtkunski, C., Swartz, M., Mshar, R., Lambert-Fair, M. A., Farrar, J. A., Glynn, M. K., Slutsker, L. (1999). A multistate outbreak of *Escherichia coli* O157, H7 infections associated with consumption of mesclun lettuce. *Arch. Intern. Med.,* 159, 1758–1764.

19. Ijzerman, M. M., Falkinham, J. O., Hagedorn, C. (1993). A liquid, colorimetric presence-absence coliphage detection method. *J. Virol. Methods,* 45, 229–234.

20. Islam, M., Doyle, M. P., Phatak, S. C., Millnerc, P., Jiangd, X. (2005). Survival of *Escherichia coli* O157, H7 in soil and on carrotsand onions grown in fields treated with contaminated manure composts or irrigation water. *Food Microbiol.,* 22(1), 63–70.

21. Jimenez, B., Asano, T. (2008). Water reclamation and reuse around the world. In: B. Jimenez and T. Asano (eds.) *Water Reuse: An International Survey of Current Practice, Issues and Needs,* IWA Publishing, London, pages 648.

22. Jimenez, B. (2003). Health risk in aquifer recharge with recycled water. In: Aertgeerts, R, Angelakis, A, eds. *State of the art report: health risk in aquifer recharge using reclaimed water.* Copenhagen, World health Organization Regional Office for Europe, pages 54–190 (Report No. EUR/03/5041122).

23. Kirkham, M. B. (1986). Problems of using wastewater on vegetable crops. *Hort. Sci.,* 21 (1), 24–27.

24. Kvanrstrom, E., Schonning, C., Carlson-Reich, M., Gustafson, M., Enockson, E. (2003). Recycling of wastewater-derived phosphorus in Swedish agriculture – A proposal. *Water Science and Technology,* 48(1), 19–25.

25. Oron, G., Goemans, M., Manor, Y., Feyen, J. (1995). Poliovirus distribution in the soil plant system under reuse of secondary wastewater. *Water Research,* 29(4), 1069–1078.

26. Rose, J. B. (2007). Water reclamation, reuse and public health. *Water Science and Technology,* 55(1–2), 275–282.

27. Scott, C. A., Faruqui, N. I., Raschid-Sally, L. (2004). Wastewater use in irrigated agriculture: Management challenges in developing countries. In: C. A. Scott, N. I. Faruqui, L. Raschid-Sally (eds) *Wastewater Use in Irrigated Agriculture: Confronting the Livelihood and Environmental Realities,* CABI Publishing, Wallingford, UK, pp 1–10.

28. Shatanawi, M. (1994). Minimizing environmental problems associated with the use of treated wastewater for irrigation in Jordan Valley: phase I. Technical Report No. 18. Water and Environ. Res., Study Center, University of Jordan, Amman, Jordan.

29. Singh, A. K., Minhas, P. S. (2011). Water management and environmental issues. In: *Souvenir, Seminar on water use in agriculture; challenges ahead.* Directorate of Water Management (ICAR), Bhubaneswar, India, 14–19 pages.

30. Strauss, M. (1985). Health aspects of night soil and sludge use in agriculture and aquaculture – Part II: Survival of excreted pathogens in excreta and faecal sludges. IRCWD News, 23, 4–9. Duebendorf, Swiss federal Institute for Environmental Science and Technology (EAWAG)/ Department of water and sanitation in developing countries (SANDEC).

31. Tripathi, V. K., Rajput, T. B. S., Patel, N. (2011). Hydraulic performance of drip irrigation system with municipal wastewater. *J. Agril. Eng.*, 48(2), 15–22.

32. Vaz da Costa Vargas, S, Bastos, R. K. X., Mara, D. D. (1996). Bacteriological aspects of wastewater irrigation. Leeds, University of Leeds, Department of Civil Engineering, Tropical Public Health Engineering (TPHE Research Monograph No. 8).

33. Water 21, (2010). *UNICEF survey finds half of water sources are polluted.* Water 21 Global News Digest Magazine, International Water Association, 27 April 2010.

34. WHO, (2006). *Guidelines for the safe use of wastewater, excreta and gray water.* Vol. II and IV. World Health Organization (WHO) Press, Geneva, Switzerland.

CHAPTER 11

BIOMETRIC RESPONSE OF EGGPLANT UNDER SUSTAINABLE MICRO IRRIGATION WITH MUNICIPAL WASTEWATER

VINOD KUMAR TRIPATHI, T. B. S. RAJPUT, NEELAM PATEL, and PRADEEP KUMAR

CONTENTS

Reprinted from Vinod Kumar Tripathi, T.B.S. Rajput, Neelam Patel, and Pradeep Kumar, 2015. Biometric response of eggplant under sustainable micro irrigation with municipal wastewater. Pages 319–330. In: *Sustainable Practices in Surface and Subsurface Micro Irrigation, volume 2.* by Megh R. Goyal. Oakville, ON, Canada: Apple Academic Press Inc.

11.1 INTRODUCTION

Worldwide, agriculture must be more proactive in managing its demand for water and improving the performance of both irrigated and rain-fed production. Availability of water for irrigation sector is declining because of increasing demand for water in domestic and industrial sector at significant rates. Production of more food to feed the burgeoning population is the big challenge. It is also vital to control the increasing prices of agricultural produce. There is a need to invest in both improved technologies and better management in order to achieve "more crops per drop." Water supply and water quality degradation are global concerns that will intensify with increasing water demand.

Worldwide, marginal-quality water is becoming increasingly important component of agricultural water supplies, particularly in water-scarce countries [18]. One of the major water resources having marginal-quality water is the municipal wastewater from urban and peri-urban areas. The wastewater has been recycled in agriculture for centuries as a means of disposal in cities such as Berlin, London, Milan and Paris [1]. However, in the recent years wastewater has gained importance in water-scarce regions. Italian legislation states that natural fresh water sources should be used as a priority for the municipal water supply, and that the recycling and reuse of water are viable alternatives for meeting industrial and agricultural needs. Wastewater reuse in agriculture requires best treatment practices, management practices and appropriate irrigation technology [8]. Wastewater treatment plants in most cities in developing countries are nonexistent or function inadequately [18]. In many cases, the quality standards for reclaimed wastewater are the same as for drinking water [7]. Therefore, wastewater in partially treated, diluted or untreated form is diverted and used by urban and peri-urban farmers to grow a range of crops [10, 15, 16].

In arid and semi arid developing countries, farmers are using municipal wastewater for irrigation by traditional surface irrigation methods (generally flood or furrow method). These methods require more water for irrigation; pose numerous problems of soil, water and environmental degradation compared to micro irrigation method. Main disadvantage of these methods are supply driven rather than crop demand driven, which cause mismatch between the need of crop and the quantity supplied [20]. Micro irrigation method may accomplish higher field level application efficiency of 80–90%, because surface runoff and deep percolation losses are minimized. Aujla et al. [4] observed a 4% higher yield of eggplant with drip irrigation compared to furrow irrigation by saving 50% water. Therefore, micro irrigation can help to give high crop yield per unit of applied water, and can allow crop cultivation in an area where available wastewater and fresh water is insufficient to irrigate with surface irrigation methods. It is prudent to reuse wastewater through subsurface micro irrigation method so that limited amount of wastewater can be applied below the ground to reduce contamination of the crop and environment. In this method, there is no risk of contamination through aerosol. Kiziloglu et al. [14], while conducting experiment with flood irrigation, concluded that primary treated wastewater can be used in sustainable agriculture in the long-term. Therefore, it is becoming increasingly important to adopt subsurface micro irrigation method [22].

Eggplant (*Solanum melongina* L.) is an extensively grown vegetable in the outskirt of Indian cities. Its annual production is 11.9 million tons that is 27.6% of the global production [11]. Al-Nakshabandi et al. [3] observed that average eggplant yield under treated effluent was twice the average eggplant production under fresh water irrigation using conventional fertilizer application in Jordan. The application of 50% water through micro irrigation can produce 4% higher eggplant yield compared to furrow irrigation with freshwater [4]. Eggplant yield with wastewater using surface irrigation methods has been studied by many researchers [3, 5, 12, 17]. Douh and Boujelben [9] and Cirelli [8] used wastewater through drip irrigation system. Most of the researchers have given emphasis on heavy metal contamination in the produce and much importance has not been given to subsurface micro irrigation. Subsurface micro irrigation with wastewater can play vital role in minimizing the contamination of the produce. According to the literature review, not enough information is available to evaluate the impact of subsurface micro irrigation. This chapter summarizes the effects of subsurface micro irrigation with wastewater on the biometric parameters of eggplant compared to surface micro irrigation.

11.2 MATERIALS AND METHODS

11.2.1 EXPERIMENTAL SITE

The experiment was conducted at the research farm of Precision Farming Development Center, Water Technology Center, Indian Agricultural Research Institute, New Delhi, India (Between latitudes 28°37′22″N and 38°39′05″N and longitudes 77°8′45″ and 77°10′24″E and AMSL 228.61 m) during November 2009 to May, 2010. January was the coldest month with a mean temperature of 14°C however; the minimum temperature dips to as low as 1°C. Frost occurs occasionally during month of December and January. The average relative humidity was 34.1 to 97.9% and average wind speed was 0.45 to 3.96 m/s.

11.2.2 SOIL CHARACTERISTICS

Soil samples were collected up to 60 cm soil depth with 15.0 cm intervals. Hydrometer method was used to determine the sand, silt and clay percentage of soil. The soil at the experimental site was deep, well-drained sandy loam soil comprising mean value 62% sand, 17% silt and 21% clay. The soil bulk density was 1.56 g.cm^{-3}. Field capacity and mean value of saturated hydraulic conductivity were 0.16 and 1.13 cm.h^{-1}, respectively.

11.2.3 EGGPLANT SEEDLINGS AND CROP PRACTICES

Seedlings for the eggplant (cv: *Supriya*) were raised in the plastic trays with the mixture of coco peat, vermiculite and perlite in the ratio of 3:2:1. Seeds were sown in November 2009 under polyhouse with partial ventilation. Light irrigation was provided frequently during warm, dry periods for adequate germination. A hand sprayer was used to spray fresh groundwater on the nursery. No wastewater was used during the growth period of nursery.

Twenty-five days old seedlings were planted in the field. Crop water requirement was met by estimating the reference evapotranspiration (ET$_0$) using the Penman-Monteith method and the crop coefficient (K$_c$) as suggested by Allen et al. [2]. The nutritional requirement (120 kg ha^{-1} N, 160 kg ha^{-1} P$_2$O$_5$ and 160 kg ha^{-1}K$_2$O) as suggested by Chadha [6] with freshwater has been suggested to decide the amount of the fertilizer to wastewater irrigated plots. Availability of nutrients in wastewater (average value: 28 mgL^{-1} N, 16 mgL^{-1} P, 28 mgL^{-1} ^{K}) was analyzed by taking the water sample during crop period. Nutrient application was done by deducting the available of nutrients in wastewater. Therefore, 64 kg ha^{-1} N, 132 kg ha^{-1} P$_2$O$_5$ and 96 kg ha^{-1} ^{K}$_2$O were applied through fertigation system.

The following treatments with different types of filter arrangement for micro irrigation systems were considered for irrigation with wastewater (WW) in this research:

- W1D1: WW through media filter and placement of drip laterals at soil surface;
- W1D2: WW through media filter and placement of drip laterals at 15 cm depth below ground surface;
- W1D3: WW through media filter and placement of drip laterals at 30 cm depth below ground surface;
- W2D1: WW through disk filter and placement of drip laterals at soil surface;
- W2D2: WW through disk filter and placement of drip laterals at 15 cm depth below ground surface;
- W2D3: WW through disk filter and placement of drip laterals at 30 cm depth below ground surface;
- W3D1: WW through media and disk filters and placement of drip laterals at soil surface;
- W3D2: WW through media and disk filters and placement of drip laterals at 15 cm depth below ground surface; and
- W3D3: WW through media and disk filters and placement of drip laterals at 30 cm depth below ground surface.

11.2.4 BIOMETRIC PARAMETERS OF EGGPLANT

The biometric parameters were leaf area index, root length, root density, yield and dry matter content were measured. All observations were made from center rows after border rows were discarded to avoid edge effects.

11.2.4.1 Leaf Area Index

Leaf area index (LAI) estimation included both assimilating area and growth. Observations for LAI were recorded during the crop season starting at 25 days after transplanting with an interval of 15 days in all treatments. LAI was determined by Canopy Analyzer (model: LAI-2000) in the experimental field. For crop production, leaf area per unit land area is more important than the leaf area of individual plants. Therefore, Leaf area index was calculated as follows:

$$LAI = \frac{LA}{LLA} \tag{1}$$

where, LAI = Leaf area index, LA = Leaf area, and LAA = Land area.

11.2.4.2 Root Length Density

Root sampling for determination of root length density was carried out during crop season starting at 25 days after transplanting with an interval of 15 days. It was done with an auger having an internal diameter 0.15 m to collect soil cores. Samples were collected at each treatment up to a depth of 45 cm. The samples were steeped and flushed prior to measure root length and root density. It was measured using root scanner (Epson Expression model LC 1600) of make (WinRHIZO 2002c). RHIZO system measured the root length by scanning the length of the root skeleton.

11.2.4.3 Yield and Dry Matter

Matured eggplant fruits were harvested manually in three stages at the interval 6–7 days as per availability of proper size. Weight of fruits was recorded for each treatment separately. Dry matter content is a measure of the quantity of total solids in fruit. It was determined by removing all the moisture from the fruit. Judgment of moisture removal was done by weighing the remaining solids with the help of digital electronic balance. Samples were kept in electric oven till it attained the constant weight. This was reported as a percentage weight of fresh fruit.

11.2.5   STATISTICAL   ANALYSIS

The experimental design was split plot randomized block design, where main plot was irrigation water at three levels and sub plot was irrigation systems, that is, placement of micro irrigation laterals at three levels (three depths). Each treatment was replicated thrice. Transformed data were analyzed by the General Linear Model (GLM) procedure of the SAS statistical software [19]. The Kolmogorov-Smirnov (K-S) statistics were used to test the goodness-of-fit of the data to normal distribution. To ensure that data came from a normal distribution, standardized skewness and kurtosis values were checked. All main-effects were compared by pair wise t-tests, equivalent to Fisher's least significant difference (LSD) test using the MEANS statement under GLM procedure with mean at $\alpha=0.05$.

11.3   RESULTS AND DISCUSSIONS

11.3.1   LEAF   AREA   INDEX

LAI was determined by leaf area meter and mean value of 20 plant sample are presented in Fig. 11.1. During initial 40–50 days after transplanting, LAI was higher under surface placed lateral treatments (W1D1, W2D1 and W3D1) in comparison to subsurface placed lateral treatments. After 25 days of transplanting, LAI under treatment W1D1 was significantly ($P<0.01$) different from W1D2 and W1D3, however, W1D2 and W1D3 were not significantly different ($P<0.05$). Similar trend was observed for water treatments (W2 and W3). After 60 days of transplanting increase rate of LAI for subsurface irrigated treatments (W1D2, W1D3, W2D2, W2D3, W3D2, and W3D3) was higher in comparison to surface irrigated treatments (W1D1, W2D1 and W3D1). After 100 days of transplanting, highest LAI of 4.23 was observed in treatment W3D2 but the lowest 3.12 was under treatment W1D1. After 115 days, almost constant value of LAI was observed in all the treatments showing no crop growth. It was also observed physically. At this stage, LAI of the surface irrigated treatments (W1D1, W2D1, W3D1) were significantly different at $P<0.01$ with subsurface irrigated treatments (W1D2, W1D3, W2D2, W2D3, W3D2 and W3D3).

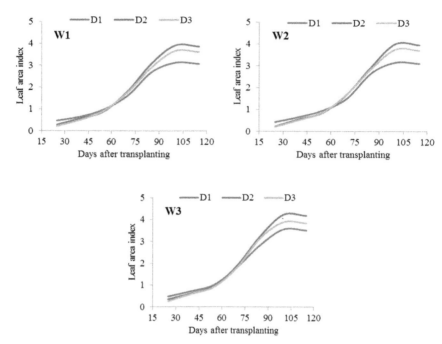

FIGURE 11.1   Temporal variation in leaf area index in different treatments of eggplant.

Subsurface treatments (W1D2, W2D2, and W3D2) at 15 cm were significantly different at $P<0.05$ compared to subsurface treatments (W1D3, W2D3, and W3D3) at 30 cm. Overall, higher LAI was observed with subsurface (15 cm) irrigated treatments. This may be due to more volume of soil under wet condition as compared with surface placed laterals.

11.3.2   ROOT LENGTH DENSITY (RLD)

Measurements of RLD and LAI were done on the same days i.e., starting from 25 days after transplanting till the last harvesting at an interval of 15 days (Fig. 11.2). During initial 40 days after transplanting, similar values of RLD were observed in all the treatments. Afterward the growth rate of roots was faster in the sequence of W1D3>W1D2>W1D1. Similar trend was observed with all the treatments with water, W2 and W3. This trend was continued till the last

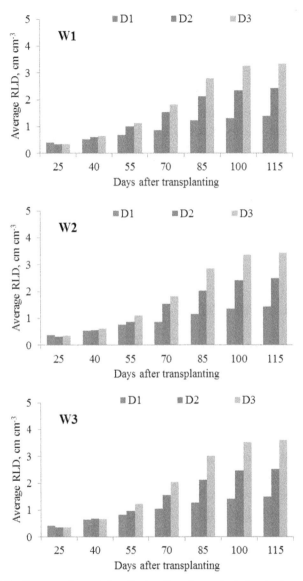

FIGURE 11.2 Increase in root length density of eggplant due to impact of lateral placement.

harvest of the crop in all the treatments. Highest RLD value of 3.6 cm cm^{-3} was observed in treatment W3D3. Under same depth of lateral placement, RLD values of 3.34 and 3.45 cm.cm^{-3} were observed in treatments W1D3 and W2D3, respectively. Minimum RLD value of 1.4 cm cm^{-3} was observed in treatment W1D1.

Higher root volume with subsurface placement of drip lateral at 30 cm (W1D3, W2D3, and W3D3) and lowest with surface place lateral (W1D1, W2D1, and W3D1) were observed visually. Based on RLD values, the treatments were significantly different ($P<0.01$) as per Student's t-test. Good relation between RLD and LAI was observed with the value of correlation coefficient 0.72. This may be due to lower flow rate in surface place lateral and smaller soil volume wetting. Yavuz et al. [23] also observed more pronounced emission uniformity reduction in surface placed laterals due to their exposure to higher temperatures in comparison with subsurface placed laterals, which achieved the best system performance in terms of emission uniformity and flow rate reduction.

11.3.3 EGGPLANT YIELD

In all the treatments, eggplant yield was higher than the national average yield of 20–25 Mgha^{-1}. Treatment wise average fruit yield was estimated by adding the weight of harvested fruit obtained in all three stages. Yield of 44.65, 45.96 and 45.32 Mgha^{-1} was recorded in treatments W1D2, W2D2, and W3D2 by subsurface placement of drip lateral at 15 cm depth, respectively. It was higher than the yield obtained in surface irrigated plots (Table 11.1). Minimum crop yield of 38.95 Mgha^{-1} was observed in the treatment W1D1 by surface placement of drip lateral. It may be due to lower flow rate in emitters causing water stress. These results agree with those obtained by Kirnak et al. [13] who demonstrated that water stress resulted in a reduction of fresh fruit yield and fruit size of eggplants. Moderate crop yield was observed in the treatments (W1D3, W2D3 and W3D3) by subsurface placement of drip lateral at 30 cm depth. The decline in crop yield for subsurface irrigation (30 cm) may be due to the fact that higher BOD$_5$ and COD values for WW reduce oxygen availability in deeper layer soil for respiration of roots. 50% decrease in soil respiration rate was observed by Singh et al. [21] in sewage water irrigated plots without drip system.

TABLE 11.1 Impact of Drip Lateral Placement on Yield and Dry Matter Content of Eggplant

Treatments	Average yield, Mg ha^{-1}	Average dry matter content, %
W1D1	38.95	8.68
W1D2	44.65	7.37
W1D3	43.26	7.92
W2D1	41.11	8.76
W2D2	45.96	7.14
W2D3	43.99	7.29
W3D1	41.46	8.51
W3D2	45.32	7.30
W3D3	44.57	7.04
LSD (P=0.05)		
Water (W)	1.19	0.46
System (D)	0.82	0.27
W×S	1.64	0.55

11.3.4 DRY MATTER (DM)

Average dry matter content for all the treatments was estimated and presented in Table 11.1. Maximum DM of 8.76% was recorded in treatment W2D1 and minimum of 7.04% with W3D3. DM in treatments with surface placed laterals was higher and significantly different than the subsurface irrigated treatments. Variation in the DM values for subsurface (30 cm) was in the range of 7.04 to 7.92% but for surface it was in the range of 8.51 to 8.76%. Lower wetting volume of soil with surface placed lateral treatments may lead to water stress. This may cause higher DM content. Kirnak et al. [13] demonstrated that water stress resulted in a reduction of fresh fruit yield and fruit size of eggplants, while it caused increases in the fruit dry matter content. An induced water shortage leads high dry matter percentages for the eggplants [8].

Statistical analysis with split plot design for fruit yield and DM content by ANOVA are presented in Table 11.2. ANOVA model for fruit yield and DM was valid with higher value of R^2 (0.86 for fruit yield and 0.94 for DM). Effect of water on yield was significant at

TABLE 11.2 Significance Level (p-Value) of the Statistical Model (Split Plot) and of Each Factor (Main Plot Water and Lateral Depth Subplot) and Interaction For Fruit Yield and Dry Matter Content

Parameter	Model	Main and interaction effect			t-grouping					
					Water (W)			Lateral depth (D)		
		W	D	W×D	W1	W2	W3	D1	D2	D3
Fruit yield	*** (R^2=0.86)	*	***	n.s.	B	A	A	B	A	A
Dry matter	*** (R^2=0.94)	***	***	***	B	A	A	A	B	C

n.s.: not significant, $P > 0.05$.
*$P<0.05$ **$P<0.01$ ***$P<0.001$.

$P<0.05$ but effect of lateral depth on yield was significant at $P<0.001$. Interaction effect was not significant on fruit yield. Higher least significant difference (LSD) values for the effect of water and lateral depth interaction can also be seen from Table 11.1. Subsurface depths D2 and D3 are in same group as per t-grouping. It may be explained that higher wetting volume of soil due to subsurface placement of lateral makes more availability of accumulated nutrients in the root zone. Individual and combined effect of water and lateral depth on DM content was significant ($P<0.001$) Different filtration system have significant role because emitter flow rate was varied with respect to time. As per t-grouping, DM was grouped separately for lateral placement depth. This shows subsurface drip can play a vital role for utilization of wastewater for irrigation.

11.4 CONCLUSIONS

The field study was conducted to evaluate eggplant performance through surface and subsurface micro irrigation system by using municipal wastewater. Higher LAI and RLD were observed in subsurface irrigated plots. Good correlation was also observed among LAI and RLD. Higher dry matter content was recorded in surface irrigated plots but higher fruit yield was recorded with subsurface irrigated plots by placing drip lateral at 15 cm depth. Irrigation with wastewater partially contributed (47%, 18%, and 40% of N, P_2O_5, and K_2O respectively) nutritional requirement of eggplant crop. With this field experiment, it can be concluded that subsurface micro irrigation might be a good alternative to use wastewater for production of eggplant. The only difficulty for subsurface micro irrigation is involvement of cost for placing drip lateral at appropriate depth. It increases the cost of cultivation. Basic information obtained about biometric response of eggplant may be used for future research.

11.5   SUMMARY

Utilization of municipal wastewater through subsurface micro irrigation reduces the environmental contamination. However, the impact of subsurface drip on biometric parameters of eggplant is a researchable issue. This study was conducted with wastewater. It was treated though three different types of filtration processes viz., media type, disk type and combined media and disk type filters. Eggplants were grown with wastewater. The root length density (RLD), leaf area index (LAI), fruit yield with their dry matter was recorded.

The data revealed that leaf area index was lower for subsurface drip during initial 55 days after transplanting but it was significantly higher in latter stage. Highest root length density 3.6 cm cm^{-3} was recorded under subsurface placement of drip lateral at 30 cm depth. Good relation between RLD and LAI was observed with the value of correlation coefficient 0.69. Highest dry matter content (8.76%) was recorded with surface placed lateral but highest fruit yield was recorded with the subsurface placed drip lateral at 15 cm depth. Subsurface micro irrigation through laterals placed at 15 cm and 30 cm depths resulted in 12% and 8.5% higher yield, respectively, in comparison to surface placed lateral. Utilization of wastewater gave the saving of 47%, 18%, and 40% of N, P_2O_5, and K_2O nutrients, respectively. Statistically the effect of filtered wastewater on the yield was significant. The findings of present study can be used for using wastewater through subsurface micro irrigation for protection of community living nearby wastewater-irrigated field.

KEYWORDS

- crop practice
- dry matter
- eggplant
- fruit yield
- leaf area index
- micro irrigation
- root length density
- statistical analysis
- subsurface drip irrigation
- wastewater

REFERENCES

1. AATSE (Australian Academy of Technological Sciences and Engineering), (2004). Water Recycling in Australia. AATSE, Victoria, Australia.
2. Allen, R. G., Pereira, L. S., Raes, D., Smith, M. (1998). *Crop Evapotranspiration. Guidelines For Computing Crop Water Requirements*. FAO Irrigation and Drainage Paper 56, FAO, Rome, Italy.
3. Al-Nakshabandi, G. A., Saqqar, M. M., Shatanawi, M. R., Fayyad, M., Al-Horani, H. (1997). Some environmental problems associated with the use of treated wastewater for irrigation in Jordan. *Agricultural Water Management*, 34(1), 81–94.
4. Aujla, M. S., Thind, H. S., Buttar, G. S. (2007). Fruit yield and water use efficiency of eggplant (*Solanummelongema* L.) as influenced by different quantities of nitrogen and water applied through drip and furrow irrigation. *Scientia Horticulturae*, 112(2), 142–148.
5. Brar, M. S., Arora, C. L. (1997). Nutrient status of cauliflower in soils irrigated by sewage and tubewell water. *Indian Journal of Horticulture*, 54(1), 80–85.
6. Chadha, K. L. (2001). *Handbook of horticulture*. Indian Council of Agricultural Research (ICAR), New Delhi, India.
7. Cirelli, G. L., Consoli, S., Di Grande, V. (2008). Long term storage of reclaimed water: the case studies in Sicily (Italy). *Desalination*, 218, 62–73.
8. Cirelli, G. L., Consoli, S., Licciardello, F., Aiello, R., Giuffrida, F., Leonardi, C. (2012). Treated municipal wastewater reuse in vegetable production. *Agricultural Water Management*, 104, 163–170.
9. Douh, B., Boujelben, A. (2010). Water saving and eggplant response to subsurface micro irrigation. *Agricultural Segment*, 1(2), AGS/1525.

10. Ensink, H. H., Mehmood, T., Vand der Hoeck, W., Raschid-Sally, L., Amerasinghe, F. P. (2004). A nation-wide assessment of wastewater use in Pakistan: an obscure activity or a vitally important one? *Water Policy*, 6, 197–206.

11. FAO, (2010). <http://www.fao.org/waicent/portal/statistics.en.asp>.

12. Irénikatché Akponikpè, P. B., Wima, K., Yacouba, H., Mermoud, A. (2011). Reuse of domestic wastewater treated in macrophyte ponds to irrigate tomato and eggplant in semiarid West-Africa: Benefits and risks. *Agricultural Water Management*, 98(5), 834–840

13. Kirnak, H., Tas, I., Kaya, C., Higgs, D. (2002). Effects of deficit irrigation on growth, yield, and fruit quality of eggplant under semiarid conditions. *Aust. J. Agric. Res.*, 53, 1367–1373.

14. Kiziloglu, F. M., Turan, M., Sahin, U., Kuslu, Y., Dursun, A. (2008). Effects of untreated and treated wastewater irrigation on some chemical properties of cauliflower (Brassica olerecea L. var. botrytis) and red cabbage (Brassica olerecea L. var. rubra) grown on calcareous soil in Turkey. *Agricultural Water Management*, 95, 716–724.

15. Lai, T. V. (2000). Perspectives of peri-urban vegetable production in Hanoi. Background paper prepared for the Action Planning Workshop of the CGIAR Strategic Initiative for Urban and Peri-urban Agriculture (SIUPA), Hanoi, 6–9 June. Convened by International Potato Center (CIP), Lima.

16. Murtaza, G., Ghafoor, A., Qadir, M., Owens, G., Aziz, M. A., Zia, M. H., Saifullah. (2010). Disposal and use of sewage on agricultural lands in Pakistan: A review. *Pedosphere*, 20(1), 23–34.

17. Papadopoulos, I., Chimonidou, D., Polycar, P., Savvides, S. (2013). Irrigation of vegetables and flowers with treated wastewater. Agricultural Research Institute, 1516 Nicosia, Cyprus. <http://om.ciheam.org/om/pdf/b56_2/00800185.pdf >.

18. Qadir, M., Wichelns, D., Raschid-Sally, L., McCornick, P. G., Drechsel, P., Bahri, A., Minhas, P. S. (2010). The challenges of wastewater irrigation in developing countries. *Agricultural Water Management*, 97(4), 561–568.

19. SAS, (2008). SAS/STAT® 9.2 User's Guide. Cary, NC: SAS Institute Inc.

20. Singh, D. K., Rajput, T. B. S. (2007). Response of lateral placement depths of subsurface micro irrigation on okra (*Abelmoschuse sculentus*). *International Journal of Plant Production*, 1, 73–84

21. Singh, P. K., Biswas, A. K., Singh, S. P. (2009). Effect of sewage water on heavy metal load on okra, spinach and cauliflower in vertisoil. *Indian J. Hort.*, 66(1), 144–146.

22. Turral, H., Svendsen, M., Faures, J. M. (2010). Investing in irrigation: Reviewing the past and looking to the future. *Agricultural Water Management*, 97(4), 551–560.

23. Yavuz, M. Y., Demirel, K., Erken, O., Bahar, E., Devecirel, M. (2010). Emitter clogging and effects on micro irrigation systems performances. *African J. Agricultural Research*, 5(7), 532–538.

CHAPTER 12

COST–BENEFIT ANALYSIS OF WASTEWATER REUSE IN MICRO IRRIGATION: A REVIEW

SUSHIL KUMAR SHUKLA and ASHUTOSH TRIPATHI

CONTENTS

12.1 INTRODUCTION

Water is one of the most important natural resource present on this holy Earth also known as blue planet. Out of total, 97.53% water is captured in ocean as saline water, and rest 2.43% water are freshwater on this earth surface, but all fresh water are not directly useful for human beings because 1.9% are locked up in ice and glaciers which is not directly useful for living beings. 0.51% fresh water is present in underground and rest 0.02% fresh water present on earth surface in river, lakes, streams etc. Living beings including humans requires water for their sustenance and performing various industrial, domestic and social activities. In spite of such a low presence of fresh water resource, human beings are recklessly exploiting the surface water as well as ground water resources without caring for the future generations to come.

Almost 70% of the fresh water is used in agricultural sector. Industrial sector has got a share of 20% and 10% of the water is utilized by residential units. Thus, it is clear that water scarcity has got direct effect on decline in agricultural productivity also [1]. We have to look for alternative sources of water for producing food grains to feed the increasing population. Wastewater reuse seems to be a better alternative. Along with industrialization and urbanization large volumes of wastewater are also being generated and their safe disposal is posing a bigger challenge to the environmentalists. The driving force behind conservation of water resources is water scarcity. Wastewater which is simply thrown out of the establishments could be diverted to agricultural fields for microirrigation. Wastewater reuse is already in practice since ancient ages. There are reports that wastewater was used back for irrigation in ancient Greece and in the Minan civilization (ca. 3000–1000 BC) [2, 3]. In the mid 1950's use of wastewater for several land application caught pace with improvement in wastewater treatment technologies. Land application became an economical substitute of discharging effluent into surface water bodies [4]. The present book chapter focuses mainly on the analyses of cost and benefit of reuse of wastewater through microirrigation.

In general, two major types of wastewater reuse have been developed and practiced throughout the world:

12.1.1 POTABLE USES

- direct, use of reclaimed water to augment drinking water supply following high levels of treatment.
- indirect after passing through the natural environment.

12.1.2 NON-POTABLE USES

- irrigated agriculture.
- use for irrigating parks, public places of forestry (fastest reuse application in Europe: Irrigation of golf courses).

- use for aquaculture.
- aquifer recharge (indirect reuse).
- or uses in industry and urban settlements.

Our main concern in current discussion is the wastewater reuse in microirrigation. The wastewater can be utilized for microirrigation, which is the backbone of developing countries like India. The wastewater being used for irrigation activity can be from different sources. The wastewater can be untreated municipal or industrial wastewater, mechanically purified wastewater or particularly or fully purified wastewater [5].

While considering wastewater reuse for agricultural activities points like advantages, disadvantages and possible harms related needs to be analyzed. The following table sums up the advantages, disadvantages and possible risks related to water conservation, diverse substances in the water and influences regarding the soil.

Wastewater is often associated with health and environmental risks and sometimes geared up supply of it to the agricultural field is also absent (Table 12.1). Thus a proper cost-benefit analysis of wastewater reuse is must prior commencement of microirrigation. Industrial wastewater has got a huge potential to be reused in irrigation of agricultural land. Apart from industries, municipal wastewater has also got tremendous usage for agricultural productivity. Among industries, wastewater from food processing industries has high value to be reused in agriculture since the main constituents are organic substances. It has been found that such wastewaters have no bad impact on the soil and crop productivity. Some of the food processing industries that have high potential wastewater reuse after

TABLE 12.1   Advantages, Disadvantages and Possible Risks of Wastewater Reuse

Advantages	Disadvantages	Risks
Enhancement of the financial efficiency of investments in wastewater disposal and irrigation	Wastewater is usually produced continuously throughout the year, whereas wastewater irrigation is mostly limited to the growing season.	Probable harm to groundwater due to heavy metal, nitrate and organic matter
Preservation of freshwater sources		
Revival of aquifers through infiltration water (natural treatment)		

TABLE 12.1 Continued.

Advantages	Disadvantages	Risks
Use of the nutrients of the wastewater (e.g., nitrogen and phosphate) ⇒ reduction of the use of synthetic fertilizer ⇒ improvement of soil properties (soil fertility; higher yields)	Several substances that can be present in wastewater in such concentrations that they are lethal for plants or lead to environmental harm	Impending damage to human health by spreading pathogenic germs
Decline of treatment costs: Soil treatment of the pre-treated wastewater via irrigation (no tertiary treatment necessary, highly reliant on the source of wastewater)		Probable harm to the soil due to heavy metal accumulation and acidification
Favorable influence of a small natural water cycle		
Fall in environmental shocks (e.g., eutrophication and minimum discharge requirements)		

treatment are: distilleries, brewery, fish processing, potato flour, sugar, starch, canning and dairy, etc.

For sustainable reuse of the wastewater two most important points needed to take care are quality of the effluent and the environmental risk associated with land application. Major sources of these wastewaters are industries and domestic sectors. These effluents contain several toxic, hazardous, organic, inorganic, degradable and non-degradable substances depending upon the type of industry from which they are produced. The main problem is disposal of such effluents into natural water bodies without being treated. Based on research it is found that sometimes domestic effluents are high potential of pollution as compared to industrial sector as they are under greater scrutiny. The domestic effluent contains a blend of chemicals at fluctuating concentrations, along with biodegradable and non-biodegradable solids. The nutrients as well as the contaminants present in the reused wastewater are absorbed by the soil and affects crop yield. Important criteria used for evaluation of the treated wastewater are as follows:

- salinity;
- heavy metals and harmful organic substances;
- pathogenic germs.

The aim of each wastewater reuse project is to protect health of public without discouraging reuse and wastewater treatment. It is the duty of regulatory bodies to maintain water quality standards in combination with requirements for treatment, sampling and monitoring. These guidelines are dependent on the kind of water use. Important water quality parameters and their significance are presented in Table 12.2.

TABLE 12.2 Physio-Chemical Parameters, Their Significance With Approximate Ranges For Treated Wastewater

Parameter	Significance	Approximate range in treated wastewater
Total Suspended Solids (TSS)	Can leads to sludge deposition and anaerobic conditions.	Less than 1–30 mg/L
Total Organic carbon (TOC)	Measures of organic carbon	1–20 mg/L
Degradable organics (COD, BOD)	Biodegradation can lead to oxygen depletion. Low to moderate concentrations of organics are required.	10–30 mg/L
Stable organics (e.g., phenols, pesticides, chlorinated hydrocarbons)	Some are toxic in the environment, accumulation processes in the soil.	
Nutrients N, P	When discharged into the aquatic environment they lead to eutrophication. In irrigation they are beneficial, nutrient source. Nitrate in excessive amounts, however, may lead to groundwater contamination.	N: 10 to 30 mg/L P: 0.1 to 30 mg/L
pH	Affects metal solubility and alkalinity and structure of soil, and plant growth.	

TABLE 12.2   Continued.

Parameter	Significance	Approximate range in treated wastewater
Heavy metals (Cd, Zn, Ni, etc.)	Accumulation processes in the soil, toxicity for plants	
Pathogenic organisms	Measure of microbial health risks due to enteric viruses, pathogenic bacteria and protozoa	Coliform organisms: <1 to 10^4/100 mL other pathogens: Controlled by treatment technology
Dissolved Inorganic (TDS, EC, SAR)	Excessive salinity may damage crops. Chloride, Sodium and Boron are toxic to some crops, extensive sodium may cause permeability problems	

The issue of microbial contamination is incredibly important in household sewage. The pathogens are not supposed to be entirely absent from industrial effluents and thus analyses of wastewater from pathogens contamination point of view are factually important. Usually the coliform bacteria are considered as indicator organism. World Bank, UNEP, UNDP and WHO in the year 1985 advocated permissible limits of coliform and *helminth eggs* to be less than 1000/100 mL and ≤1, respectively. The shielding of public health, especially that of workers and consumers, is one of the most important part of any water reuse project. To accomplish the objective, it is most significant to counteract or eradicate any pathogenic life form that may be present in the wastewater. For irrigation of non-food crops waste water can be reused after secondary treatment but for other applications, further disinfection by methods like chlorination or ozonation may be necessary. A case study of biological treatment of distillery effluent and its suitability for reuse in microirrigation is shown in the present discussion.

Distillery is one of the highly polluting and growth-oriented industries in India with respect to the extent of water pollution and the quantity of wastewater generated. In Distillery industries fresh water used in manufacturing of alcohol and generates huge amount of waste water which contains large amount of BOD, COD, TSS, important nutrient such as nitrate, phosphate, potassium etc.

If we develop cost effective treatment process for treatment of spent wash and utilize treated wastewater for microirrigation, it surely could be one of the promising techniques for conservation of freshwater and reduction of treatment cost. This chapter presents cost-benefit analysis of wastewater reuse in micro irrigation.

12.2   MATERIALS AND METHODS

12.2.1   WASTEWATER SOURCE

The anaerobically biodigested distillery effluent (ABDE) was obtained from a cane molasses-based Lord's Distillery Ltd. Nandganj, Ghazipur, U.P. India and stored at low temperature (3–4°C) in a deep freezer. DSW and BDE was characterized and analyzed for pH, total solids (TS), chemical oxygen demand (COD), biological oxygen demand (BOD), based on the standard methods for examination of water and wastewater [13] as shown in Table 12.3.

TABLE 12.3   Composition of Anaerobically Biodigested Distillery Effluent Obtained From Lords Distillery, Nandganj, Ghazipur, UP

Parameter	Composition
Color	Dark Brown
pH	7.65
Alkalinity (mg/L)	2300
Total solids (mg/L)	51100
Total dissolved solids (mg/L)	44600
Total Suspended solids (mg/L)	6500
BOD (mg/L)	6200
COD (mg/L)	42500

12.2.2 COAGULANT TREATMENT

Biodigested distillery effluent (BDE) was pretreated with coagulants such as Aluminum Chloride ($AlCl_3$), Ferric chloride ($FeCl_3.6H_2O$) and Potash Alum ($K_2SO_4.Al_2 (SO_4)_3.24H_2O$) with the optimum coagulant dose of 8 g, 3 g and 3.5 g in 100 ml respectively. pH of BDE was adjusted with 35% HCl. Coagulation was done in jar test apparatus (VELP Scientifica, Model JLT6, France). Coagulants were added to the effluent and flash-mixed for 2 min at 100 rpm and, thereafter, slowly mixed at 30 rpm for 30 min. The supernatant obtained after 1 h settling was used for fungal treatment.

12.2.3 MICROORGANISM AND INOCULUMS

The decolorization of BDE was carried out using *Aspergillus niger* ATCC No. 26550 and NCIM No. 684. The strain was obtained from NCL, Pune, It was maintained on PDA incubated at 30°C for 4–5 days and stored under refrigeration at 4–6°C [11].

12.2.3.1 Flask Cultures

Screening and optimization studies were performed in 250 mL shake flasks with 100 mL of BDE after coagulation. The different concentration of nutrients such as carbon, nitrogen, $MgSO_4$ sources is added. After sterilization in autoclave, flasks were inoculated with fungal pellets (10%). Flasks were then, kept inside the shaker at 180 rpm at 30°C. Samples were withdrawn every 24 h for observing maximum decolorization.

12.2.4 ADSORBENTS

Commercial activated carbon, ash from Naphtha cracking plant (available in Research Laboratory of the Department, procured from FCI Ltd Gorakhpur) and acid treated bagasse (Lord's Distillery Ltd, Nandganj, Ghazipur, UP, India) were selected as adsorbents. The activated carbon and naphtha ash were used as such after grinding and sieving to get required sizes (75, 66 and 53 μm, respectively).

The bagasse (53 μm) was acid treated using hydrochloric acid, sulphuric acid, nitric acid and phosphoric acid. The activation was carried out separately with sulphuric acid, phosphoric acid, nitric acid and hydrochloric acid. Bagasse was treated with acids in a ratio of 4:3. The acidified mixture was kept in an air oven maintained at 150 ±5°C for 12 hrs. The resulting char was washed with distilled water and then soaked overnight in 1% sodium carbonate solution to remove the residual acid. The soaked material was again washed with distilled water the following day and dried at 105°C for 24 hrs. Preliminary investigation revealed the superiority of phosphoric treated bagasse over other acid treated bagasse, so for further experimentation only phosphoric treated bagasse was used as adsorbent.

12.2.4.1 Adsorption Experiments

Batch adsorption studies were performed for the treatment of BDE using activated carbon, naphtha ash and acid treated bagasse. The experiments were carried out by taking varying amounts (1–7 g/L) of adsorbents in 100 ml sample of biodegraded BDE in 250 ml conical flasks mounted in a variable speed (50–200 RPM). The pH adjustment was carried out by addition of 0.1 N HCl and 0.1 N NaOH prior to the start of the adsorption experiments. Effect of temperature on adsorption was evaluated in the range of 25 to 45°C. The range of temperature was selected on the basis of exit temperature of the effluent coming out of the anaerobic digester of a distillery and the average ambient temperature prevalent in India.

The extent of adsorption was determined at different time intervals till the attainment of saturation. The suspension was taken out at the end of the reaction time, filtered in vacuum and the supernatant liquid was analyzed spectrophotometrically to estimate the residual adsorbate concentration.

12.2.5   DECOLORIZATION ASSAY

Samples amounting 5 ml were taken from shake flasks and centrifuged at 10,000 rpm for 10 minutes. The supernatant after centrifugation was diluted 10 times and used for color reduction measurement. The absorbance was measured at 475 nm using Systronics make double beam spectrophotometer (Model 2202). The decolorization yield was expressed as the percentage decrease in absorbance at 475 nm related to the initial absorbance at the same wavelength. All experiments were performed in triplicates and samples were withdrawn at regular interval of time (24 hr.) for decolorization measurements.

% decolorization = [(initial OD – final OD)/initial OD] × 100   (1)

12.3 RESULTS AND DISCUSSION

Batch studies were performed using *Aspergillus niger* for biodegradation of biodigested distillery effluent (BDE) procured from Lord's distillery Ltd. Nandganj, Ghazipur. The primary investigation revealed requirement of excessive dilution resulting in additional wastage of water. The maximum percentage decolorization thus obtained with *Aspergillus niger* was 63.5% at excessive dilution of 100%. Treating the BDE with different coagulants removed the excessive need of dilution. Depending upon its cost effectiveness and degradation ability Alum has been selected as the best coagulant. The filtrate after coagulation with alum was used for aerobic decolorization of BDE. The characteristics of BDE before pretreatment with alum are given in Table 12.4. Color removal of 92.5% and COD reduction of 78.5% was observed after alum treatment. These were due to precipitation of hydrocolloids and the dissolved matter.

12.3.1 AEROBIC DEGRADATION OF COAGULATED BDE

The filtrate of coagulated BDE was used for biological study. Various parameters influencing decolorization using *Aspergillus niger* were optimized. Throughout the experiments, small (2–6 mm diameter), compact and uniform pellets of the fungus were used. Best color removal was observed after 3–4 days in each case. The experiments were repeated thrice to validate the results. The remaining color and organic load have been reduced by aerobic degradation employing *Aspergillus niger*. The total decolorization obtained after complete treatment (fungal and coagulation)

TABLE 12.4   Typical Characteristics of BDE Before and After Coagulation

Parameters	BDE	Treated BDE		
		Potash Alum	**Aluminum Chloride**	**Ferric Chloride**
COD (mg/L)	42500	9450	14110	17425
BOD (mg/L)	6200	1705	2790	3410
Color	Blackish brown	Light yellow	Light brown	Greenish brown
pH	7.6	4.8	3.9	4.2

was 97.2%, and COD reduction was 96.2%, which indicates fungal treatment after pre-treatment with alum is a viable option for the treatment of anaerobically biodigested distillery Effluent (BDE) [15].

12.3.2   DECOLORIZATION OF BIODEGRADED BDE BY ADSORPTION

The aerobic biodegradation of coagulated BDE is unable to remove all the colors from BDE and produced water dose not meet the requirements of discharge on land and surface water-bodies. Therefore a further treatment is required before its disposal. In view of its simplicity and effectiveness, adsorption can be successfully employed as a polishing treatment process for decolorizing the effluent from biodegradation unit. Attempts were made to study the effectiveness of various adsorbents and optimize the relevant process parameters. Activated carbon, bagasse and naphtha ash were used as adsorbents. These results are presented and discussed in this section.

12.3.2.1   Characterization of Adsorbents

The physical and chemical characteristics of screened adsorbents are listed in Table 12.5. The fixed carbon content viz. the carbonaceous residue determined by subtracting the percentages of moisture, volatile matter, and ash was highest in naphtha ash and lowest in bagasse. Naphtha ash showed the highest bulk density among the three carbons.

TABLE 12.5 Characteristics of Adsorbents

Parameters	Adsorbent Samples		
	Bagasse (75 μm)	Activated carbon (75 μm)	Naphtha Ash (75 μm)
Moisture	15.4	10.6	7.8
Ash	10.5	4.1	2.6
Volatile matter	43.1	24.5	21.2
Fixed carbon	24.5	62.3	74.3
Bulk density (g/ml)	0.14	0.39	0.46
Surface area (m²/g)	926	1356	1422
pH	4.3	8.1	5.8

This is desirable, especially in column operations since adsorbent with high bulk density holds more adsorbate per unit volume and regeneration frequency is reduced [12]. Also, higher bulk density imparts more mechanical strength [16]. Apart from low fixed carbon, bagasse has the highest moisture among the three samples. High moisture content is not desirable as it dilutes the adsorption capacity of the adsorbent and, thus larger dosages would be required [16]. Furthermore, the highest ash and volatile matter contents was present in bagasse. Since ash is the residue that remains when the carbonaceous material is burned, the ash comprises inorganic constituents (mainly minerals such as silica, aluminum, iron, magnesium and calcium), associated with carbonaceous adsorbent. Bagasse had the lowest surface area (926 m²/g), which is considerably lower than surface area of the activated carbon and naphtha ash. In general, higher surface area is desirable since it increases the likelihood of adsorption on the outer surface by providing a very large interfacial area for adsorption [14]. In addition, there is also the possibility of intra-particle diffusion from the outer surface into the pores of the materials. That is why; color removal was significantly lower in case of bagasse than that obtained with commercial activated carbon and naphtha ash. This directly correlated to the comparatively lower surface area of bagasse due to two possible factors. First by low temperature acid carbonization is not followed by an additional activation step. Second, low temperature carbon is micro-porous with a low degree of meso-porosity [8]. This is a limitation since larger mean pore radius i.e., more mesopores would translate into higher adsorption capacity for larger molecules like melanoidins [6].

12.3.2.2   Effects of Adsorbent Dose

The effect of naphtha ash dose having different particle size, that is, 75, 66, and 53 µm was evaluated for the removal of color. Maximum color removal of 98% was obtained at the dose of 5 g/L of 53 µm naphtha ash. Decreased color removal efficiency with increasing particle size obtained through the experiment was due to decreasing surface area available for decolorization. Thus naphtha ash of 53 µm was used for further studies using a dose of 5 g/L. Finally it had been possible to conclude that through treatment of BDE using coagulation, aerobic biodegradation and adsorption nearly 99% decolorization and 98% COD reduction obtained which makes it fit to be used in microirrigation.

12.3.3   INTEGRATING PLANNING APPROACH OF AGRICULTURAL REUSE PROJECTS

One of the most significant wastewater reuse option is microirrigation in agriculture. It is a wastewater reuse option, which has to be included with wastewater reclamation, in an overall planning effort for public health protection, environmental pollution control, and water resources management [3]. It motivates the planning and management of agricultural reuse projects wants to consider institutional and legal, socio-economic, financial, environmental, technical and psychological aspects [10]. To get public approval and keenness to implement reuse projects, regulatory settings of directives can act as a tool. The alternative formulation and analyses should consider the following major feasibility criteria:

- Engineering feasibility.
- Economic feasibility.
- Financial feasibility: Development of a construction financing plan and revenue program.
- Institutional feasibility: Formal discussion with suppliers, wholesalers, retailers, and users of reclaimed water with the objective to reach an consent on legal and operational responsibilities.

- Environmental impact.
- Social impact and public acceptance.
- Market feasibility.
- Integrated approach involves the necessity of harmonization and collaboration between wastewater treatment agencies and the reclaimed water users. The proactive and participatory approaches between the water users associations need to be promoted. This might ensure safe and effective uses of effluent as well increase the reuse rate through reuse based on the demand [5].

12.3.4   ECONOMIC CONSIDERATIONS

To rationalize the benefits from implementation of wastewater reuse, a well-designed integrated planning process is obligatory. Description of the project, cost estimation, and identification of a potential reclaimed water market are essential ingredients of planning for wastewater reuse.

The social and economic benefits related to wastewater reuse for microirrigation needs to be assessed properly. Two terms economic and financial analyses are most often used in cost-benefit analyses for reuse of wastewater through microirrigation. The economic analyses is done to comprehend the social impact caused by wastewater reuse projects, whereas financial analyses are targeted on the local ability to raise money from project revenues, governmental grants, and loans to pay for the project.

A cost-benefit analysis of reuse operations is valuable. A water reuse project generates fiscal and non-fiscal benefits. Encouraging aspects related are:

- significance of water and nutrients;
- upgrading in the environment, for example, quality of receiving bodies;
- enhancement in public health;
- settlement for wastewater agencies and local authorities;
- decline of effluent discharge and preservation of discharge capacity;
- abolition of certain treatment processes to meet mass limits;
- trade of recycled water.

12.3.5 POSSIBLE UNENTHUSIASTIC ASPECTS

The unenthusiastic threats are aquifer pollution mainly by pathogens and organic trace elements, health risk due to infected crops, storage and transportation overheads, treatment costs, etc.

12.3.6 PUBLIC APPROVAL

The public acceptance is essential for initiating, implementing and sustaining long-term wastewater reuse program. Therefore, for sustainable water recycling scheme, proper understanding of the social and cultural aspects of water reuse is essential. In absence of public rejection a reuse project may not be successful. Even for non-potable reuse purposes, the public attitude plays an important role, including the perception of water quality, willingness to pay or to accept any wastewater reuse project [13].

By working on the public as well as on the institutional approval, it comes out that wastewater reuse has different driving forces. First it is an additional water supply in water sparse regions and second it can be a practical option to the disposal of treated effluents in surface water bodies. The use of alternative resolutions to the expulsion of wastewater in susceptible areas, where advanced tertiary treatment is not affordable, needs to be encouraged.

Keeping in mind, public acceptance and health issues a risk assessment must be part of the planning process. For example a watchful consideration of the degree of potential health risks involved in wastewater reuse for microirrigation is necessary. The amount of risks then might be weighted against necessity and derived benefits of the water reuse in order to make a sound conclusion on the project.

12.4 CONCLUSIONS

This review on cost-benefit analyses of wastewater reuse in agriculture through microirrigation shows that an incorporated planning approach, considering monetary as well as environmental and health issues, related to water reuse is essential to guaranty a triumph. Moreover it has been shown that the issue of wastewater reclamation is discussed and implemented all over the world. A case study of the distillery effluent from Lord's distillery Ltd. Nandganj, Ghazipur, U.P., India is also discussed. The satisfactory results obtained made it fit for reuse in microirrigation.

The adjustment of wastewater reuses for microirrigation to the local conditions should increase the profit and decrease the health dangers. Besides this will outcome in a higher public acceptance, which is crucial for execution of reuse project. Wastewater reuse in agriculture has been revealed as one significant management subject for sustainable use of the restricted freshwater reserve, next to demand oriented water distribution because of the potential financial and environmental profit. It is indispensable and important to commence and sustain wastewater reuse projects on a global scale, in view of the fact that our population and the food demand is mounting gradually, whereas water availability is limited.

12.5 SUMMARY

The present review includes reuse of wastewater in various sectors and its cost-benefit analyses focusing mainly on agriculture. Industrial as well as household effluents are two important sources for reuse of wastewater. While considering wastewater reuse for agricultural activities factors like advantages, disadvantages and potential harms related needs to be analyzed. A proper cost-benefit analysis of wastewater reuse is must prior commencement of microirrigation. For sustainable reuse of the wastewater two mainly vital points needed to take concern are quality of the effluent and the environmental threat associated with land use. Based on research it is found that sometimes domestic effluents are high prospective of pollution as evaluated to industrial sector as they are under greater inspection. The domestic effluent contains intermingle of chemicals at fluctuating concentrations, along with biodegradable and non-biodegradable solids. The endeavor of each wastewater reuse project is to safeguard wellbeing of public without dispiriting reuse and wastewater treatment. Parameters like Total Suspended Solids (TSS), Total

Organic carbon (TOC), Degradable organics (COD, BOD), Stable organics (e.g., phenols, pesticides, chlorinated hydrocarbons), pH, Heavy metals (Cd, Zn, Ni, etc.), Pathogenic organisms, Dissolved Inorganic (TDS, EC, SAR) etc. needs to be within the permissible level for safe reuse of wastewater.

A detailed case study of decolorization of Biodigested Distillery effluent (BDE) obtained from Nandganj distillery plant located at Ghazipur district in Uttar Pradesh was explained in the discussion. It was concluded that through treatment of BDE using coagulation, aerobic biodegradation and adsorption nearly 99% decolorization and 98% COD reduction was obtained which makes it fit to be used in microirrigation. Fungus *Aspergillus niger* was used as a microbial source for aerobic biodegradation and naphtha ash at dose of 5g/L was employed as adsorbent for the treatment of distillery effluent rendering it fit for microirrigation.

It is a wastewater reuse option, which has to be included with wastewater reclamation, in an overall planning effort for public health protection, environmental pollution control, and water resources management. It motivates the planning and management of agricultural reuse projects wants to consider institutional and legal, socio-economic, financial, environmental, technical and psychological aspects. Integrated approach involves the obligation of coordination and cooperation between wastewater treatment agencies and the reclaimed water users. The proactive and participatory approaches between the water users associations need to be promoted. This might ensure safe and effective uses of effluent as well increase the reuse rate through reuse based on the demand. The social and economic benefits related to wastewater reuse for microirrigation needs to be assessed properly. Two terms economic and financial analyses are most often used in cost-benefit analyses for reuse of wastewater through microirrigation. The economic analyses is done to comprehend the social impact caused by wastewater reuse projects, whereas financial analyses are targeted on the local ability to raise money from project revenues, governmental grants, and loans to pay for the project. The public approval is indispensable for initiating, implementing and sustaining long-term wastewater reuse program. Risk assessment considering public acceptance and health issues ought to be ingredient of the integrated planning process of reuse of wastewater.

KEYWORDS

- adsorption
- adsorbent dose
- aerobic degradation
- agriculture
- alum
- coagulant
- cost-benefit analysis
- decolorization
- distillery
- effluent
- economy
- industry
- integrated
- microirrigation
- reuse
- wastewater

ACKNOWLEDGEMENTS

The authors acknowledge: University Grants Commission (UGC, India) for the financial support; Combined laboratory facilities of Department of Chemical Engineering, School of Biochemical Engineering; and Collaborative and continuous support of Central University of Jharkhand and Amity University, Noida.

REFERENCES

1. Angelakis, A. N., Bontoux, L. (2000). Wastewater reclamation and reuse in Eureau countries. Water Policy Journal.
2. APHA-AWWA-WPCF, (1989). *Standard Methods for the Examination of Water and Waste Water*. 17th ed., Washington, DC.

3. Asano T., Levine A. D. (1996). Wastewater reclamation, recycling and reuse: Past, present, and future. *Water Science and Technology*, 33(10–11), 1–14.
4. Asano T. (1998). Wastewater reclamation and reuse. *Water Quality Management Library*, Volume 10, Technomic Publishing Company, Lancaster, PA, USA.
5. Bahri, A. (1999). Agricultural reuse of wastewater and global water management. *Water Science and Technology*, 40(4–5), 339–346.
6. Castro, J. B., Bonelli, P. R., Cerrella, for example, Cukierman, A. L. (2000). Phosphoric acid activation of agricultural residues and bagasse from sugar cane: Influence of the experimental conditions on adsorption characteristics of activated carbons. *Ind. Eng. Chem. Res.*, 39(11), 4166–4172.
7. Donta, A. A. (1997). *Der Boden als Bioreaktor bei der Aufbringung von Abwasser auf landwirtschaftlich genutzte Flächen*, Veröffentlichungen des Institutes für Siedlungswasser-wirtschaft und Abfalltechnik der Universität Hannover, Heft 100.
8. Girgis, B. S., Khalil, L. B., TawWk, T. A. M. (1994). Activated carbon from sugar cane bagasse by carbonization in the presence of inorganic acids. *J. Chem. Technol. Biotechnol.*, 61(1), 87–92.
9. HRH The Prince of Orange, (2002). No Water No Future: A Water Focus for Johannesburg. Contribution to the Panel of the UN Secretary General in preparation for Johannesburg Summit, <http://www.nowaternofuture.org/pdf/NW_NF_August2002.pdf>
10. Lazarova, V., Levine B., Sack J., Cireli, G., Salgot, M. (2000). Role of water reuse in enhancement of integrated water management in Europe and Mediterranean countries. 3rd International Symposium on Wastewater Reclamation, Recycling and Reuse; France 3.-7. July 2000.
11. Miranda, P. M., G. G. Benito, N. S. Cristobal, C. H. Nieto, (1996). Color elimination from molasses wastewater by *Aspergillus niger*. Biores. *Technol.*, 57, 229–235.
12. Ng, C., Losso, J. N., Marshall, W. E., Rao, R. M. (2002). Physical and chemical properties of selected agricultural byproduct-based activated carbons and their ability to adsorb geosmin. *Bioresource Technol.*, 84, 177–185.
13. Shahalam, A. B. M., Mansour, A. R. (1989). Modeling health risks associated with wastewater reuse in irrigation. *Journal of Environmental Science and Health*, A24(2), 147–166.
14. Shukla, A., Zhang, Y.-H., Dubey, P., Margrave, J. L., Shukla, S. S. (2002). The role of sawdust in the removal of unwanted materials from water. *J. Hazard. Mater.*, B95, 137–152.
15. Shukla, S. K., Mishra, P. K., Srivastava, K. K., Srivastava, P. (2010). Treatment of anaerobically digested distillery effluent by *Aspergillus niger*. The IUP *J. Chemical Engineering*, II(2), 7–18.
16. Sivabalan, R., Rengaraj, S., Arabindoo, B., Murugesan, V. (2002). Preparation and characterization of activated carbon from casurina seed and pinnaie seed coat. In: Trivedy, R. K. (Ed.), *Industry and Environment*. Daya Publishing House, New Delhi, pages 181–185.

APENDIX I

Point Source Releasing Industrial Effluent In Water Body.

A flood irrigated wheat crop (Source: http://en.wikipedia.org).

CHAPTER 13

EVALUATION OF MICRO IRRIGATION WITH MUNICIPAL WASTEWATER

VINOD KUMAR TRIPATHI, T. B. S. RAJPUT, NEELAM PATEL, and LATA

CONTENTS

13.1 INTRODUCTION

There is a great challenge ahead to produce more food for increasing population using from less and less water, because the demand of domestic and industrial water consumption is increasing. Agriculture sector is in

Reprinted from Vinod Kumar Tripathi, T.B.S. Rajput, Neelam Patel, and Lata, 2015. Evaluation of micro irrigation with municipal wastewater. Chapter 8, In: *Sustainable Micro Irrigation Management for Trees and Vines, Volume 3* by Megh R. Goyal, Oakville, ON, Canada: Apple Academic Press Inc.

completion with allocation of fresh water. Contrarily, increasing urbanization is resulting in increasing domestic wastewater (WW) generation.
Currently, partially treated and untreated WW is discharged into rivers or
lands causing various environmental concerns. On the other hand, WW
is beneficial, if it is scientifically used for irrigation as it can act as an
important source of water and nutrient [19]. Although WW has been used
to irrigate crops, rangelands, forests, parks and golf courses in many parts
of the world [1], yet unrestricted irrigation may expose the public to a variety of pathogens such as bacteria, viruses, protozoa, or helminthes and
exposure to heavy metals. The factors that influence the use of WW for
irrigation are: the degree of wastewater treatment, the crop type and its use
(e.g., human consumption or not, consumption after cooking, animal consumption fresh or sun-dried, etc.), the degree of contact with WW, and the
irrigation method. Therefore, it is preferable to have the irrigation method
having specific characteristics to minimize the various risks namely plant
toxicity due to direct contact between leaves and water; salt accumulation
in the root zone; health hazards related to aerosol spraying and direct contact with irrigators and product consumers; water body contamination due
to excessive water loss by runoff and percolation [18]. In this sense, use of
WW to agricultural crops through micro irrigation system is the safest way
to manage WW resource [4].

Micro irrigation system applies precise amount of water to the crop at
the right time and ensure its uniform distribution in the field. Although, it
is the most effective method for WW reuse, yet the suspended solids and
organic matter contained in WW can lead to a high risk of system failure
due to clogging of the drippers and inadequate filtering systems. These
risks depend on the level of treatment, the WW has undergone. Tertiary
treatment and chlorination have been found to be effective to reduce clogging caused by bacteria and algae, but, in most arid and semiarid developing countries and in small communities, extremely stringent quality
standards would lead to unsustainable costs [8]. In micro irrigation system, quality of water, emitter characteristics and filter efficacy would play
a key role in minimize clogging but other factors being same, most important feature for success with WW is filtration [15, 17].

In India mostly gravel media filter, screen filter and disk filters are used
to clean the water for micro irrigation system. Capra and Scicolone [7]

indicate that screen filters are not suitable for use with WW, with the exception of diluted and settled WW. They also observed almost similar performance in disk and gravel media filter with treated municipal WW. Besides, many researchers have conducted studies on WW using micro irrigation mostly by surface placement of lateral and mostly in laboratories [7, 9, 13, 22]. In India, subsurface micro irrigation has not been evaluated using WW. Therefore, there is a need to develop the methodology for using untreated WW through micro irrigation on sustainable basis. Therefore, this chapter discusses the research studies in realistic field situations using surface and subsurface drip systems with three kinds of filters to develop guidelines for using wastewater in micro irrigation.

13.2 MATERIALS AND METHODS

13.2.1 WATER RESOURCES

The field experiments were conducted at Precision Farming Development Centre of Water Technology Centre, IARI, Pusa, New Delhi during 2008–09 and 2009–10. Randomized block statistical design was used in field experiments. WW was collected from the drain passing through Indian Agricultural Research Institute (IARI). Water samples were analyzed for pH, electrical conductivity (EC), total Solids (dissolved and undissolved), turbidity, calcium, magnesium, carbonate, bicarbonate, total Coliform and *E. coli* according to the standard methods [2].

13.2.2 EXPERIMENTAL SET-UP

Micro irrigation system was installed for WW and GW separately (Fig. 13.1). In-line lateral (J-Turbo Line) with 40 cm emitter spacing was laid on the ground for surface drip and was buried at a depth of 15 cm from ground surface for subsurface drip irrigation. System included sand media filter (F1), disc filter (F2), and screen filter (F3). WW was allowed to pass through filter F1 and F2 singly as well as in combination of both the filters (F3). GW was also passed through combination of filters for comparison. Main line was connected to submains for each of the plots through a gate valve.

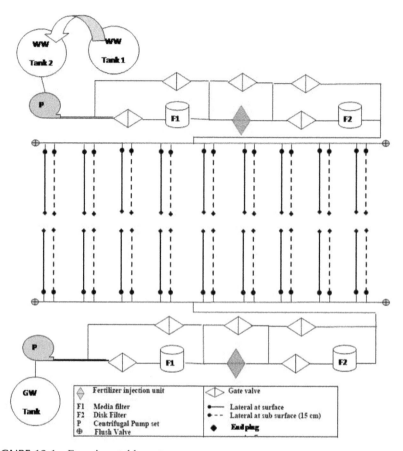

FIGURE 13.1   Experimental layout.

13.2.3   OPERATIONAL PROCEDURE

WW collected from the drain was stored in the tank one for settlement for 24 h so that all the suspended foreign particles were settled. After that the settled WW was transferred to tank two. This step was important to improve the quality of WW by reducing the suspended particles and to avoid frequent chocking of filters. Water from tank two was fed to the filtration system and then allowed to pass through emitters. The pump was turned on, and emitters were allowed to operate for approximately 2 min to allow air to escape. The water collection period was set at 5 min.

Quantity of flow of water from drip emitter was collected in containers at 98.06 kPa pressure and was repeated for three times. The flow rate was estimated by dividing total volume collected to the time of collection. The measurement was taken from randomly located sampling emitters to evaluate the performance evaluation of micro irrigation. Discharge from SDI laterals was measured by excavating the soil around the buried drip laterals so that an emitter is visible with sufficient space below it for placement of the container to collect discharged water from it as suggested by Camp et al. [6]; and Magwenzi, [14]. Performance of system was evaluated at normal operating pressure to discharge sufficient water for infiltration and to avoid ponding near the emitter. As per manufactures recommendation, operating pressure of 98.06 kPa was considered appropriate. To achieve accurate pressure, emitter level measurement was done at the lateral with digital pressure gauge having the least count of 0.01 kPa.

13.2.4 PARAMETERS FOR EVALUATION OF PERFORMANCE OF MICRO IRRIGATION

The parameters to evaluate the performance of the sustainable micro irrigation system were: Head-discharge relationship of emitters; irrigation uniformity, discharge variation, coefficient of variation and uniformity coefficient.

13.2.4.1 Head-Discharge Relationship of Emitters

A numerical description of pressure flow characteristics for a given emitter device is based on flow rate versus pressure curve described below:

$$q = CH^x \tag{1}$$

where, q = emitter flow rate (m³ s⁻¹); C = dimensional emitter coefficient that accounts effects of real discharge (l s⁻¹); H = pressure head in the lateral at the location of emitters (m); and x = exponent characteristic of the emitter (dimensionless). The exponent x indicates the flow regime and emitter type and typically ranges between 0.0 and 1.0. This exponent is a measure of flow rate sensitivity to pressure change. A higher value of x indicates higher sensitivity. The emitter exponent x and constant value C were derived using a linear regression equation: $(\text{Log } q) = (\text{Log } C + x \text{ Log } H)$, or $Y = mx + C$.

13.2.4.2   Coefficient of Variation

The coefficient of variation (CV_q) of the emitter discharge in the lateral was calculated [5, 23] using the following relation:

$$CV_q = \frac{SD}{q} 100 \qquad (2)$$

where, SD = standard deviation of emitter discharge (lph); and q = mean discharge in the same lateral (lph). Minimum CV_q was observed at 98.06 kPa pressure. Therefore, it was selected as the operating pressure for evaluation of clogging.

13.2.4.3   Emitter Flow Rate (% of Initial)

The emitter flow rate (% of initial) (R) is defined in Eq. (3):

$$R = \frac{q}{q_{ini}} 100 \qquad (3)$$

where, q = the mean emitter discharges of each lateral (lph); and q_{ini} = corresponding mean discharge (lph) of new emitters at the same operating pressure of 98.06 kPa.

13.2.4.4   Uniformity Coefficient

Uniformity coefficient, UC, is defined by Christiansen [10] as follows:

$$UC = 100 \left[1 - \frac{\frac{1}{n} \sum_{i=1}^{n} |q_i - q|}{q} \right] \qquad (4)$$

where, q_i = the measured discharge of emitter i (lph); q = the mean discharge at drip lateral (lph); and n = the total number of emitters to be evaluated.

13.2.4.5 Variation in Flow Rate (q_{var})

Emitter flow rate variation, q_{var} [23] was with the following equation:

$$q_{var} = \frac{q_{max} - q_{min}}{q_{max}} \tag{5}$$

where, q_{max} = maximum flow rate (lph); and q_{min} = minimum flow rate (lph).

13.2.5 STATISTICAL ANALYSIS

Statistical analysis was carried out using the GLM procedure of the SAS statistical package (SAS Institute, Cary, NC, USA). The model used for analysis of variance (ANOVA) included water from different filters and placement of lateral as fixed effect and interaction between filtered water and depth of emitter. The ANOVA was performed at probabilities of 0.05 or less level of significance to determine whether significant differences existed among treatment means.

13.3 RESULTS AND DISCUSSIONS

13.3.1 CHARACTERIZATION OF THE WASTE WATER

The physical, chemical and biological characteristics of WW and GW are presented in Table 1. It was observed that EC values for WW were lower than those for groundwater (GW). The EC values for GW varied from 1.89 to 2.58 dS m^{-1} with an average of 2.16 dS m^{-1}, and for WW it was in the range of 1.63 to 1.90 dS m^{-1} with a mean of 1.74 dS m^{-1}. Lower EC values indicate that salt content in the WW did not contribute much in chemically induced emitter clogging. Variation in pH values for WW was 6.60 to 7.32 with an average of 6.87 that was lower than GW (mean value 7.40 with range 6.95 to 8.57) indicating slight acidic nature of WW in comparison to GW. The pH may not have direct impact on clogging but it can accelerate

TABLE 13.1 Physicochemical and Biological Properties of Water Used For Irrigation

Properties	Units	Wastewater	Groundwater
		Mean ± SD	Mean ± SD
EC	dS m^{-1}	1.48±0.23	2.17±0.25
pH	—	7.33±0.35	7.4±0.43
Total Solids	mgL^{-1}	849.8±148.5	967.4±212.6
Turbidity	NTU	46.25±10.23	1.50±0.52
Ca	mgL^{-1}	82.16±19.99	44.58±8.27
Mg	mgL^{-1}	33.61±5.5	35.28±5.81
CO$_3$	mgL^{-1}	119.5±41.69	58.0±8.23
HCO$_3$	mgL^{-1}	415.27±69.7	364.33±70.7
Total coliforms	MPN mL^{-1}	41.7.7 172,437	nd

nd = not detected.

the chemical reactions or biological growth involved in clogging [12, 16]. Variation in values of total solids for WW was 733 to 1297 mg.L^{-1} with an average value of 989 mg.L^{-1} but for GW it was in the range of 800 to 1533 mg.L^{-1} with a mean of 967 mg.L^{-1}. Total solids (1533 mg.L^{-1}) were highest for GW in the month of May.

Turbidity for WW was always high and in the range of 33 to 68 NTU with a mean value of 55 NTU but GW had negligible turbidity levels with maximum of only 2 NTU. The WW contained surface runoff and foreign particles from anthropogenic pollution. Variation in calcium content of WW was from 68 to 136 mg.L^{-1} with a mean value of 94.6 mg.L^{-1} but for GW it was in the range of 36 to 66 mg.L^{-1} with an average of 45 mg.L^{-1}. Variation in magnesium content of WW was 25 to 38 mg.L^{-1} with mean value of 32 mg.L^{-1} but for GW it was in the range of 23 to 42 mg.L^{-1} with mean value of 36 mg.L^{-1}. Carbonate content of GW was in the range of 48 to 78 mg.L^{-1} with mean value of 58 mg.L^{-1} but for WW it was in the range of 12 to 78 mg.L^{-1} with an average value of 43.5 mg.L^{-1}. The variation in bicarbonate content of WW was 440 to 610 mg.L^{-1} with an average of 516 mg.L^{-1} but for GW it was in the range of 264 to 496 mg.L^{-1} with a mean of 364 mg.L^{-1}. Presence of carbonate for WW was less than GW but the range for WW was higher. It may be due to the reason that carbonate get converted into bicarbonate with the availability of other ions and variation in temperature. This also gives an indication of the presence of magnesium carbonate in GW and calcium carbonate in WW. Microbial contamination as indicated by total coliforms (mean value 2.0×10^7) was observed for WW only. Based on these quality parameters, it was concluded that clogging problem can be encountered more in WW than GW.

13.3.2 HYDRAULIC CHARACTERISTICS OF EMITTER

Coefficient for Q-H equation (Eq. (1)) decreased with the time of operation of emitters in all filtration systems as a result of partial clogging (Table 13.2). Theoretically the exponent for the emitter was 0.5, which comes under category of completely turbulent hydraulic regime [11]. In normal pressure range, exponent was more than 0.5 for gravel media filtered WW and less than 0.5 in case of disk filter. Performance of exponent was close to 0.5 in combination filter for both WW and groundwater. The coefficient of regression (R^2) was 0.99 in most of the situations indicating that the Q-H equation described the flow-pressure relationship precisely.

TABLE 13.2 Q-H Relationships For Emitter Under Different Filtration System With Wastewater and Groundwater

Filter	Placement of lateral	Stage	Coefficient, c	Exponent, x	R^2
Gravel media (F1)(Wastewater)	Surface	Beginning	3.768	0.521	0.99
		Middle	3.599	0.511	0.99
		End	3.484	0.534	0.99
	Subsurface	Beginning	3.768	0.521	0.98
		Middle	3.520	0.533	0.98
		End	3.455	0.533	0.99
Disk (F2) (Wastewater)	Surface	Beginning	3.538	0.485	0.99
		Middle	3.435	0.489	0.99
		End	3.368	0.486	0.99
	Subsurface	Beginning	3.538	0.485	0.99
		Middle	3.417	0.488	0.99
		End	3.345	0.485	0.99

TABLE 13.2   Continued

Filter	Placement of lateral	Stage	Coefficient, c	Exponent, x	R^2
Combination of F1 and F2 (with Wastewater)	Surface	Beginning	3.548	0.494	0.99
		Middle	3.476	0.492	0.99
		End	3.403	0.494	0.99
	Subsurface	Beginning	3.548	0.494	0.99
		Middle	3.465	0.494	0.99
		End	3.388	0.497	0.99
Combination of F1 and F2 (with Groundwater)	Surface	Beginning	3.548	0.494	0.99
		Middle	3.446	0.494	0.99
		End	3.350	0.493	0.99
	Subsurface	Beginning	3.548	0.494	0.99
		Middle	3.431	0.494	0.99
		End	3.350	0.493	0.99

13.3.3   COEFFICIENT OF VARIATION OF EMITTER DISCHARGE (CV_q)

The CV_q of the discharge for different filters and for the combination of both filters are presented in Fig. 13.2. After one year, maximum CV_q of 3.49 and 4.57% was observed with WW in surface and subsurface drip irrigation systems, respectively. As shown in Fig. 13.2 after one year in all filter condition, CV_q was less than 5%. Hence the performance can be rated as excellent [3]. After two years of experimentation, maximum variation of 10.16% was observed with disk filter in subsurface drip system. The performance of combination with filter for WW and GW was excellent with only 4% of variation in surface drip system. Maximum deviation of 6.46% was observed in subsurface drip system with both filter combination in GW. The results indicate that one-year operation of the emitters did not cause much variation but continuous two years of operation caused significant variation in emitter discharge. This is also supported by the computation of the standard error, which was lower in first year but was significantly higher in second year under all filters situation with both types of water. Coefficient of variation in subsurface condition was always poor than surface.

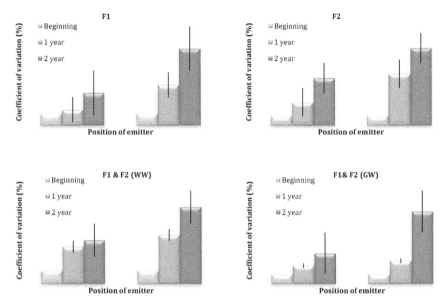

FIGURE 13.2   Coefficient of variation in emitter discharge under different filtration systems for wastewater and groundwater at 98.06 kPa pressure.

13.3.4   EMITTER FLOW RATE VARIATION

The study on flow rate variation was carried out with pressure variation so that flow rate reduction can be explained by clogging of emitters alone. Maximum reduction in flow rate was observed with gravel media filter (F1) and minimum with combination of both filters under WW. The clogging due to disk filter (F2) remained in between these two values (Fig. 13.3).

The results of the statistical analysis revealed that after 2 years of experiment, there was significant effect of filter, emitter placement and their interaction on the discharge of drip emitters (Table 13.3). In the beginning of experiment there was no significant effect of emitter placement and their interaction with filtered water because emitters were new and there was no clogging. After continuous use, clogging takes place and effect of different filtration system start showing up in the discharge of emitters. At the end of one-year effect of filtration system was significant but effect of emitter placement was not significant. Both were significant after two-year use. These results prove that clogging is a dynamic phenomenon over time [21].

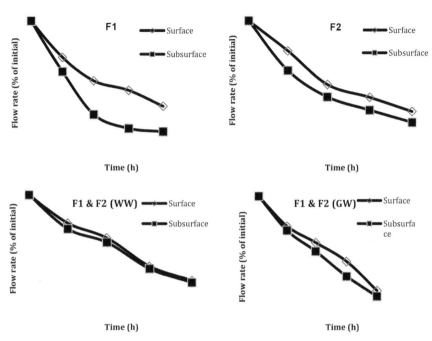

FIGURE 13.3 Emitter flow rate (% of initial flow rate) under different filtration system for wastewater and groundwater at 98.06 kPa pressure.

TABLE 13.3 Significance Level (P-Value) of the Statistical Model and of Each Factor and Interaction For Emitter Flow Rate

Parameter	Time		
	Beginning	**1 year**	**2 year**
Model	*** (R^2=0.97)	** (R^2=0.87)	*** (R^2=0.93)
Filter (F)	n.s.	**	***
Emitter placement (EP)	n.s.	n.s.	*
F × EP	n.s.	**	***

n.s.: not significant, $P > 0.05$; *: $P < 0.05$; **: $P < 0.01$; ***: $P < 0.001$.

13.3.5 UNIFORMITY OF WATER APPLICATION

Variations in uniformity coefficient and flow rate are presented in Table 13.4. Least variation in flow rate with maximum uniformity was observed at the beginning of experiment. The variation in flow rate increased with the operation of drip system and maximum variation with minimum uniformity coefficient

TABLE 13.4 Uniformity Coefficient and Variation in Flow Rate (q_{var}) Resulting From the Performance Evaluation of Micro Irrigation System

Filter	Depth of placement of lateral	Uniformity Coefficient			Variation in flow rate (q_{var})		
		Beginning	1 year	2 year	Beginning	1 year	2 year
Gravel media (F1)	Surface	98.59	96.26	94.84	0.049	0.106	0.171
	15 cm	98.56	95.24	92.05	0.049	0.120	0.219
Disk (F2)	Surface	98.89	96.52	94.73	0.032	0.090	0.117
	15 cm	98.91	95.00	93.15	0.032	0.106	0.181
Combination of F1 and F2	Surface	99.01	96.67	95.27	0.048	0.092	0.158
	15 cm	99.05	95.70	94.17	0.048	0.108	0.189
Combination of F1 and F2 with GW	Surface	99.07	97.29	95.55	0.048	0.102	0.131
	15 cm	99.02	96.02	93.50	0.048	0.100	0.160

were reached at the end of two years of experimentation. Performance of filter combination for both types of water could be rated as good [20]. After two years minimum uniformity coefficient under both filter combinations was 94.17 and 93.50 with WW and GW, respectively, for subsurface drip. As per general criteria for q_{var} values of 0.10 or less are desirable and 0.1 to 0.2 is acceptable and greater than 0.2 being unacceptable. Two out of three filtration systems (F2 and F3) gave variation in flow rate under acceptable limit.

13.4 CONCLUSIONS

The hydraulic performance of the drip emitters revealed that for continuous use of WW, filtration with a combination of gravel and disk filter would be most appropriate strategy against emitter clogging. It resulted in a better emitter discharge exponent, a reasonably good coefficient of variation and uniformity coefficient.

13.5 SUMMARY

Generation of WW in huge amounts is putting a lot of pressure to irrigation engineers for its safe reuse in agriculture. Though WW supports major and minor nutritional requirements of crops, but the presence of microbial contaminants and toxic elements in WW, limits its use. Utilization of WW for irrigation through micro irrigation system is the best choice to reduce the chances of contamination due to restricted quantity of application. Since clogging is the main problem associated with WW utilization through micro irrigation system, its remediation is required for enhanced utilization of WW through micro irrigation system. Physical and chemical characteristics of WW were determined and compared with GW.

While higher EC, pH, Mg, and carbonate were observed in GW but WW contained higher turbidity, total solids, HCO_3, and Ca. The population of total coliforms (2.72×10^4 to 5.2×10^7) and *E. coli.* (1.8×10^3 to 2.64×10^6) were detected in WW. The hydraulic performance of drip emitters was studied for two years (2009 and 2010) with municipal WW and groundwater (GW) using gravel media (F1), disk filter (F2) and combination of gravel and disk filters (F3). Filtration using F3 gave emitter discharge exponent close to 0.5 with R^2 value of 0.99. Emitter flow rate decreased with the increase in time of operation of the system. Coefficient of variation less than 4% with WW and GW showed excellent performance in surface placed drip lateral after two years of operation. After filtration with F3, coefficient of variation (CV_q) of 4.0% with WW and 6.46% with GW was observed under subsurface (15 cm deep) placement of lateral.

ACKNOWLEDGEMENTS

Authors are thankful to the National Committee on Plasticulture Applications in Horticulture (NCPAH), Department of Agriculture and Cooperation, Ministry of Agriculture, Government of India for providing the necessary funds to conduct this research.

KEYWORDS

- **coefficient of variation**
- **coliforms**
- **crop growth**

- dripper
- emitter
- emitter discharge
- emitter discharge exponent
- emitter flow rate
- emitter hydraulics
- emitter placement
- filter
- ground water, GW
- head-discharge relationship
- hydraulic coefficient
- micro irrigation
- National Committee on the Plasticulture Applications in Horticulture India, NCPAH
- uniformity coefficient
- waste water, WW
- water characterization
- World Health Organization, WHO

REFERENCES

1. Al-Jamal, M. S., Sammis, T. W., Mexal, J. G., Picchioni, G. A., Zachritz, W. H. (2002). A growth-irrigation scheduling model for wastewater use in forest production. *Agricultural Water Management*, 56, 57–79.
2. APHA, (2005). *Standard methods for the examination of water and wastewater* (21st ed.). Washington, DC: American Public Health Association (APHA).
3. ASABE, (2003). Design and installation of micro irrigation systems. In: EP405.1, *ASABE standards, 50th edition.* American Society of Agricultural and Biological Engineers: St. Joseph, Michigan.
4. Ayers, R. S., Westcot, D. W. (1991). *Water quality for agriculture. FAO Irrigation and Drainage paper* 29, Rome: FAO.
5. Bralts, F. V., Kesner, D. C. (1983). Micro irrigation field uniformity estimation. *Trans. ASAE*, 26, 1369–1374.
6. Camp, C. R., Sadler, E. J., Busscher, W. J. (1997). A comparison of uniformity measures for micro irrigation systems. *Trans. ASAE*, 40(4), 1013–1020.

7. Capra, A., Scicolone, B. (2004). Emitter and filter test for wastewater reuse by micro irrigation. *Agricultural Water Management*, 68(2), 135–149.
8. Capra, A., Scicolone, B. (2007). Recycling of poor quality urban wastewater by micro irrigation systems. *Journal of Cleaner Production*, 15, 1529–1534.
9. Cararo, D. C., Botrel, T. A., Hills, D. J., Leverenz, H. L. (2006). Analysis of clogging in drip emitters during wastewater irrigation. *Applied Engineering in Agriculture*, ASABE, 22(2), 251–257.
10. Christiansen, J. E. (1942). Hydraulics of sprinkler systems for irrigation. *Trans ASCE*, 107, 221–239.
11. Cuenca, R. H. (1989). *Irrigation system design: an engineering approach*. Prentice-Hall: Englewood Cliffs, New Jersey, pages 317–350.
12. Dehghanisanij, H., Yamamoto, T., Rasiah, V., Utsunomiya, J., Inoue, M. (2004). Impact of biological clogging agents on filter and emitter discharge characteristics of micro irrigation system. *Irrigation and Drainage*, 53, 363–373.
13. Liu, H., Huang, G. (2009). Laboratory experiment on drip emitter clogging with fresh water and treated sewage effluent. *Agricultural Water Management*, 96, 745–756.
14. Magwenzi, O. (2001). Efficiency of subsurface micro irrigation in commercial sugarcane field in Swaziland. <http://www.sasa.org.za/sasex/about/agronomy/aapdfs/ magwenzi.pdf>, pages 1–4.
15. McDonald, D. R., Lau, L. S., Wu, I. P., Gee, H. K., Young, S. C. H. (1984). *Improved emitter and network system design for reuse of wastewater in micro irrigation. Technical Report no 163*, Water Resources Research Centre, University of Hawaii at Manoa, Honolulu.
16. Nakayama, F. S., Bucks, D. A. (1991). Water quality for drip/trickle irrigation: a review. *Irrigation Science*, 12, 187–192.
17. Oron, G., Shelef, G., Turzynski, B. (1979). Trickle irrigation using treated wastewaters. *J. Irrig. Drain. Div.*, 105(IR2), 175–186.
18. Pereira, L. S., Oweis, T., Zairi, A. (2002). Irrigation management under water scarcity. *Agricultural Water Management*, 57, 175–206.
19. Pescod, M. D. (1992). *Wastewater treatment and use in agriculture*. FAO Irrigation and Drainage paper 47, Rome: FAO.
20. Puig-Bargues, J., Arbat, G., Barragan, J., Ramirez de Cartagena, F. (2005). Hydraulic performance of micro irrigation subunits using WWTP effluents. *Agricultural Water Management*, 77(1–3), 249–262.
21. Ravina, I., Paz, E., Sofer, Z., Marcu, A., Shisha, A., Sagi, G. (1992). Control of emitter clogging in micro irrigation with reclaimed wastewater. *Irrigation Science*, 13(3), 129–139.
22. Rowan, M., Manci, K., Tuovinen, O. H. (2004). Clogging incidence of micro irrigation emitters distributing effluents of different levels of treatments. Conference proceeding on On-Site wastewater Treatment, Sacramento, California, USA, 21–24 March, pp. 84–91.
23. Wu, I. P., Howell, T. A., Hiler, E. A. (1979). Hydraulic design of micro irrigation systems. *Hawaii Agric. Exp. Stn. Tech. Bull. 105*, Honolulu.

CHAPTER 14

PERFORMANCE OF MICRO IRRIGATED POTATO WITH TREATED WASTEWATER

H. A. A. MANSOUR and CS. GYURICZA

CONTENTS

14.1 INTRODUCTION

The automatic subsurface drip irrigation system may increase water use efficiency (WUE) due to reduced evaporation, drip irrigation is an application of water only near the root zone and the entire area is not wetted [6–8].

Modified and printed from Mansour, H. A. and Gyuricza, Cs., 2013. Impact of water quality and automation controller drip irrigation systems on potato growth and yield. *International Journal of Advanced Research*, 1(9):235–244. Open access article at: www.journalijar.com."

Sammis [12] compared sprinkler, surface drip, subsurface drip, and furrow irrigation systems for the production of potato and lettuce in New Mexico. Subsurface automatic drip irrigation (SDI) at suction of 20 kPa was among the most efficient irrigation systems. Shae et al. [13] studied four options for management of automatic drip irrigation of potatoes in North Dakota. Automation of the irrigation based on a soil water tension irrigation criterion of 30 kPa had relatively high WUE. Smajstrla et al. [17] compared automated SDI irrigation with the conventional semi-closed seepage sub-irrigation in Florida. The conventional irrigation system is under criticism because of surface runoff and nutrient contamination of adjoining waterways. The SDI system required more electrical energy but used 36% less water to achieve the same potato yield. Steyn et al. [18] examined irrigation-scheduling options for drip-irrigated potatoes.

Irrigation requirements vary with locations, soil types, and cultural practices. Under conditions of limited water supply, higher benefits may be achieved by adopting suitable irrigation and planting techniques [14]. Furrow and sprinkler irrigation methods are widely used in early potato production in Mediterranean cropping systems. Onder et al. [11] stated that automatic drip irrigation has not been widely used in potato production, because of higher initial cost of installation. In recent years, cost of installation has relatively decreased with improvement of technology. Also, the use of automation in drip irrigation has increased in most crop commodities, mainly for vegetables and fruits, to improve WUE and nutrition supply in Mediterranean cropping systems. Nowadays, subsurface automatic drip irrigation is also under evaluation to improve WUE, since water is getting scarcer and more valuable year by year. Therefore, the automation of drip irrigation will possibly be increased in early potato production in Mediterranean region. In the last few years, increasing number of UK growers have been carrying out small on-farm trials of trickle irrigation on potatoes [21]. There have been many reports on the effects of water stress and irrigation regimes on potato crop in Mediterranean-type environments as well as other parts of the world. However, not enough information is available on the WUE, growth, and yield of potato crop with on farm automatic surface drip and subsurface automatic drip irrigation systems.

For sprinkler-irrigated potato, extensive work has been done on potato response to N fertilizer and N losses, but relatively few studies have studied potato N fertilization and losses under automatic drip irrigation. Sprinkler irrigation at different irrigation criteria was compared to surface automatic drip and buried automatic drip irrigation (with a range of fertilization treatments), for potato yield in Minnesota [20]. Less water was required using either automatic drip irrigation system. Surface automatic drip and buried automatic drip systems were among the most productive systems for total and marketable yields. Furthermore, automatic drip irrigation or sprinkler irrigation (at relatively dry soil criteria) reduced nitrate leaching in potato field compared to normal sprinkler irrigation [19]. Neibling and Brooks [10] reported that reduced nitrogen rates did not affect potato yield, when irrigated with a subsurface automatic drip irrigation.

Simonne et al. [16] showed that automatic drip irrigation had potential as an economical viable potato production method in the southeastern United States. Optimized irrigation rates of 99 to 86% of the water were used in their irrigation model. Zartman et al. [22] examined tape depth installation and emitter spacing on tuber yield and grade of Norgold Russet potato in Lubbock, Texas. Tape depth or emitter spacing did not influence potato yield, but the proportion of misshaped tubers were greater when the tape was buried at 0.2 m than with shallower placement. Soil temperature was greater with the tape at 0.2 m than at 0.1 m or 0.025 m. DeTar et al. [2] found that automatic drip tape depths of 0.08 m (above the seed piece) and 0.46 m (below the seed piece) performed better than intermediate and greater depths. Fabeiro et al. [5] used 10 automatic drip irrigation treatments to examine the effects of the timing of irrigation deficits on potato yield and WUE in Spain. Irrigation deficits occurring during mid- and late-season tuber bulking were particularly damaging to yield. High yield combined with high WUE was observed when irrigation deficits restricted to early in the season. Shock et al. [15] investigated the performance of 'Umatilla Russet' under automatic drip irrigation in silt loam soil. The factors considered in the study were tape placement (one tape per row or one tape per two rows) and four soil water tension levels for automatically starting irrigation (15, 30, 45, and 60 kPa). They concluded that drip tape placement had a significant effect on every variable except total marketable yield and bud-end fry color for which interactions of irrigation criteria with tape number were significant. Tape placement and irrigation criterion interacted to influence total yield, total marketable potatoes, and US No. 2 yield. Results indicated potato should be irrigated at a suction of 30 kPa in silt loam soil to apply 2.5 mm water at each irrigation episode. The irrigation criterion only influenced the total US No.1 and over-340-g tuber size categories. Potato cultivars gave very different performance under automatic drip irrigation [4].

This chapter discusses performance of growth parameters, yield and water use efficiency of potato (*Solanum tuberosum* L. cv. Diamond) under Egyptian conditions under drip irrigation systems, that are operated by an automatic controller, using freshwater (FW) and agricultural treated wastewater (TAWW).

14.2 MATERIALS AND METHODS

During 2011 and 2012, field experiments were conducted in sandy soil at the Experimental Farm of National Research Center (NRC), El-Nubaria, El-Behira Governorate. Physical, chemical and hydraulic properties of the soil were determined according to standard methods. A compete randomized split-plot design was used to evaluate the following treatments:

- Installation depths of lateral drip lines:
    1. surface drip irrigation system with automatic controller at ground surface (DS);
    2. subsurface drip irrigation system with automatic controller 15 cm depth (DSS15);
    3. subsurface drip irrigation system with automatic controller at 30 cm depth (DSS30).

- Two irrigation water sources:
    1. freshwater (FW); and
    2. treated agricultural waste water (TAWW).

All treatments were replicated three times. All plots were irrigated at an interval of 3–4 days. Amount of irrigation water was measured with a flow-meter at the entrance of a submain to each plot. Irrigation operation was ended two weeks before initiation of harvesting of tubers. Leaf area (cm²) was measured by digital planomater. The LAI (Leaf area index) was calculated by dividing the total leaf area with the corresponding land area.

Each subplot consisted of six rows 4.2 m in width and 10.0 m in length (42 m²). The rows were spaced 0.70 m. On 5 March of 2011 and 25 February of 2012, the medium-early potato cultivar Diamond was hand planted at a row spacing of 0.25 m and 0.10 m depth. Plots were fertilized with 90 kg/ha of N, P, K before planting and an additional nitrogen dose of 60 kg per ha was side-dressed at the beginning of tuber bulking [11]. At preplant, drip lateral lines were installed at 0.15 and 0.30 m depths [1] in treatments DSS15 and DSS30. Cumulative evapo-transpiration (ET) and irrigation water requirement (IWR) of potato during growing season were estimated as 83.2 and 76.1 m³/ha using methods by Doorenbos and Pruitt [3].

The automatic controller system (Fig. 14.1) for drip irrigation has been described in detail by Mansour [8]. The soil moisture and temperatures were measured. The entire field was divided into small sections so that each section had one moisture sensor (densiometer) and one temperature sensor (RTD PT100) as shown in Fig. 14.2. These sensors were buried in the ground at the recommended depth. Once the soil reached the adequate

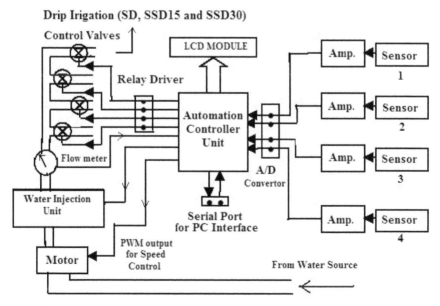

FIGURE 14.1 Automatic controller unit for drip irrigation [8].

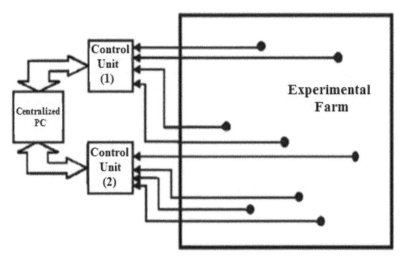

FIGURE 14.2   Layout for automatic controller unit in the field [8].

moisture level, the sensor sent a signal to the micro controller to turn off the relays, which control the valves of surface and subsurface automatic drip irrigation systems.

Surface automatic drip irrigation lines of 16 mm in diameter were installed after the emergence. The drippers were built-in line type (GR) with a flow rate of 4 lph at 1.0 bar. The DSS15 and DSS30 treatments received 1000 lph of water per 100 m of hose at 1.0 bar pressure. Dripper spacing was 0.20 m in DSS15 and DSS30 drip methods. One drip line for each crop line was used in the experiment. The middle four rows in each plot were hand harvested on 14 June of 2011 and 9 June of 2012, respectively.

The number of tubers per plant, tuber yield per plant, and mean tuber weight (g) were determined from 10 randomly selected plants in each subplot. After harvesting, tubers of each plot were graded into three size categories (>45, 28–45 and <28 mm unmarketable) and were weighed. The potato yield of unmarketable tubers was generally very low and was excluded from the data [9]. This classification has also been used by the Egyptian growers that processed the early potato.

The data were subjected to analysis of variance (ANOVA) using M-Stat. The significant differences between treatments were determined using least significant difference.

14.3   RESULTS AND DISCUSSION

14.3.1   GROWTH PARAMETERS

Data tabulated in Table 14.1 is plotted in Figs. 14.3–14.6, for effects of fresh water (FW), treated agricultural wastewater (TAWW), and three automatic drip irrigation systems on number of plants per square meter, number of branches and plant length.

Number of plants per square meter is shown in Table 14.1 and Fig. 14.3. During the first season of 2011, with FW, the highest value was 24.7 under DS, followed by 24 under SSD15. The lowest value was 23.2 under SSD30. Whereas with TAWW, the highest value was 25.5 under DS, followed by 23.8 under SSD15. The lowest value was 23.8 under SSD30.

TABLE 14.1   Effects of Water Quality, Automatic Surface and Subsurface Drip Irrigation Systems on Vegetative Growth of Potato

Year	Water quality	Irrigation system	Number of Plants/m2	Number of branches	Plant length, cm
2011	FW	DS	24.7	3.3	45.3
		DSS15	24.0	3.5	45.6
		DSS30	23.2	3.6	46.3
	TAWW	DS	25.5	3.7	46.3
		DSS15	24.4	3.8	46.5
		DSS30	23.8	3.9	46.6
		Mean	24.3	3.6	46.1
2012	FW	DS	25.6	3.4	45.2
		DSS15	24.8	3.6	45.4
		DSS30	24.5	3.7	46.8
	TAWW	DS	26.2	3.8	46.4
		DSS15	25.3	3.9	46.8
		DSS30	24.8	4.2	47.3
		Mean	25.2	3.8	46.3
	LSD at P = 0.05		0.07	0.06	0.04
	Interactions		0.06	0.04	0.05

DS: Surface drip, DSS15: subsurface drip with lateral at 15 cm, DSS30: sub-surface drip with lateral 30 cm. FW: Fresh water and TAWW: treated Agricultural wastewater.

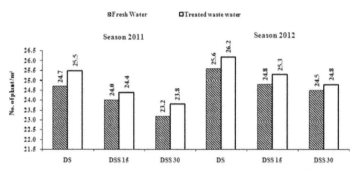

FIGURE 14.3 Effect of water quality and three drip irrigation systems on number of plants per square meter of potato.

The differences were significant at 5% level in number of plants per square meter among irrigation systems. During the second season of 2012, with FW, the highest value was 26.6 under DS, followed by 24.8 under SSD15. The lowest value was 24.5 under SSD30. Whereas with TAWW, the highest value was 26.2 under DS, followed by 25.3 under SSD15. The lowest value was 24.8 under SSD30. The differences were significant at 5% level in number of plants per square meter among irrigation systems.

The increase in number of plants per square meter under FW during the 2nd season of 2012 were 3.5, 3.2 and 5.3% in each of three irrigation systems. While with TAWW, values were 2.7, 3.6 and 4.0% compared to season 2011 under DS, DSS15, and DSS30, respectively.

With FW, the mean of number of branches (Fig. 14.4) was 3.3, 3.5, and 3.6 (season 2011) and 3.4, 3.6 and 3.7 (season 2012) under DS, DSS15, and DSS30, respectively. In both seasons, the difference in leaf dry weight between any two irrigation systems were significant at 5% level. During

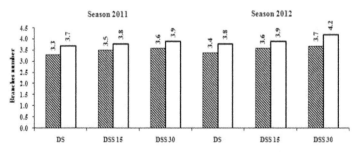

FIGURE 14.4 Effects of water quality and three drip systems on number of branches.

season 2012, the increases in number of branches with FW was 2.9, 2.8 and 2.7%. Whereas with TAWW, values were 2.6, 2.6 and 7.1% compared to season 2011 under DS, DSS15, and DSS30, respectively.

Plant length under FW (Fig. 14.5) was 45.3, 45.6, 46.3 in 2011 and 45.2, 45.4, 46.8 cm in 2012 for each irrigation system. Whereas under TAWW (Fig. 14.5), values were 46.3, 46.5, 46.6 in 2011 and 46.4, 46.8, 47.3 in 2012 by using DS, DSS15, and DSS30. The increase in mean of plant length (cm) was 0.4% in 2nd season compared to 1st season. The differences in plant length factors and treatments under study were significant at 5% level in both seasons.

The production yield of potato in the both seasons is shown in Fig. 14.6. Under FW during the first season, the highest value by using DSS30 was 10.7 ton/ha, followed by 9.3 ton/ha in DSS15, while the lowest value of

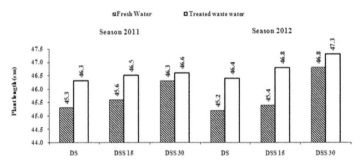

FIGURE 14.5   Effects of water quality and three drip systems on plant length.

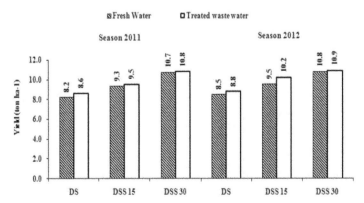

FIGURE 14.6   Effects of water quality and three drip systems on potato yield.

potato yield was 8.2 ton/ha under DS. On the other hand during the second season, potato yield was highest (10.5 ton/ha) under DSS30, followed by 9.5 ton/ha under DSS15, while the lowest value of potato yield was 8.5 ton/ha under DS.

Under TAWW during the first season, the highest value by using DSS30 was 10.8 ton/ha, followed by 9.5 ton/ha in DSS15, while the lowest value of potato yield was 8.6 ton/ha under DS. On the other hand during the second season, potato yield was highest (10.9 ton/ha) under DSS30, followed by 10.2 ton/ha under DSS15, while the lowest value of potato yield was 8.8 ton/ha under DS.

The differences in yield between factors and any two-irrigation systems were significant at 5% level during the two seasons. The results are in agreement with Simonne et al. [16]. The data shows that increase in 2nd season compared to the 1st season for potato yield was maximum with FW under DS (3.5%), followed by DSS15 (2.2%), while the minimum percentage of increase in potato yield (0.9%) was under DS30. On the other hand under TAWW, values of increase were 2.3, 6.9 and 0.9% under DS, DSS15 and DSS30 treatments, respectively.

14.3.2 WATER USE EFFICIENCY

Data in Table 14.2 is plotted in Fig. 14.7 for two water qualities and three drip irrigation methods, during the two seasons. During the 1st season, treatment DSS30 gave the highest value of WUE 0.129 with FW and 0.130 ton/m^3 with TAWW, followed 0.112 with FW and 0.114 ton/m^3 with TAWW under DSS15. While the lowest value of WUE was 0.099 with FW and 0.103 ton/m3 with TAWW, respectively, when DS treatment was used.

For the 2nd season, DSS30 has the highest values with FW and TAWW (0.130 and 0.131ton/m^3). Under DSS15, those values were 0.114 and 0.123 ton/m^3 with DS and TAWW, respectively. While the lowest values under FW and TAWW were 0.102 and 0.106 ton/m^3 using DS. Based on the interactions, the LSD values at 5% level were significant among three irrigation treatments in both seasons, respectively. For both water quality treatments, the increase in WUE during the 2nd season was 2.9, 1.8 0.8% compared to 2.3, 7.3, 0.8%, in 1st season, for each of irrigation method, respectively.

TABLE 14.2 Effects of Water Quality and Three Drip Irrigation Methods on Yield and WUE of Potato

Year	Water quality	Irrigation system	Water amount (m³/ha)	Yield (ton/ha)	WUE (ton/m³)
2011	FW	DS	83.2	8.2	0.099
		DSS15		9.3	0.112
		DSS30		10.7	0.129
	TAWW	DS		8.6	0.103
		DSS15		9.5	0.114
		DSS30		10.8	0.130
		Mean		9.5	0.115
2012	FW	DS		8.5	0.102
		DSS15		9.5	0.114
		DSS30		10.8	0.130
	TAWW	DS		8.8	0.106
		DSS15		10.2	0.123
		DSS30		10.9	0.131
		Mean		9.8	0.118
	LSD at P = 0.05			0.08	0.001
	Interactions			0.05	0.001

WUE: water use efficiency, DS: Surface drip, DSS15: subsurface drip with lateral at 15 cm, DSS30: sub-surface drip with lateral 30 cm. FW: Fresh water and TAWW: treated Agricultural wastewater.

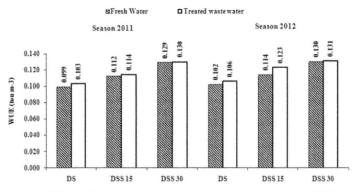

FIGURE 14.7   Effects of water quality and three irrigation methods on WUE.

14.4   CONCLUSIONS

Three irrigation systems positively affected the growth parameters and WUE of potato. Under sandy soil conditions, the laterals at deeper depths lead to increasing the ability of soil to retain more moisture than the nearest layer above it. This is favorable for growth of potato plant under drought conditions. It can be recommended to use DSS30 with TAWW, which has more nutrient content.

14.5   SUMMARY

Potato (*Solanum tuberosum* L.) response to two water qualities and three automatic drip irrigation methods was studied for two growing seasons 2011 and 2012. Field experiments were conducted in sandy soil at the Experimental Farm of National Research Center (NRC), El-Nubaria, El-Behira Governorate. It can be concluded that subsurface automatic drip irrigation system with laterals at 30 cm depth (DSS30) using TAWW gave best potato productivity.

KEYWORDS

- automatic controller drip irrigation
- Automation
- Egypt
- freshwater, FW
- potato
- sandy soil
- subsurface drip irrigation, DSS
- surface drip irrigation, DS
- treated agricultural wastewater, TAWW
- water use efficiency, WUE

REFERENCES

1. Camp, C. R. (1988). Subsurface drip irrigation: a review. *Trans. ASAE*, 41(5), 1353–1367.
2. DeTar, W. R., Browne, G. T., Phene, C. J., Sanden, B. L. (1996). Real-time irrigation scheduling of potatoes with sprinkler and subsurface drips systems. In: *Proc Int. Conf. on Evapotranspiration and Irrigation Scheduling*, eds. C. R. Camp, E. J. Sadler, and R. E. Yoder, Pages 812–824. ASAE, St. Joseph, Michigan.
3. Doorenbos, J., Pruitt, W. O. (1977). *Crop Water Requirements*. Irrigation and Drainage Report 24. Food and Agricultural Organization of the United Nations, Rome, Italy.
4. Eldredge, E. P., Shock, C. C., Saunders, L. D. (2003). Early and late harvest potato cultivar response to drip irrigation. In: Yada, R. Y. (ed.) *Potatoes – Healthy Food for Humanity*. *Acta Horticulturae*, 619, 233–239.
5. Fabeiro, C., Martin de Santa Olalla, F., de Juan, J. A. (2001). Yield and size of deficit irrigated potatoes. *Agricultural Water Management*, 48, 255–266.
6. Goyal, Megh R. (2012). *Management of Drip/Trickle or Micro Irrigation*. Oakville, ON, Canada: Apple Academic Press Inc.,
7. Goyal, Megh R. (2015). *Research Advances in Sustainable Micro Irrigation, volumes 1 to 10*. Oakville, ON, Canada: Apple Academic Press Inc.,
8. Goyal, Megh R., H. A. A. Mansour (Eds.), (2015). *Closed circuit trickle irrigation design: Theory and Applications, volumes 7*. Oakville, ON, Canada: Apple Academic Press Inc.,
9. Islam, T., Sarker, H., Alam, J., Rashid, H. U. (1990). Water use yield relationships of irrigated potato. *Agric. Water Management*, 18, 173–179.
10. Neibling, H., Brooks, R. (1995). Nitrogen management impacts on nitrate leaching under potato. *Environ. Qual.*, 29, 251–261.
11. Onder S., Caliskan, M. E., Onder, D., Caliskan, S. (2005). Different irrigation methods and water stress effects on potato yield and yield components. *Agricultural Water Management*, 73, 73–86.
12. Sammis, T. W. (1980). Comparison of sprinkler, trickle, subsurface and furrow irrigation methods for row crops. *Agron. J.*, 72(5), 701–704.
13. Shae, J. B., Steele, D. D., Gregory, B. L. (1999). Irrigation scheduling methods for potatoes in the Northern Great Plains. *Trans. American Society of Agricultural Engineers*, 42, 351–360.
14. Sharma, S. K., Dixit, R. S., Tripathi, H. P. (1993). Water management in potato (*Solanum tuberosum*). *Indian J. Agron.*, 38, 68–73.
15. Shock, C. C., Eldredge, E. P., Saunders, D. (2002). Drip irrigation management factors for Umatilla Russet potato production. In: Malheur Experiment Station Annual Report, pages 157- 169. Special Report 1038, Oregon State University.
16. Simonne, E., Ouakrim, N., Caylor, A. (2002). Evaluation of an irrigation scheduling model for drip-irrigated potato in Southeastern United States. *HortScience*, 37(1), 104–107.
17. Smajstrla, A. G., Locascio, S. J., Weingartner, D. P., Hensel, D. R. (2000). Subsurface drip irrigation for water table control and potato production. *Appl. Eng. Agric.*, 16, 225–229.

18. Steyn, J. M., Du Plessis, H. F., Fourie, P., Ross, T. (2000). Irrigation scheduling of drip irrigated potatoes. *Micro irrigation technology for developing agriculture.* 6th International Micro-irrigation Congress. South Africa. October 22–27.
19. Waddell, J. T., Gupta, S. C., Moncrief, J. F., Rosen, C. J., Steele. D. D. (1999). Irrigation and nitrogen management effect on potato yield, tuber quality, and nitrogen uptake. *Agron. J.*, 91, 991–997.
20. Waddell, J. T., Gupta, S. C., Moncrief, J. F., Rosen, C. J., Steele. D. D. (2000). Irrigation and nitrogen management impacts on nitrate leaching under potato. *Environ. Qual.*, 29, 251–261.
21. Weatherhead, E. K., Knox, J. W. (1997). Drip irrigation revisited. *Irrigation News*, 25, 11–18.
22. Zartman, R. E., Rosado – Carpio, L., Ramsey, R. H. (1992). Influence of trickle irrigation emitter placement on yield and grade distribution of potatoes. *HortTechnology*, 2, 387–391.

CHAPTER 15

EFFECTS OF TREATED WASTEWATER ON WATER USE EFFICIENCY OF POTATO

H. A. MANSOUR, M. S. GABALLAH, M. ABD EL-HADY and EBTISAM I. ELDARDIRY

CONTENTS

15.1 INTRODUCTION

The Increase for water demand in the world (especially in arid and semi-arid regions such as Saudi Arabia) resulted in searching for effective ways

---

In this chapter: One feddan (Egyptian unit of area) = 4200 m².

Modified and printed from H. A. Mansour, M. S. Gaballah, M. Abd El-Hady and Ebtisam I. Eldardiry, 2014. Influence of different localized irrigation systems and treated agricultural wastewater on distribution uniformities, potato growth, tuber yield and water use efficiency. *International Journal of Advanced Research*, 2(2):143–150. Open access article at: www.journalijar.com.

to use of water resources rationality by farm. Since, Egypt lacks water for agriculture, therefore it was necessary to use alternative systems of modern irrigation to contribute to the provision of water for irrigation in arid regions.

The potato (*Solanum tuberosum* L.) is one the most important vegetable crops in the world in terms of production and cultivated area, as well as one of the most widely consumed vegetable crop by humans. It is an important food source of nutrients [8]. In arid and semiarid regions, potato is sensitive to water stress and irrigation has become an essential component of potato production in comparison with other crops [29]. Shock et al. [25] stated that potato can tolerate water deficit before tuber-set without reduction in tuber quality under some water stress conditions. Potato may be quite sensitive to drought [28] as it needs frequent irrigations for suitable growth and optimum yield [11, 31]. Doorenbos and Kassam [1] have reported that initial vegetative stage is not sensitive to the moisture stress. In contrast, Hassan et al. [7] found that the tuber formation stage was more sensitive than bulking and tuber enlargement stages. Thornton [27] and Shock [24] found that all growing stages of potato, especially tuber formation stage, are very sensitive to water deficit. Whereas, Wright and Stark [29] found that some stress can be tolerated during early vegetative growth and late tuber bulking stage under water deficit conditions. Irrigation management plays a key role in soil organic matter turnover [21].

Michael [13] reported that drip irrigation has several advantages compared to other irrigation systems, such as: increased crop yield, water and energy savings, increased water and fertilizer use efficiencies, tolerance to windy atmospheric conditions, decreased labor cost, protection from pests and diseases, applicable on slopping lands, suitability with different types of soils and improved the salinity conditions. Goyal et al. [4–6] has discussed in detail drip irrigation technology and its application in various situations. Yildirim and Korukcu [30] indicated that drip irrigation gave higher crop yield and adequate soil moisture in the active root zone with minimum water losses. Sharma [22] found that drip irrigation saved 80% of water compared to furrow irrigation system. Singh et al. [23] found that potato yield was 88.20 ton/ha with drip irrigation compared to

76.17 ton/ha with furrow irrigation and 84.2 ton/ha with sprinkler irrigation. Ibragimov et al. [9] reported that yield was increased by 18–42%
and water use efficiency (WUE) was increased by 35 to 103% under drip
irrigation system. Tagar et al. [26] found that drip irrigation saved 56.4%
of water and gave 22% more yield compared to furrow irrigation method.
Moreover they found that WUE was 4.87 under drip irrigation compared
to 1.66 in furrow irrigation.

Katerji et al. [10] stated that stomatal conductance was clearly different
between loamy and clayey soils. These trends were the same in wheat, but
much less clear than in potatoes. According to both parameters, potatoes
are more sensitive than wheat to water stress caused by soil or by salinity.
WUE is defined as the tuber yield per unit of water consumption [2]. Miller
and Donahue [14] reported that potato may respond differently in puddled
low-land rice fields. Soil compaction may affect root bulking. Irrigation
can loosen the soil and can improve the root bulking. Rashidi and Gholami
[18] illustrated that WUE of potato in Iran ranged from 1.92 to 5.25 kg
per m^3. They added that few numbers of irrigations can reduce soil compaction. Potato responds very well to fertilizer application. Nagaz et al.
[15] found that WUE varied around 8–14 kg per m^3 for planted potato.
Wright and Stark [29] reported that the WUE ranged from 0.05 to 0.1 kg
per ha-m^3. Sharma [22] stated that the high wind velocity had no effect on
drip irrigation system, because water is applied directly to the root zone.

This chapter discusses benefits of reusing agricultural drainage water
as an alternative to fresh water in irrigated potato under dry conditions
of Egypt. The chapter also presents effects of water quality and subsurface drip irrigation on: performance parameters of potato and water use
efficiency.

15.2 MATERIALS AND METHODS

During 2012 and 2013, field experiments were conducted in sandy soil at
the Experimental Farm of National Research Center (NRC), El-Nubaria,
El-Behira Governorate, Egypt. Physical, chemical and hydraulic properties of the soil were determined according to standard methods [3, 12, 19].

A compete randomized split-plot design was used to evaluate the following treatments:

- Two irrigation methods: Bubbler and drip irrigation systems.
- Two irrigation water sources: Freshwater (FW) and treated agricultural waste water (TAWW).
- Three irrigation depths based on percentage of crop evapotranspiration (ET), %: 50, 75, and 100.

15.2.1 DISTRIBUTION UNIFORMITY OF LOCALIZED IRRIGATION SYSTEMS

For every plot, the distribution uniformity (DU) of irrigation was estimated along the lateral drip lines of automatic drip irrigation system at an operating pressure of 1.0 bar by using 20 collection cans and the following equation:

$$DU = (qm/qa) \times 100 \tag{1}$$

where: DU = distribution uniformity, %; qm = the average flow rate of the emitters in the lowest quartile, lph; and qa = the average flow rate of all emitters under test, lph.

During the growing season, potential evapotranspiration (ETo) and irrigation water requirement (IWR) of potato per season were estimated as 83.2 and 76.1 m^3/ha at the experimental site, respectively [2].

15.2.2 CROP ESTABLISHMENT

The potato seeds were planted at a row spacing of 70 cm on April 9 during each season using seeding rate of 2000 kg/ha. The fertilizer formula was 60:35:45 (%) of N-P-K. Fertilizers N, P and K were applied @ 145, 80, 125 kg/ha, respectively. To prevent any possible water deficit stress during the vegetative growth stage, irrigation was applied at 9 and 13 days after sowing. All plots were irrigated at an irrigation interval of 3–4 days. Amount of irrigation water was measured at the entrance of each line of drip irrigation by a flow-meter. Irrigation operation was stopped two weeks before initiation of harvesting of potato. Leaf area (cm^2) was calculated by digital planometer. Leaf area index (LAI) was calculated by dividing the total leaf area with the corresponding land area:

$$LAI = (total\ leaf\ area/unit\ land\ area) \qquad (2)$$

Treatment means were compared using analysis of variance (ANOVA) and the least significant at $P = 5\%$.

15.3 RESULTS AND DISCUSSION

15.3.1 BUBBLER IRRIGATION: DISTRIBUTION UNIFORMITY

Distribution uniformity (DU) was used as an index to validate the irrigation systems. The DU estimates the homogeneity of distribution of irrigation water. Table 15.1 shows the volume of water that was collected in each can that was randomly located in the field below the emission point. Collection time was 5 minutes and flow rate was calculated in lph. The average of lowest five observations was = 1.230 lph and the average of water applications was 1.292 lph. Distribution uniformity for the bubbler irrigation system is shown in Fig. 15.1. Emission uniformity of bubbler irrigation was 95.20% that is acceptable.

TABLE 15.1 Water Volume in Each Can Under Bubbler and Drip Irrigation Systems

Can number	Bubbler irrigation system	Drip irrigation system	Can number	Bubbler irrigation system	Drip irrigation system
	Water volume (lph)			Water volume (lph)	
1	1.22	0.23	11	1.28	0.24
2	1.22	0.24	12	1.29	0.25
3	1.23	0.24	13	1.29	0.25
4	1.24	0.24	14	1.30	0.25
5	1.24	0.24	15	1.31	0.25
6	1.25	0.24	16	1.31	0.25
7	1.25	0.24	17	1.32	0.25
8	1.26	0.24	18	1.32	0.26
9	1.27	0.24	19	1.33	0.26
10	1.27	0.24	20	1.33	0.26

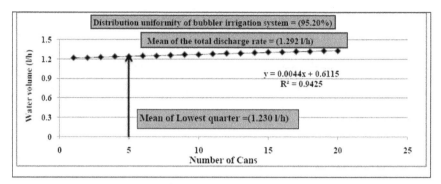

FIGURE 15.1   Distribution uniformity for bubbler irrigation system.

15.3.2 DRIP IRRIGATION: DISTRIBUTION UNIFORMITY

Table 15.1 shows the volume of water that was collected in each can that was randomly located under the built-in-GR dripper. Figure 15.2 shows that average of the lowest quarter was 0.238 lph and average of all observations was 0.248 lph. Distribution uniformity for drip irrigation is shown in Fig. 15.2. Emission uniformity of drip irrigation was 95.97% that was acceptable.

15.3.3 PERFORMANCE OF POTATO

Table 15.2 shows the effects of irrigation methods, water quality and irrigation depths on plant length in cm, leaf area index (LAI) and number

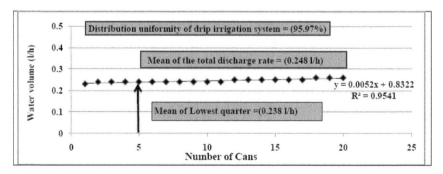

Fig. (2). Distribution uniformity of drip irrigation system.

FIGURE 15.2   Distribution uniformity for drip irrigation system.

TABLE 15.2 Effects of Localized Irrigation Systems, Water Quality and Irrigation Rates on Performance of Potato

Localized irrigation system (I)	Water quality (II)	Water applied (% ET) (III)	Plant length (cm)	LAI	Average number of branches
		%	cm	—	Nos.
Bubbler	FW	50	0.51a	2.1a	4.1a
		75	0.55b	2.2b	3.7b
		100	0.62c	3.3c	3.3c
	TAWW	50	0.53d	2.5d	4.3d
		75	0.58e	3.2e	4.2e
		100	0.65f	3.4f	4.6f
	FW	50	0.52g	2.3g	4.4g
		75	0.57h	3.3h	4.0h
		100	0.64i	3.4i	3.5i
	TAWW	50	0.54j	3.2j	4.6j
		75	0.58k	3.3k	4.5k
		100	0.67L	3.5L	4.1L
LSD at P = 0.05			0.03	0.1	0.1
Interactions					
I × II			0.01	0.1	0.2
I × III			0.02	0.2	0.1
II × III			0.01	0.1	0.1

FW= Fresh water, TAWW= Treated wastewater; ET= Evapotranspiration, LAI = Leaf area index.

of branches. The highest values of plant length were 0.54, 0.58, 0.67 cm in I50, I75, I100 with TAWW under drip irrigation system compared to 0.53, 0.58, 0.64 cm in I50, I75, I100 with TAWW under bubbler irrigation system. Whereas, the lowest values were with FW under bubbler irrigation system (0.53, 0.58, 0.65 cm), followed by FW under drip irrigation system (0.52, 0.57; 0.64 cm) for applied water treatments (50, 75, 100%), respectively.

The values of LAI took the same trend as of plant length. Highest values of LAI were with TAWW under drip and bubbler irrigation systems (3.2, 3.3 3.5) and (2.5, 3.2, 3.4), respectively. In contrast, the lowest values of

LAI were with FW under bubbler and drip irrigation systems (2.3, 3.3, 3.4) and (2.1, 2.2, 3.3) for three water treatments (50, 75, 100%), respectively.

The average number of branches took the same trend as of plant length and LAI. The average of branches number values were highest with TAWW under drip irrigation system (4.6, 4.5, 4.1), followed by TAWW under bubbler irrigation system (4.3, 4.2, 4.6). Whereas, the lowest values were with FW under bubbler irrigation system (4.1, 3.7, 3.3), followed by FW with Drip irrigation system (4.4, 4.0, 3.5) under three water treatments (50, 75, 100%), respectively.

According to LDS values for plant length and average branches number, the differences were significant at 5% level among all values (Table 15.1). The differences were significant at 5% level among LAI values exception under some similar and non-similar water treatments. Concerning the vegetative characteristics of potato under study, the interactions, between the main factors I, II and III (I x II, I x III, II x III), were significant at 1% level. The results in this chapter are in agreement with those by Nedunchezhiyan et al. [17] and Roy et al. [20].

Table 15.3 shows the effects of irrigation systems, water quality and three irrigation rates on tuber yield and water use efficiency (WUE). The values of tuber yield took the same trend as of vegetative growth parameters. The highest values of tuber yield were with TAWW under drip and bubbler irrigation systems (9.2, 11.12, 11.8 ton/ha) and (9.17, 10.8, 11.6 ton/ha), respectively. While the lowest values of tuber yield were with FW under bubbler and drip irrigation systems (8.6, 10.6, 11.3 ton/ha) and (8.9, 10.7, 11.5 ton/ha) for three water treatments of (50, 75, 100%), respectively.

15.3.4 WATER USE EFFICIENCY

According to data in Table 15.3, the WUE took the same trend as potato vegetative growth and tuber yield. WUE values were highest with TAWW under drip irrigation system (0.24, 0.19, 0.14 ton/m³), followed by TAWW under bubbler irrigation system (0.22, 0.19, 0.14 ton/m³). Whereas, the lowest values of WUE were with FW with bubbler irrigation system (0.21, 0.18, 0.14 ton/m³), followed by FW under drip irrigation system (0.23, 0.18, 0.13 ton/m³) for three water treatments (50, 75, 100%), respectively. This finding is inconsistent with Nagaz et al. [15], who reported that the

TABLE 15.3   Effect of Localized Irrigation System, Water Quality and Different
Irrigation Rates, on Tuber Yield and Water Use Efficiency of Potato

Localized irrigation system (I)	Water quality (II)	Water applied (% ET) (III)	Water applied	Tuber yield	WUE
		%	m³/ha	tons/ha	tons/m³
Bubbler	FW	50	41.6	8.6a	0.21a
		75	58.2	10.6b	0.18b
		100	83.2	11.3c	0.14c
	TAWW	50	41.6	9.2d	0.22d
		75	58.2	10.8e	0.19e
		100	83.2	11.5f	0.14f
	FW	50	41.6	8.9g	0.23g
		75	58.2	10.7h	0.18h
		100	83.2	11.5i	0.13i
	TAWW	50	41.6	9.2j	0.24j
		75	58.2	11.2k	0.19k
		100	83.2	11.8L	0.14L
LSD at P = 0.05			–	2.2	0.02
Interactions					
I × II			–	1.2	0.01
I × III			–	1.1	0.01
II × III			–	1.3	0.02

FW= Fresh water, TWW= Treated wastewater; ET= Evapotranspiration.

range of WUE was from 44.1 to 63.4 kg/ha-mm and from 8 to 14 kg/m³,
respectively.

According to LSD values in Table 15.3 for potato yield and WUE, the
differences were significant at 5% level among all values. Also the interactions among the factors I, II and III (I x II, I x III, II x III) were significant
at 5% level. The data agrees with Nasseri and Bahramloo [16].

15.4   CONCLUSIONS

The irrigation methods, treated agricultural wastewater, and irrigation rate
of 100%ET were on positively correlated with potato vegetative growth,

tuber yield and WUE. This can be attributed to the improvement of soil hydro-physical properties by using of drip irrigation system compared to bubbler irrigation system; and the soluble nutrients in the treated agricultural waste water compared to fresh water. Estimated field uniformity distribution of bubbler and drip irrigation systems were 95.20 % and 95.97 %, respectively.

15.5 SUMMARY

This chapter discusses benefits of reusing agricultural drainage water as an alternative to fresh water in irrigated potato under dry conditions of Egypt. The chapter also presents effects of water quality and subsurface drip irrigation on: performance parameters of potato and water use efficiency. During 2012 and 2013, field experiments were conducted in sandy soil at the Experimental Farm of National Research Center (NRC), El-Noubaria Governor, Egypt. A compete randomized split-plot design was used to evaluate the following treatments:

- Two irrigation methods: Bubbler and drip irrigation systems.
- Two irrigation water sources: Freshwater (FW) and treated agricultural waste water (TAWW).
- Three irrigation depths based on percentage of crop evapotranspiration (ET), %: 50, 75, and 100

Field uniformity distribution of bubbler and drip irrigation systems were 95.20% and 95.97%, respectively. The parameters of vegetative growth (plant length, LAI, number of branches) under Irrigation system, treated agricultural waste water (TAWW) were increased by 6.0%, 14.6% and 15.3% compared to bubbler irrigation system and fresh water (FW), respectively. Yield and WUE under drip irrigation system and TAWW were increased by 5.4% and 5.7% compared to bubbler irrigation system and fresh water (FW), respectively. Plant length and leaf are index (LAI) values were highest under the control treatment (100%ETo water application). Vegetative growth, tuber yield and WUE can be arranged in ascending order: 50% < 75% < 100% due to amount of water. The effects of drip irrigation system and TAWW were positive on vegetative growth parameters (yield, and WUE). This can be attributed to the improvement of soil physical characteristics under drip irrigation system and the soluble nutrients in the TAWW relative to fresh water.

KEYWORDS

- **bubbler irrigation**
- **crop water requirement**
- **deficit irrigation**
- **distribution uniformity, DU**
- **drip irrigation**
- **Egypt**
- **freshwater, FW**
- **localized irrigation**
- **potato**
- **treated agricultural waste water, TAWW**
- **tuber yield**
- **vegetative growth**
- **water evapotranspiration, ET**
- **water use efficiency, WUE**

REFERENCES

1. Doorenbos, J., A. H. Kassam, (1979). Yield response to water. Irrigation and Drainage Manual, page 33. Food and Agricultural Organization of the United Nations, Rome, Italy.
2. Doorenbos, J., W. O. Pruitt, (1977). *Crop Water Requirements*. Food and Agricultural Organization of the United Nations, Rome, Italy.
3. Gee, G. W., J. W. Bauder, (1986). Particle size analysis. Pages 383–412, In: *Methods of Soil Analysis, Part 1*. ASA and SSSA, Madison, WI.
4. Goyal, Megh R. (2012). *Management of Drip/Trickle or Micro Irrigation*. Oakville, ON, Canada: Apple Academic Press Inc.,
5. Goyal, Megh R. (2015). *Research Advances in Sustainable Micro Irrigation, volumes 1 to 10*. Oakville, ON, Canada: Apple Academic Press Inc.,

6. Goyal, Megh R., H. A. A. Mansour (Eds.), (2015). *Closed circuit trickle irrigation design: Theory and Applications, volumes 7.* Oakville, ON, Canada: Apple Academic Press Inc.,

7. Hassan, A. A., A. A. Sarkar, M. H. Ali, N. N. Karim, (2002). Effect of deficit irrigation at different growth stage on the yield of potato. *Pakistan J. Biol. Sci.,* 5, 128–134.

8. Hassan, M. A. (1999). In vitro production of potato (*Solanum tuberosum*) with micro-tubes: The influence of jasmonic acid, silver nitrate and some physical factors. *Annual Agric. J. of Moshtohr Univ.,* 37(1), 433–448.

9. Ibragimov, N., Evtt, S. R. Esanbekov, Y., Kamilov, B. S., Mirzaev, L., Lamers, J. P. A. (2007). Water use efficiency of irrigated cotton in uzbekistan under drip and furrow irrigation. *Agricultural Water Management,* 90(1/2), 335–238.

10. Katerji, N., van Hoorn, J. W., Hamdy, A., Bouzid, N., EI-Sayed Marous, S., Mastrorilli, M. (1992). Effect of salinity on water stress, growth, and yield of broadbeans. *Agric. Water Manage.,* 21, 107–117.

11. Kiziloglu, F. M., U. Sahin, T. Tune and S. Diler, (2006). The effect of deficit irrigation on potato evapotranspiration and tuber yield under cool season and semiarid climatic conditions. *J. Agron.,* 5, 284–288.

12. Klute, A., C. Dirksen, (1986). Hydraulic conductivity and diffusivity: Laboratory methods. Pages 687–734, In: *Methods of soil analysis. Part 1.* ASA and SSSA, Madison, WI.

13. Michael, A. M. (2008). *Irrigation Theory and Practice.* Vikas Publishing House Pvt. Ltd., Delhi, India.

14. Miller, R. W., R. L. Donahue, (1992). *Soils: An Introduction to Soils and Plant Growth.* Prentice Hall of India, New Delhi, India.

15. Nagaz, K., M. M. Masmoudi and N. B. Mechlia, (2007). Soil salinity and yield of drip-irrigated potato under different irrigation regimes with saline water in arid conditions of Southern Tunisia. *J. Agron.,* 6, 324–330.

16. Nasseri, A., R. Bahramloo, (2009). Potato cultivar Marfuna yield and water use efficiency responses to early-season water stress. *International Journal of Agriculture & Biology,* 11(2), 201–204.

17. Nedunchezhiyan, M., G. Byju, R. C. Ray, (2012). Effect of irrigation, and nutrient levels on growth and yield of sweet potato in rice fallow. *International Scholarly Research Network (ISRN) Agronomy,* pages 1–13.

18. Rashidi, M., M. Gholami, (2008). Review of crop water productivity values for tomato, potato, Melon, watermelon and cantaloupe in Iran. *Int. J. Agric. Biol.,* 10, 432–436.

19. Rebecca, B. (2004). Soil Survey Laboratory Methods Manual. USDA Soil Survey Laboratory Investigations Report No. 42. Room 152, 100 Centennial Mall North, Lincoln, NE 68508- (3866). (402) 437–5006.

20. Roy Chowdhury, S., R. Singh, D. K. Kundu, E. Antony, A. K. Thakur, and H. N. Verma, (2002). Growth, dry-matter partitioning and yield of sweet potato (*Ipomoea batatas* L.) as influenced by soil mechanical impedance and mineral nutrition under different irrigation regimes. *Advances in Horticultural Science,* 16(1), 25–29.

21. Shepherd, M., Hatley, D., Gosling, P. (2002). Assessing soil structure in organically farmed soils. In: *Proceedings of UK Organic Research 2002,* the COR Conference,

Aberystwyth (ed. J. Powel), pp. 143–144. Organic Centre Wales, Institute of Rural Studies, University of Wales, Aberystwyth.

22. Sharma, B. R. (2001). Availability, status and development and opportunities for augmentation of groundwater resources in India. *Proceeding ICAR-IWMI Policy Dialogue on Ground Water Management*, November 6- 7 at CSSRI, Karnal, pp. 1–18.

23. Singh, N., Sood M. C., Lal, S. S. (2005). Evaluation of potato based cropping sequences under drip, sprinkler and furrow methods of irrigation. *Potato Journal*, 32(¾), 175–176.

24. Shock, C. C. (2004). Efficient irrigation scheduling. Malheur Experiment Station, Oregon State University, Oregon, USA.

25. Shock, C. C., J. C. Zalewski, T. D. Stieber and D. S. Burnett, (1992). Impact of early-season water deficits on Russet Burbank plant development, tuber yield and quality. *American Potato J.,* 69, 793–803.

26. Tagar, A., F. A. Chandio, I. A. Mari, and B. Wagan, (2012). Comparative study of drip and furrow irrigation methods at farmer's field in Umarkot. *World Academy of Science – Engineering and Technology*, 69, 863–867

27. Thornton, M. K. (2002). *Effects of Heat and Water Stress on the Physiology of Potatoes*. Idaho Potato Conference, Idaho.

28. Van Loon, C. D. (1981). The effect of water stress on potato growth, development and yield. *American Potato J.,* 58, 51–69

29. Wright, J. L., J. C. Stark, (1990). Potato. In: Stewart, B. A., D. R. Nielson (eds.), *Irrigation of Agricultural Crops*, pp: 859–889. American Society of Agronomy, Crop Science Society of America, Soil Science Society of America, Madison, USA.

30. Yildirim, O., A. Korukcu, (2000). Comparison of drip, sprinkler and surface irrigation systems in orchards. Faculty of Agriculture, University of Ankara, Ankara Turkey. 47p.

31. Yuan, B. Z., S. Nishiyama, Y. Kang, (2003). Effects of different irrigation regimes on the growth and yield of drip-irrigated potato. *Agric. Water Manag.,* 63, 153–167

CHAPTER 16

SOIL EVAPORATION FROM DRIP IRRIGATED MAIZE

HOSSEIN DEHGHANISANIJ and HANIEH KOSARI

CONTENTS

16.1   INTRODUCTION

Development of reliable methods of measuring evapotranspiration components plays an important role in crop growth modeling and irrigation management. Except initial crop growth stage, soil evaporation includes smaller proportion of evapotranspiration (ET) and since it is not directly related to crop yield, it is not considered. However, it is an important loss in initial crop growth stage.

Various methods are available to measure soil evaporation (E_{soil}). In the past, soil evaporation has been measured by water balance using micro-lysimeters [4, 5]. However, there are limitations of using micro-lysimeters, since the soil inside the micro-lysimeter is hydraulically isolated and it may dry differently from the undisturbed around soil [1]. Also when evaporation is small, micro-lysimeters do not give correct results [7]. In an alternate method, soil evaporation is measured indirectly by measuring both ET and crop transpiration and E_{soil} is calculated by subtracting these two values. With these methods, precision of evaporation is dependent on the precision of ET and crop transpiration. Furthermore, the correlation may not exist between the calculated and actual soil evaporation due to temporal and spatial variations of measurements.

Bowen ratio energy balance (BREB) is one of the simplest and most applicable methods of latent heat flux measurement. This method has been widely used under different conditions and the results have shown that it is one of the reliable methods of ET measurement. In a research by Ashktorab et al. [1], soil evaporation was measured by BREB from a bare soil. Results indicated a good correlation between soil evaporation measured by BREB and micro-lysimeter method. Therefore, they suggested measuring soil evaporation under crop canopy [1]. Other researchers used their method to measure E_{soil} under maize canopy [13] and tomato [2] and indicated reliable results. Therefore under limitations with other methods, the BREB method for E_{soil} measurement has been widely accepted.

Micro irrigation system can reduce E_{soil}. However, proper design and adequate management must be implemented to avail the advantages offered by BREB method. This issue becomes more important due to high initial investment of drip irrigation system. Therefore, a detailed research can provide adequate information.

This chapter discusses the research results to measure soil evaporation in surface and subsurface drip irrigation by Bowen ratio energy balance (BREB) method.

16.2 THEORETICAL ASPECTS OF ENERGY BALANCE AT SOIL SURFACE/PLANT CANOPY

Governing equations for energy balance are given below:

$$R_{ns} - \lambda E_s - H_s - G = 0 \tag{1}$$

$$R_{ns} = R_n \exp[- 0.622LAI + 0.055LAI^2] \tag{2}$$

$$\beta_s = [H_s / (\lambda E_s)], \text{ or} \tag{3}$$

$$\beta_s = \gamma[\Delta T / (\Delta e)] \tag{4}$$

$$\gamma = [C_p P]/[\varepsilon L_v] \tag{5}$$

$$\lambda E_s = [R_{ns} - G] / [1 + \beta_s], \text{ or} \tag{6}$$

$$H = \beta_s [R_{ns} - G] / [1 + \beta_s] \tag{7}$$

$$R_{nc} = \lambda E_s + H_c, \text{ plant canopy} \tag{8}$$

$$R_{nc} = R_{n-} R_{ns}, \text{ plant canopy} \tag{9}$$

$$\lambda E_c = \lambda E - \lambda E_s, \text{ plant canopy} \tag{10}$$

In Eqs. (1)–(10): Subscript c refers to plant canopy and subscript s refers to soil surface; Rns is the net radiation reaching the soil surface; λEs is the soil surface latent heat flux; Hs is sensible heat flux; G is soil heat flux (all units of w.m^{-2}); LAI = Leaf area index; ΔT is air temperature; and Δe is vapor pressure differences between the two measurement levels; Cp is the specific heat of air at constant pressure (1.01 kJ.kg^{-1}.c^{-1}); c = units of temperature in °c; P is atmospheric pressure (kPa); ε is the ratio between the molecular weights of water vapor and air (= 0.622); β_s = Bowen ratio at the soil surface; and Lv is latent heat of vaporization (kJ.kg^{-1}). Psychrometric constant (γ) for the experiment site was 0.058 (kPa.c^{-1}).

Energy balance at soil surface can be expressed as shown in Eq. (1). In equation (1), the convention used for the signs of the energy fluxes was Rn positive downward, and G is positive when it is conducted downward from the surface, λE and H are positive upward. R$_{ns}$ was determined by the empirical Eq. (2) with Rn and LAI values [9, 10, 13].

Partitioning of energy between λEs and Hs is determined by the BREB [1, 3, 11–13] by Eq. (3). Assuming equality of eddy transfer coefficients for sensible heat and water vapor in the averaging period and measuring air temperature and vapor pressure gradients between the two levels, the

Bowen ratio (βs) is calculated by Eq. (4). γ is psychrometric constant is calculated by the Eq. (5). Using Eqs. (4) and (5) and measurements of air temperature and vapor pressure gradients near the soil surface, Bowen ratio at soil surface was determined. By solving Eqs. (1) and (3) simultaneously, latent heat flux from the soil surface was determined by Eq. (6). The Eqs. (8)–(10) are for plant canopy. The Eqs. (1)–(7) are for soil surface.

16.3   MATERIAL AND METHODS

16.3.1   EXPERIMENTAL SITE

This research was conducted at Experimental Station of Agricultural Engineering Research Institute (AERI), Karaj-Iran (located at 35°21′ N, 51°38′ E, 1312.5 m above sea level). The field was prepared for planting in spring. Results from soil analysis up to 80 cm depth showed that the soil texture was loam (47% sand, 44% silt, 9% clay) with ECe = 1.7. Maize crop (var. Double Cross 370) was planted on 15 June 2009. The crop was planted at 0.75 m row width and north-south orientation. The experiment site was a 40×60 m² (Fig. 16.1).

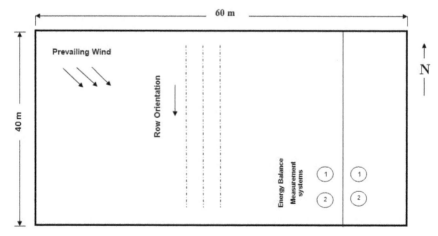

FIGURE 16.1   Schematic diagram showing field position and location of energy balance measurement systems, 15 June of 2009.

The field was bordered by irrigated maize plots except no crop on western side. Irrigation water was from a well with good water quality as indicated by chemical analysis of water. A drip-tape irrigation system with 0.30 m dripper spacing was used to apply irrigation. Drip tapes were positioned 15 cm below the soil surface in subsurface drip irrigation (SDI) in 14 rows on eastern part of the field. Depth of positioning of laterals was selected based on previous research studies in this area. For the rest of the field, drip tapes were placed on soil surface nearest to the plant row.

Crop water requirement (CWR) was estimated based on longtime meteorological data (average of 1988 to 2008) and by Penman Monteith method in FAO 56 [2]. From the early crop growth period, 20% of irrigation based on 3 day intervals was applied to prevent water stress. Recommended nitrate fertilizer (@ 400 kg/ha) was distributed during the crop growth period and closely to crop establishment through the irrigation system (fertigation). At 41–44 and 59–62 day after emergence (DAE), leaf area index (LAI) was measured on 41, 44, 59, and 62 DAE. During each sampling, 3–5 plants were selected randomly and the whole leaf area of a plant was measured with leaf area meter (Area Measurement system, DELA-T Devices, England) in the laboratory. Then, LAI was calculated by multiplying the average plant leaf area by plant density. LAI values for the days between the days of measurement were obtained by linear interpolation [6]. Automatic weather station was established in the field simultaneously at the start of experiment. Hourly average values of solar radiation (Rs), air temperature, relative humidity and wind speed were measured and logged on continuously.

16.3.2 ENERGY BALANCE MEASUREMENTS

From 41–44 to 59–62 DAE, soil evaporation (Es) was determined by measuring all energy fluxes at soil surface simultaneously in two irrigation systems by two repetitions which were separated by 5 m distance. Energy balance equipment's consisted of a net radiometer (CNR1, Kipp & Zonen), two soil heat flux plates (MF-180M, EKO Japan) and two handmade thermocouple ventilated psychrometers for Bowen ratio measurement at soil surface. The details of constructed psychrometers have been described in Kosari [2014]. Measuring systems in subsurface drip irrigation were placed 8.5 m from the east edge of the field as the system number 2 was positioned 5 m from the south edge to maximize the fetch to height ratio when prevailing wind (northwestern to south eastern) were present (Fig. 16.1). This value was greater than the minimum ratio reported by Ashktorab et al. [8] for measuring Bowen ratio during the experiment period. Measurement equipment's in each measurement system were installed on a tall rod. Two ventilated psychrometers used for measuring temperature and water vapor gradients at soil surface were fixed 0.1 m apart on the rod as the lowest one was positioned 0.05 m above the soil surface [1, 13].

Net radiometer was also installed at 1 m above crop canopy to measure the total net radiation available at field level. Soil heat flux was measured with two soil heat flux plates positioned 0.02 m below the soil surface, one in plant row and the other in plant row aisle. All data were measured every minute by a CR23X data-logger connected to an AM16/32 multiplexer (Campbell Scientific, Inc., UT) and averaged at 30 min intervals.

16.4   RESULTS AND DISCUSSION

16.4.1   CLIMATIC PARAMETERS

Table 16.1 shows daytime average values of meteorological parameters measured by the automatic weather station during crop growth. Parameters were maximum, minimum and average air temperature, relative humidity, and wind speed. Plants on days 41- 44 DAE were in developing stage and on days 59–62 DAE were in mid-season stage. Irrigation applications were on 41, 44, 59 and 63 DAE, with last irrigation at 4 days interval, because of technical problems. During the growth period, the 42 and 60 DAE received maximum and minimum solar radiation, respectively.

16.4.2   ENERGY BALANCE MEASUREMENTS

16.4.2.1   Soil Surface

Table 16.2 shows daytime average of energy balance measurements (w/m²) at soil surface for two irrigation systems. Net radiation values ranged from

TABLE 16.1 Daytime Average Values of Climatic Data During the Maize Growth Period

Growth Stage	DAE	Air Temperature (°c)			Relative Humidity (%)			Wind Speed	Solar Radiation
	Days	Avg.	Min.	Max.	Avg.	Min.	Max.	m/s	MJ/m²/day
Development	41	16.67	34.44	24.10	18.92	78.16	53.02	2.90	45.17
	42	16.51	34.07	24.12	15.64	83.07	52.64	3.38	46.82
	43	16.10	33.30	23.39	26.80	77.82	54.09	2.95	46.21
	44	16.56	32.89	23.79	28.66	74.72	52.16	2.92	44.31
	59	19.91	37.38	28.30	12.42	65.51	36.40	1.48	40.97
Mid-season	60	19.96	35.64	27.42	17.78	65.98	41.67	2.33	40.14
	61	19.09	35.57	27.00	13.29	72.40	41.68	2.05	45.18
	62	16.00	34.45	24.24	13.06	59.69	38.43	1.90	45.07

304 to 333 (w/m²). It was considered correct, because all effective parameters for net radiation (climate conditions, soil texture, crop, soil surface color, etc.) were the same in two irrigation systems. Therefore, net radiation reaching the soil surface (Rns) was the same in both systems, because leaf area index (LAI) measurements were same in two irrigation systems. Since there was a short irrigation interval in drip irrigation, the crop root zone area and soil surface were wet. Therefore, there were no significant changes in net radiation before and after irrigation. Net radiation measurements also showed that Rns values decreased with crop growth and LAI increased.

Latent heat flux at soil surface accounted for about 32 to 60 (w/m²) in surface drip irrigation and 40 to 73 (w/m²) in subsurface drip irrigation. As shown in Table 16.2, soil latent heat flux accounted for a large portion of the net radiation as is expected under non-stressed conditions [7]. The $\lambda E_s / R_{ns}$ varied from 41 to 63% in surface drip irrigation and 57 to 71 in subsurface drip irrigation system.

Soil heat flux (G) varied from 25 to 47 in surface drip irrigation and 17 to 25 (w/m²) in subsurface drip irrigation, which is equal to 36 to 53 % and 18 to 31% of net radiation reaching the soil surface in surface and subsurface drip irrigation, respectively. Soil surface energy balance measurements showed that R_{ns} was partitioned primarily between soil heat flux and latent heat flux, and there was very little sensible heat flux.

16.4.2.2 Maize Field

Table 16.3 shows daytime average of energy balance measurements (w/m²) in the maize field for two irrigation systems. The G variation was 17 to 25 in SDI and 25 to 47 W/m² in DI. The λEs accounted for about 40 to 73 (W/m²) in SDI, where maximum value of λEs was 60 W/m² for surface DI. Accordingly, λEs/Rns and G/Rns ratios ranged from 56 to 71 % and 18 to 31 %, respectively. The G in SDI was much less compared to that in DI during crop developing stage than mid-season stage. The λEs was larger in SDI compared to that in DI. These results contributed to higher possible potential for T in SDI during crop developing stage (Table 16.3).

TABLE 16.2 Daytime Average Energy Fluxes At Soil Surface in Surface and Sub-Surface Drip Irrigation

Daytime average energy fluxes at soil surface					
G/R_{ns}	$\lambda E_s/R_{ns}$	λE_s	G	R_{ns}	DAE
%		w/m^2			Days
Sub-surface drip irrigation					
21	59	63	23	107	41
18	71	73	19	104	42
26	56	54	25	96	43
26	66	58	23	88	44
29	71	50	20	70	59
25	64	43	17	67	60
28	69	49	20	71	61
31	57	40	22	70	62
Surface drip irrigation					
44	51	55	47	107	41
43	58	60	44	104	42
48	41	39	46	96	43
53	44	39	47	88	44
49	46	32	34	70	59
45	52	35	30	67	60
39	59	42	28	71	61
36	63	44	25	70	62

16.4.2.3 Plant Canopy

Table 16.4 shows daytime average of energy balance measurements (w/m^2) at the maize canopy, for two irrigation systems. Daytime average of energy balance measurements at Maize canopy showed that λEc increased during the crop development from 134 to 227 (W/m^2), which resulted in $\lambda Ec/Rnc$ ratio of 58 to 90% in SDI. These ratios showed that canopy latent heat fluxes were often lower than the available energy, which resulted

TABLE 16.3 Daytime Average Energy Fluxes in the Maize Field

DAE	R_n	Subsurface drip irrigation (SDI)				Surface drip irrigation (DI)			
		G	λE	λE/R_n	G/R_n	G	λE	λE/R_n	G/R_n
		W/m²			%	W/m2			%
41	322	23	–	–	7	47	197	61	14
42	333	19	207	62	6	44	266	80	13
43	329	25	212	64	8	46	255	78	14
44	320	23	242	76	7	47	243	76	15
59	315	20	247	78	6	34	296	94	11
60	304	17	235	77	6	38	253	83	10
61	326	20	248	76	6	28	271	83	9
62	322	22	267	83	7	25	245	76	8

TABLE 16.4 Daytime Average Energy Fluxes At the Maize Canopy

DAE	R_{nc}	Subsurface drip irrigation (SDI)		Surface drip irrigation (DI)	
		λE_c	λE_c/R_{nc}	λE_c	λE_c/R_{nc}
	W/m²		%	W/m²	%
41	215	–	–	142	66
42	230	134	58	205	89
43	233	158	68	216	93
44	232	184	79	204	88
59	245	197	80	264	108
60	237	192	81	218	92
61	255	199	78	229	90
62	252	227	90	201	80

some sensible heat flux conducting away from canopy level. In DI, λEc increased during the crop development from 142 to 264, which was less compared to that for SDI (Table 16.4).

16.4.3 DIURNAL ENERGY BALANCE PATTERN FOR A SAMPLE DAY

Despite discussion on daytime averages of energy balance components, evaluation of its diurnal pattern yields useful information. The Fig. 16.2 indicates diurnal trends of the energy balance components at soil surface for both irrigation systems on 60th DAE. This day was selected because it is a representative of a cloudy day. Average air temperature and relative humidity was 27.4°C and 41.6%, respectively. Daytime average of net radiation available at maize field was 304 w/m², which was the smallest value in the measurement period. Maximum Rn and G were 674 and 140 w/m², which occurred some minutes before and after 13:00 PM, respectively. As shown in Fig. 16.2, variation of Rn values is not symmetrically, because a bell shape curve signifies that there was some cloud cover in sky during the day.

As shown in Fig. 16.3, both methods have the same trend and a good correlation ($R^2 = 0.92$ and 0.95 for DAE 60 and 61, respectively). Therefore evapotranspiration by BREB showed 9% variations compared to P-M method, which is acceptable. Positive value of MBE parameter shows overestimation of P-M method compared to BREB. These differences can be attributed to different measurements of effective and required parameters in both methods. In other words, BREB evapotranspiration was obtained by direct measurement of required parameters at field level and soil surface, while in P-M method maize evapotranspiration was obtained

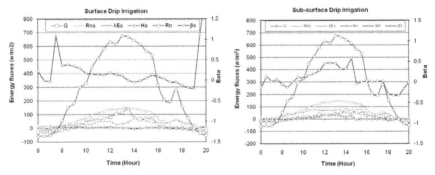

FIGURE 16.2 Diurnal trends of energy balance components at soil surface in surface (top) and sub-surface drip irrigation (bottom).

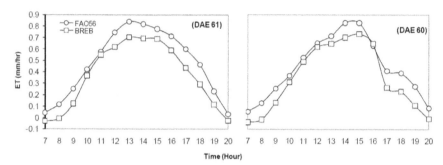

FIGURE 16.3   Diurnal trend of evapotranspiration by BREB and P-M method (FAO56) for 60 and 61 DAE.

by estimation of reference evapotranspiration and crop coefficient. Other researchers have found a good correlation between reference evapotranspiration by BREB and Penman method on irrigated grass.

Available energy (= R_n – G) and λE in maize field for the 60th DAE is shown in Fig. 16.4. The linear regression lines between λE and Rn-G were obtained with high values of $R^2 = 0.99$ for both SDI and DI. Based on the slope of the lines in Fig. 16.4, there was no significant reduction in available energy between the maize field and soil surface for two irrigation systems. Accordingly, SDI aimed at reducing soil evaporation compared to DI, is not effective when soil surface is covered by canopy (LAI = 3.5). During the daytime at 60th DAE, only 22% of net radiation reached the soil surface. Most of the energy was split between λEs and G and the

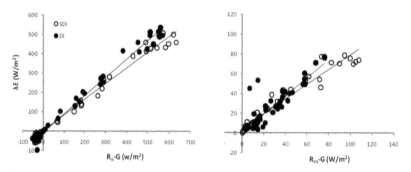

FIGURE 16.4   Available energy (= Rn-G) and latent heat flux (λE) from the maize field and soil (=Rns-G and λEs) on 60th DAE for surface (DI) and subsurface (SDI) drip irrigation.

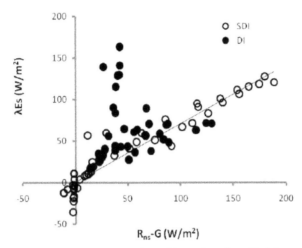

FIGURE 16.5   Available energy (= Rns-G) and latent heat flux (λEs) from soil surface in 41st DAE for surface (DI) and subsurface (SDI) drip irrigation.

value of Hs was small. The λEs was less than available energy except in the afternoon suggesting that the soil surface was absorbing energy from within – canopy air stream, which provided energy for λEs. Similar results have been reported by Ham et al. [7] for soil surface energy balance relationships. Daytime average of λEs was about 0.94% of (= Rns-G) and only about 6% of (= Rns-G) was used as sensible heat. Zeggef et al. [13] reported that Rns-G was split between λEs (0.52%) and Hs (48%) at maize field (LAI = 1); and βs ranged from -0.2 to 1.2 but it typically ranged between -0.2 to 0.5 on 60th day.

Available energy (= R_{ns} – G) and $λE_s$ from the soil surface for 41st DAE is shown in Fig. 16.5. The 41st DAE presented crop-developing stage, when LAI was about 2.20. Accordingly, there was a wide scattering for available energy under DI, which might be due to non-uniform wetting of soil under DI. However, the conditions under SDI were uniform. Linear regression lines between λEs and Rns-G are shown by solid lines.

16.5   CONCLUSIONS

The Bowen ratio energy balance (BREB) method can be used for partitioning ET for surface drip irrigation (DI) and subsurface drip irrigation

(SDI) systems. Partitioning ET can provide us useful information for better irrigation management during crop growth stages and development of new irrigation techniques. Partitioning ET and measurement of energy balance in maize field, plant canopy and soil surface by BREB showed that soil had major impact on the energy balance between the soil and canopy, when soil surface is not fully covered by crop canopy. In crop developing stage, energy balance of maize field was different in DI and SDI. These differences can be contributed to pronounced differences between the irrigation systems in early crop development stage, when soil surface is not covered fully by crop canopy.

Daytime soil heat flux values were greater in DI ($25–47$ W/m^2) than in SDI ($17–25$ W/m^2). It may be due to heat convection in surface drip irrigation while moving down the water from the surface and higher water temperature when drip tapes were positioned on the ground. Therefore, available energy for soil evaporation (= Rns-G) was lower in surface drip irrigation. The λEs accounted for about 41 to 63% of Rns in surface drip irrigation compared to 56 to 71% in subsurface drip irrigation. It was observed that the ground in both surface and subsurface drip irrigation became wet but reverse direction of moisture movement in subsurface system, and it may have contributed to more evaporation in subsurface drip irrigation. Based on the results of this study, it is recommended to give more consideration to the depth of installation of drip laterals in subsurface drip irrigation system.

16.6 SUMMARY

Evaporation reduction is one of the advantages of drip irrigation. A research was conducted at Experimental Station of AERI, Karaj-Iran to measure soil surface evaporation in maize field by BREB method in surface drip irrigation and subsurface drip irrigation systems. In surface drip irrigation, the drip tapes were placed nearest to the crop row along the crop row; and in subsurface drip system, drip tapes were placed 0.15 m below soil surface under the crop rows. Four components of soil surface energy balance (net radiation reaching soil surface (R_{ns}), soil surface heat flux (G), sensible heat flux (H_s) and soil latent heat flux (λE_s) were calculated and discussed for two irrigation systems. Daytime average of energy balance components (w/m^2) and soil surface evaporation (mm/day) were calculated in these irrigation systems.

During measurement period, net radiation values ranged from 304 to 333 w/m^2, which caused net radiation reaching the soil surface in the range of 67–107 w/m^2 in both systems. As expected, R_{ns} values were decreased with crop growth and leaf area index (LAI) increased later in crop development period. Soil heat flux accounted for about 36–53% of R_{ns} in surface drip irrigation and about 17–25% in subsurface drip irrigation. Daytime soil heat flux values were greater in surface drip irrigation. The λE_s accounted for about 41–63% of R_{ns} in surface drip irrigation and was about 56–71% in subsurface drip irrigation. It was observed that the ground in both systems became wet but reverses direction of moving moisture in subsurface system due to more evaporation in subsurface drip irrigation. Accordingly in subsurface drip irrigation, depth of installation of emitter lateral line is of utmost importance.

ACKNOWLEDGEMENTS

Authors acknowledge financial support by Agricultural Engineering Research Institute (AERI), Iran; and research guidance by Dr. Majid Liaghat and Dr. Farhad Mirzaei, Department of Irrigation and Reclamation Engineering, Agricultural and Natural Resources Campus, University of Tehran, Iran.

In this chapter, Tables 16.3 and 16.4 and Figs. 16.3–16.5 are from, "*H. Dehghanisanij and H. Kosari, 2011. Evapotranspiration Partitioning in Surface and Subsurface Drip Irrigation Systems. Chapter 11, pages 211–224. In: Evapotranspiration – From Measurements to Agricultural and Environmental Applications, Dr. Giacomo Gerosa (Ed.).* Open Access Article: http://www.intechopen.com/books/evapotranspiration-from-measurements-to-agricultural-and-environmental-applications/ evapotranspiration-partitioning-in-surface-and-subsurface-drip-irrigation-systems."

KEYWORDS

- air temperature
- bare soil
- Bowen ratio, β
- Bowen ratio energy balance, BREB
- Brazil
- canopy temperature
- crop growth
- days after emergence, DAE
- drip irrigation
- emitter
- energy balance
- evaporation, E
- evapotranspiration, ET
- heat flux
- Iran
- lateral line
- latent heat flux
- leaf area index, LAI
- Lysimeter
- maize
- micro Bowen ratio
- net radiation
- pineapple
- psychrometric constant, γ
- relative humidity
- soil evaporation, E_{soil}
- soil heat flux
- subsurface drip irrigation, SDI
- surface drip irrigation
- transpiration, T
- tropical environment
- vapor pressure gradient

REFERENCES

1. Ashktorab, H., Pruitt, W. O., Paw, U. K. T., George, W. V. (1989). Energy balance determination close to the soil surface using a micro Bowen ratio system. *Agric. For. Meteorol.*, 46, 259–274.
2. Ashktorab, H., Pruitt, W. O., Paw, U. K. T. (1994). Partitioning of evapotranspiration using lysimeter and micro Bowen-ratio system. *ASCE J. of Irrigation and Drainage*, 120(2), 450–464.
3. Azevedo, P. V., Souza, C. B., Silva, B. B., Silva, V. P. R. (2007). Water requirement of pineapple crop grown in a tropical environment, Brazil. *Agric. Water Manage.*, 88, 201–208.
4. Boast, C. W., Robertson, T. M. (1982). A micro lysimeter method for determining evaporation from a bare soil: Description and laboratory evaluation. *Soil Sci. Soc. Am. J.*, 46, 689–696.
5. Conaway, J., Van Bavel, C. H. M. (1967). Evaporation from a wet soil surface calculated from radiometrically determined surface temperatures. *J. Appl. Meteorol.*, 6, 650–655.
6. Gardiol, J. M., L. A. Serio, and A. I. Della Maggiora, 2003, Modeling evapotranspiration of corn (*Zea mays*) under different plant densities. *J. Hydrol.*, 271, 188–196.
7. Ham, J. M., Heilman, J. L., Lascano, R. J. (1990). Determination of soil water evaporation and transpiration from energy balance and stem flow measurements. *Agric. For. Meteorol.*, 59, 287–301
8. Heilman, J., L., C., L. Brittin, and C. M. U. Neale, (1989). Fetch requirements of Bowen ratio measurements of latent and sensible heat fluxes. *Agric. For. Meteorol.*, 44, 261–273.
9. Jara, J., Stockle, C. O., Kjelgard, J. (1998). Measurement of evapotranspiration and its components in a corn (*Zea Mays L.*) field. *Agric. For. Meteorol.*, 92, 131–145.
10. Kato, T., R. Kimura, and M. Kamichica, (2004). Estimation of evapotranspiration, transpiration ratio and water use efficiency from a sparse canopy using a compartment model. *Agric. Water Manage.*, 65, 173–191.
11. Prueger, J. H., Hatfield, J. K., J. L. Pikul, (1997). Bowen ratio comparisons with lysimetric evapotranspiration. *Agron. J.*, 89, 730–736.
12. Rosenberg, N. J., Blad, B. L., Verma, S. B. (1983). *Microclimate: The biological environment*. Wiley, New York, 495 pages.
13. Zeggaf, T. A., Takeuchi, S., Dehghanisanij, H., Anyoji, H., Yano, T. (2008). A Bowen ratio technique for partitioning energy fluxes between maize transpiration and soil surface evaporation. *Agron. J.*, 100, 1–9.

FURTHER READING

1. Algozin, K. A., V. F. Bralts and J. T. Ritchie, (2001). Irrigation scheduling for a sandy soil using mobile frequency domain reflectometry with a checkbook method. *Journal of Soil and Water Conservation*, 56(2), 97–100
2. Allen, R. G., Pereira, L. S., Raes, D., Smith, M. (1998). Crop evapotranspiration guidelines for computing crop requirements. Irrigation and drainage paper 56. FAO. Rome, Italy. Pages 300.

3. Al-Shrouf, A. (2005). The safe use of marginal quality water in agriculture, challenges and future alternative, I: Saline water. Paper presented at The Sixth Annual UAE University Research Conference at the United Arab Emirates University. April 24–26, Al-Ain.UAE.

4. Al-Shrouf, A. (2008). Irrigation requirements of date palm (Phoenix dactylifera) in UAE conditions. Paper presented at the ninth Annual UAE University Research Conference at the United Arab Emirates University. April 21–23, Al-Ain, UAE.

5. American Society of Agricultural Engineers (ASABE), (1985). *Design, installation and performance of trickle irrigation systems.* ASAE standard EP 405, St. Joseph, Michigan, pages 507–510.

6. Aragüés R., V. Urdanoz, M. Çetin, C. Kirda, H. Daghari, W. Ltifi, M. Lahlou, A. Douaik. Soil salinity related to physical soil characteristics and irrigation management in four Mediterranean irrigation districts. Agricultural Water Management, Volume 98, Issue 6, April 2011, Pages 959–966.

7. Ashktorab, H., Pruitt, W. O., Paw, U. K. T. (1994). Partitioning of evapotranspiration using lysimeter and Micro-Bowen Ratio system. ASCE J. of Irrigation and Drainage, 120(2, March/April), 450–464.

8. Ashktorab, H., Pruitt, W. O., Paw, U. K. T., George, W. V. (1989). Energy balance determination close to the soil surface using a Micro-Bowen Ratio system. Agricultural and Forest Meteorology Journal, 46, 259–274.

9. Boast, C. W., Robertson T. M. (1982). A micro-lysimeter method for determining evaporation from a bare soil: Description and laboratory evaluation. Soil Sci. Soc. Am. J., 46, 689–696.

10. Bowen, I. S. (1926). The ratio of heat losses by conduction and by evaporation from any water surface. In: Rosenberg, N. J., Blad, B. L., Verma, S. B. (1983). Microclimate: The biological environment. Wiley, New York, 495 pages.

11. Brady, N. C., R. R. Weil, (2007). *The nature and properties of soils* (14th ed.). Prentice Hall, Upper Saddle River, NJ.

12. Bralts, V. F. (1986). Operational principles-field performance and evaluation. In: *Trickle irrigation for crop production.* Amsterdam, Elsevier, pages 216–240.

13. Bureau of Agriculture and Rural Development (BoARD), (2008). A survey conducted in the annual report of the District (wukro) office of Agriculture and Rural Development.

14. Cardenas, B., Lailhacar, and M. D. Dukes, (2010). Precision of soil moisture sensor irrigation controllers under field conditions. *Agricultural Water Management*, 97(5), 666–672.

15. Chávez, J. L., Pierce, F. J., Elliott, T. V., Evans, R. G. (2009). A remote irrigation monitoring and control system for continuous move systems. Part A: description and development. *Precision Agriculture,* 11(1), 1–10.

16. Clesceri L. S., Greenberg, A. E., Eaton, A. D. (1998). *Method 2540D: Standard Methods for the Examination of Water and Wastewater.* 20th Edition. American Public Health Association, Washington DC.

17. Damas, M., Prados, A. M., Gómez, F., Olivares, G. (2001). HidroBus System: Fieldbus for integrated management of extensive areas of irrigation land. Microprocessors Mi-crosyst, 25, 177–184.

18. De Lange, M. (1998). Promotion of low cost and water saving technologies for small-scale irrigation. South Africa: MBB Consulting Engineers.

19. De Tar, W. R. (2004). Using a subsurface drip irrigation system to measure crop water use. Irri. Sci., 3, 111–122.
20. Dehghanisanij, H., H. Kosari, (2011). Evapotranspiration Partitioning in Surface and Subsurface Drip Irrigation Systems. Chapter 11, pages 211–224. In: Evapotranspiration – From Measurements to Agricultural and Environmental Applications, Dr. Giacomo Gerosa (Ed.). Open Access Article from: <http://www.intechopen.com/books/evapotranspiration-from-measurements-to-agricultural-and-environmental-applications/evapotranspiration-partitioning-in-surface-and-subsurface-drip-irrigation-systems>.
21. Dehghanisanij, H., Yamamoto, T., Rasiah, V. (2004). Assessment of evapotranspiration estimation models for use in semi-arid environment. Agricultural Water Management Journal, 64, 91–106.
22. Delgado, Jorge A., Peter M. Groffman, Mark A. Nearing, Tom Goddard, Don Reicosky, Rattan Lal, Newell R. Kitchen, Charles W. Rice, Dan Towery, and Paul Salon, (2008). New technology to increase irrigation efficiency. *Journal of Soil and Water Conservation,* 63(1), 11A.
23. Doorenbos, J., Pruitt, W. O. (1977). *Crop Water Requirements.* FAO Irrigation and Drainage Paper 24, pages 144.
24. FAO, (1984). Localized Irrigation: Design, installation, operation and evaluation. Irrigation and Drainage Paper 36, FAO, Rome.
25. FAO, (1980). Localized Irrigation: Design, installation, operation and evaluation. Irrigation and Drainage Paper 36, FAO, Rome.
26. FAO, (1998). Institution and technical operations in the development and management of small-scale irrigation. Pages 21–38. Proceedings of the third session of the multilateral cooperation workshops for Sustainable Agriculture, Forest and Fisheries Development, Tokyo, Japan, 1995, FAO Water Paper 17, Rome.
27. Federal Democratic Republic of Ethiopia Population Censes Commission (F. D. R. E. P. C. C.), (2008). Population and housing census summary and statistical report. Pages 54.
28. Folhes, M. T., C. D. Rennó, and J. V. Soares, (2009). Remote sensing for irrigation water management in the semi-arid Northeast of Brazil. *Agricultural Water Management,* 96(10), 1398–1408.
29. Fritschen, L. J., Shaw, R. H. (1961). Transpiration and evaporation of corn as related to meteorological factors. Agron. J., 53, 71–74.
30. Gardiol, J. M., Serio, L. A., Della Maggiora, A. I. (2003). Modelling evapotranspiration of corn (Zea mays) under different plant densities. Journal of Hydrology, 271, 188–196.
31. Goyal, Megh R. (2014). Research Advances in Micro Irrigation. Book series by Apple Academic Press Inc.,
32. Goyal, Megh R., Eric W. Harmsen, (2014). *Evapotranspiration: Principles and Applications for Water Management.* Apple Academic Press Inc.,
33. Haftay, Abrha, (2009). Crop water fertilizer interaction and physico-chemical properties of the irrigated soil. Post graduate studies (unpublished) by Mekelle University, Mekelle, Ethiopia.
34. Ham, J. M., Heilman, J. L., Lascano, R. J. (1990). Determination of soil water evaporation and transpiration from energy balance and stem flow measurements. Agric. For. Meteorol., 59, 287–301.

35. Ham, J. M., Heilman, J. L., Lascano, R. J. (1991). Soil and canopy energy balances of a row crop at partial cover. Agron. J., 83, 744–753.
36. Haman, D. Z., Smajstrla, A. G., 2002). Scheduling tips for drip irrigation of vegetable. Publication No: AE259. Florida Extension Service, University of Florida.
37. Haman, D. Z., Smajstrla, A. G., Zazueta F. S. (1987). Water quality problems affecting micro irrigation in Florida. Agricultural Engineering Extension Report 87–2. IFAS, University of Florida
38. Hanson, B. A., Fauton, D. W., May, D. (1995). Drip irrigation of row crops: An overview. *Irrigation Science*, 45(3), 8–11.
39. Hanson, B. R., Schwankl, L. J., Schulbach, K. F., Pettygrove, G. S. (1997). A comparison of furrow, surface drip, and subsurface drip irrigation on lettuce yield and applied water. Agric. Water Manag., 33, 139–157.
40. Hargreaves, G. H., Merkley, G. P. (1998). *Irrigation Fundamentals*. Water Resources Publications, Colorado, USA.
41. Harrold, L. L., Peters, D. B., Driebelbis, F. R., Mc-Guiness, J. L. (1959). Transpiration evaluation of corn grown on a plastic-covered lysimeter. Soil Sci. Soc. of Am. Proc. 23, 174–178.
42. Heilman, J. L., C. L., Brittin and Neale, C. M. U. (1989). Fetch requirements of Bowen ratio measurements of latent and sensible heat fluxes. Agric. For. Meteorol., 44, 261–273.
43. Integrated Food Security Program (IFSP), (2005). A study conducted in the five year development plan of the drought-prone areas of Tigray regional state districts. Mekelle, Tigray, Ethiopia.
44. Isaya, V. S. (2001). Drip Irrigation: Options for smallholder farmers in Eastern and Southern Africa. Regional Land Management unit (RELMA/SIDA), technical and book series 24, Nairobi, Kenya.
45. Jara, J., Stockle, C. O., Kjelgard, J. (1998). Measurement of evapotranspiration and its components in a corn (Zea Mays L.) field. Agric. For. Meteorol., 92, 131–145.
46. Kato, T., R. Kimura and Kamichica, M. (2004). Estimation of evapotranspiration, transpiration ratio and water use efficiency from a sparse canopy using a compartment model. Agric. Water Manage., 65, 173–191.
47. Keller, J., Keller, A. A. (2003). Affordable drip irrigation systems for small farms in developing countries. Proceedings of the irrigation Association Annual Meeting in San Diego CA, 18–20 November. Falls Church, Virginia, Irrigation Association.
48. Kim, Y., Evans, R. G. (2009). Software design for wireless sensor-based site-specific irrigation. *Computers and Electronic in Agriculture*, 66(2), 159–165.
49. Kim, Y., Evans, R. G., Iversen, W. M. (2008). Remote sensing and control of an irrigation system using a distributed wireless sensor network. *IEEE Transaction on Instrumentation and Measurement*, 57(7), 1379–1387.
50. Kirsten, U., Sygna, L., O'brien, K. (2008). Identifying sustainable path ways for climate adoption and poverty reduction. Pages 44.
51. Kosari, H., Dehghanisanij, H., Mirzaei, F., Liaghat, A. M. (2010). Evapotranspiration partitioning using the Bowen Ratio Energy Balance method in a sub-surface drip irrigation. Journal of Agricultural Engineering Research, 11(3), 71–86.
52. Kruse, for example, (1978). Describing irrigation efficiency and uniformity. *J. Irrig., Drain Div., ASCE*, 104(IR1), 35–41.

53. Malik, R. S., Kumar, K., Bandore, A. R. (1994). Effects of drip irrigation levels on yield and water use efficiency of pea. *Journal of Indian Society Soil Science*, 44(3), 508–509.
54. McCready, M. S., M. D. Dukes, and G. L. Miller, (2009). Water conservation potential of smart irrigation controllers on St. Augustine grass. *Agricultural Water Management*, 96(11), 1623–1632.
55. Meixian, Liu, Jing-song Yang, Xiao-ming Li, Mei Yu, and Jin Wang, (2012). Effects of irrigation water quality and drip tape arrangement on soil salinity, soil moisture distribution, and cotton yield (*Gossypium hirsutum L.*) under mulched drip irrigation in Xinjiang, China. *Journal of Integrative Agriculture*, 11(3), 502–511.
56. Neal, J. S., S. R. Murphy, S. Harden, W. J. Fulkerso, (2012). Differences in soil water content between perennial and annual forages and crops grown under deficit irrigation and used by the dairy industry. *Field Crops Research*, 137, 148–162.
57. Nigussie Haregeweyn, Abraha Gebrekiros, Atsushi Tsunkeawa, Mitsuru Tsubo, Derege Meshesha and Eyasu Yazew (2011). Performance Assessment and Adoption Status of Family Drip Irrigation System in Tigray State, Northern Ethiopia, Water Conservation, Dr. Manoj Jha (Ed.), ISBN: 978–953–307–960–8, InTech, Available from: http://www.intechopen.com/books/water-conservation/performance-assessment-and-adoption-status-of-family-drip-irrigation-system-in-tigray-state-northern
58. Nolz, R., G. Kammerer, P. Cepuder, (2013). Calibrating soil water potential sensors integrated into a wireless monitoring network. *Agricultural Water Management*, 116(1), 12–20.
59. Ortega, F., Samuel, O., Richard, H., Cuenca, M., English, J. (1995). Hourly grass evapotranspiration in modified maritime environment. Journal of Irrigation and Drainage, ASCE, 121(6), 369–373.
60. Padhi, J., R. K. Misra, and J. O. Payero, (2012). Estimation of soil water deficit in an irrigated cotton field with infrared thermography. *Field Crops Research*, 126(14), 45–55.
61. Peters, D. B., Russell, M. B. (1959). Relative water losses by evaporation and transpiration in field corn. Soil Sci. Soc. Am. J., 23, 170–173.
62. Phene, C. J., Davis, K. R., Hutmacher, R. B., Bar-Yosef, B., Meek, D. W., Misaki, J. (1991). Effect of high frequency surface and subsurface drip irrigation on root distribution of sweet corn. Irri. Sci., 12, 135–140.
63. Polak, P., Sivanappan, R. K. (2004). The potential contribution of low-cost drip irrigation to the improvement of irrigation productivity in India. Indian water resources management sector review, report on the irrigation sector. The World Bank in cooperation with the Ministry of Water Resources, Government of India, pages 121–123.
64. Rangaswamy, R. (1995). *Agricultural statistics*. New Age International Publishers. Pages 105–110.
65. Richard, E. (1993). Plant expert systems in agriculture and resource management. *Technological Forecasting and Social Change*, 43(3–4), 241–225.
66. Ritchie, J. T. (1971). Dryland evaporation flux in a sub-humid climate: I. Micrometeorological influences. Agron. J., 63, 51–55.
67. Sakellariou-Makrantonaki, M., Kalfountzos, D., Vyrlas, P. (2002). Water saving and yield increase of sugar beet with subsurface drip irrigation. Global Nest: The Int. J., 4(2–3), 85–91.

68. Sakuratani, T. (1987). Studies on evapotranspiration from crops. II. Separate estimation of transpiration and evaporation from a soybean field without water shortage. Agric. For. Meteorol., 59, 287–301.

69. Sánchez, N., J. Martínez-Fernández, J. González-Piqueras, M. P. González-Dugo, G. Baroncini-Turrichia, E. Torres, A. Calera, and C. Pérez-Gutiérrez, (2012). Water balance at plot scale for soil moisture estimation using vegetation parameters. *Agricultural and Forest Meteorology*, Pages 1–9.

70. Senn, A. A., Cornish, P. S. (2000). An example of adoption of reduced cultivation in Sydney:vegetable growers. In: Soil 2000, New Horizons for a new century. Australian and New Zealand Joint Soils Conference. Volume 2, Oral papers. Eds. Adams J. A., Metherell, A. K. December 3–8, Lincoln University. New Zealand Society of Soil Science. Pages 263–264.

71. Sepaskhah, A. R., Ilampour, S. (1995). Effects of soil moisture on evapotranspiration partitioning. Agric. Water Manage., 28, 311–323.

72. Seyed, Hamid Ahmadi, Finn Plauborg, Mathias N. Andersen, Ali Reza Sepaskhah, Christian R. Jensen, and Søren Hansen, (2011). Effects of irrigation strategies and soils on field grown potatoes: Root distribution. *Agricultural Water Management*, 98(8), 1280–1290.

73. Seyed, Hamid Ahmadi, Mathias N. Andersen, Finn Plauborg, Rolf T. Poulsen, Christian R. Jensen, Ali Reza Sepaskhah, and Søren Hansen, (2010). Effects of irrigation strategies and soils on field grown potatoes: Yield and water productivity. *Agricultural Water Management*, 97(11), 1923–1930

74. Shaw, R. H. (1959). Water use from plastic-covered and uncovered corn plots. Agron. J. 51, 172–173.

75. Shawcroft, R. W., Gardner, M. H. (1983). Direct evaporation from soil under a row crop canopy. Agric. Meteorol., 28, 229–238.

76. Shock, C. C., A. B. Pereira, B. R. Hanson, M. D. Cahn, (2007). Vegetable irrigation. Pages 535–606. In: R. Lescano and R. Sojka (eds.) *Irrigation of agricultural crops*. Agron. Monogr. 30, ASA, CSSA, and SSSA, Madison, WI.

77. Smajstrla, A. G., Boman, B. J., Pitts, D. J., Fzueta, F. S. (2002). Field evaluation of micro irrigation water application uniformity. Fla. Coop. Ext. Ser. Bul.265. Univ. of Fla.

78. Steduto, P., Hsiao, T. C. (1998). Maize canopies under two soil water regimes: I, Diurnal patterns of energy balance, carbon dioxide flux, and canopy conductance. Agric. For. Meteorol., 89, 169–184.

79. Tedeschi, A., R. Dell'Aquila, (2005). Effects of irrigation with saline waters, at different concentrations, on soil physical and chemical characteristics. *Agricultural Water Management*, 77(1–3), 308–322.

80. Thompson, R. B., M. Gallardo, L. C. Valdez, M. D. Fernández, (2007). Using plant water status to define threshold values for irrigation management of vegetable crops using soil moisture sensors. *Agricultural Water Management*, 88(1–3), 147–158.

81. Vellidis, G., Tucker, M., Perry, C., Wen, C., Bednarz, C. (2008). A real-time wireless smart sensor array for scheduling irrigation. *Comput. Electron. Agric.*, 61, 44–50.

82. Wang, N., Zhang, N., Wang, M. (2006). Wireless sensors in agriculture and food industry-recent development and future perspective. *Computers and electronics in Agriculture*, 50(1), 1–14.

83. Whiting, D., Card, A., Wilson, C. Moravec, C., Reeder, J. (2011). *Managing Soil Tilth, Texture, Structure and Pore Space*. Colorado Master Gardner Program, Colorado State University Extension. CMG Garden Notes #213.

84. Wu, I. P. (1983). A unit-plot for drip irrigation lateral and sub-main design. ASAE Paper, St.

85. Yunusa, I. A. M., Walker, R. R., Liu, P. (2004). Evapotranspiration components from energy balance, sap flow and micro-lysimetry techniques for an irrigated vineyard in inland Australia. Agric. For. Meteorol., 127, 93–107.

86. Zeggaf, T. A., Takeuchi, S., Dehghanisanij, H., Anyoji, H., Yano, T. (2008). A Bowen ratio technique for partitioning energy fluxes between maize transpiration and soil surface evaporation. *Agron. J.,* 100, 1–9.

APPENDICES

APPENDIX A

Conversion SI and Non-SI Units

To convert the Column 1 in the Column 2, Multiply by	Column 1 Unit SI	Column 2 Unit Non-SI	To convert the Column 2 in the Column 1 Multiply by
Linear			
0.621 _____	kilometer, km (10^3 m)	miles, mi _____	1.609
1.094 _____	meter, m	yard, yd _____	0.914
3.28 _____	meter, m	feet, ft _____	0.304
3.94×10^{-2} _____	millimeter, mm (10^{-3})	inch, in _____	25.4
Squares			
2.47 _____	hectare, he	acre _____	0.405
2.47 _____	square kilometer, km^2	acre _____	4.05×10^{-3}
0.386 _____	square kilometer, km^2	square mile, mi^2 _____	2.590
2.47×10^{-4} _____	square meter, m^2	acre _____	4.05×10^{-3}
10.76 _____	square meter, m^2	square feet, ft^2 _____	9.29×10^{-2}
1.55×10^{-3} _____	mm^2	square inch, in^2 ———	645
Cubics			
9.73×10^{-3} _____	cubic meter, m^3	inch-acre _____	102.8
35.3 _____	cubic meter, m^3	cubic-feet, ft^3 _____	2.83×10^{-2}
6.10×10^4 _____	cubic meter, m^3	cubic inch, in^3 _____	1.64×10^{-5}
2.84×10^{-2} _____	liter, L (10^{-3} m^3)	bushel, bu _____	35.24
1.057 _____	liter, L	liquid quarts, qt _____	0.946
3.53×10^{-2} _____	liter, L	cubic feet, ft^3 _____	28.3
0.265 _____	liter, L	gallon _____	3.78
33.78 _____	liter, L	fluid ounce, oz _____	2.96×10^{-2}
2.11 _____	liter, L	fluid dot, dt _____	0.473

(Modified and reprinted with permission from: Goyal, Megh R., (2012). Appendices. Pages 317–332. In: *Management of Drip/Trickle or Micro Irrigation* edited by Megh R. Goyal. New Jersey, USA : Apple Academic Press Inc.)

Conversion SI and Non-SI Units (Continued)

To convert the Column 1 in the Column 2, Multiply by	Column 1 Unit SI	Column 2 Unit Non-SI	To convert the Column 2 in the Column 1 Multiply by
Weight			
2.20×10^{-3}	gram, g (10^{-3} kg)	pound,	454
3.52×10^{-2}	gram, g (10^{-3} kg)	ounce, oz	28.4
2.205	kilogram, kg	pound, lb	0.454
10^{-2}	kilogram, kg	quintal (metric), q	100
1.10×10^{-3}	kilogram, kg	ton (2000 lbs), ton	907
1.102	mega gram, mg	ton (US), ton	0.907
1.102	metric ton, t	ton (US), ton	0.907
Yield and Rate			
0.893	kilogram per hectare	pound per acre	1.12
7.77×10^{-2}	kilogram per cubic meter	pound per fanega	12.87
1.49×10^{-2}	kilogram per hectare	pound per acre, 60 lb	67.19
1.59×10^{-2}	kilogram per hectare	pound per acre, 56 lb	62.71
1.86×10^{-2}	kilogram per hectare	pound per acre, 48 lb	53.75
0.107	liter per hectare	galloon per acre	9.35
893	ton per hectare	pound per acre	1.12×10^{-3}
893	mega gram per hectare	pound per acre	1.12×10^{-3}
0.446	ton per hectare	ton (2000 lb) per acre	2.24
2.24	meter per second	mile per hour	0.447
Specific Surface			
10	square meter per kilogram	square centimeter per gram	0.1
10^3	square meter per kilogram	square millimeter per gram	10^{-3}
Pressure			
9.90	megapascal, MPa	atmosphere	0.101
10	megapascal	bar	0.1
1.0	megagram per cubic meter	gram per cubic centimeter	1.00
2.09×10^{-2}	pascal, Pa	pound per square feet	47.9

Conversion SI and Non-SI Units (Continued)

To convert the Column 1 in the Column 2, Multiply by	Column 1 Unit SI	Column 2 Unit Non-SI	To convert the Column 2 in the Column 1 Multiply by
1.45×10^{-4} _____	pascal, Pa	pound per square inch _____	6.90×10^3
Temperature			
1.00 (K-273) _____	Kelvin, K	centigrade, °C _____	1.00 (C + 273)
(1.8 C + 32) _____	centigrade, °C	Fahrenheit, °F _____	(F–32)/1.8
Energy			
9.52×10^{-4} _____	Joule J	BTU _____	1.05×10^3
0.239 _____	Joule, J	calories, cal _____	4.19
0.735 _____	Joule, J	feet-pound _____	1.36
2.387×10^5 _____	Joule per square meter	calories per square centimeter _____	4.19×10^4
10^5 _____	Newton, N	dynes _____	10^{-5}
Water Requirements			
9.73×10^{-3} _____	cubic meter	inch acre _____	102.8
9.81×10^{-3} _____	cubic meter per hour	cubic feet per second _____	101.9
4.40 _____	cubic meter per hour	galloon (US) per minute _____	0.227
8.11 _____	hectare-meter	acre-feet _____	0.123
97.28 _____	hectare-meter	acre-inch _____	1.03×10^{-2}
8.1×10^{-2} _____	hectare centimeter	acre-feet _____	12.33
Concentration			
1 _____	centimol per kilogram	milliequivalents per 100 grams _____	1
0.1 _____	gram per kilogram	percents _____	10
1 _____	milligram per kilogram	parts per million _____	1
Nutrients for Plants			
2.29 _____	P	P_2O_5 _____	0.437
1.20 _____	K	K_2O _____	0.830
1.39 _____	Ca	CaO _____	0.715
1.66 _____	Mg	MgO _____	0.602

Conversion SI and Non-SI Units (Continued)

To convert the Column 1 in the Column 2, Multiply by	Column 1 Unit SI	Column 2 Unit Non-SI	To convert the Column 2 in the Column 1 Multiply by

Nutrient Equivalents

Column A	Column B	Conversion A to B	Equivalent B to A
N	NH_3	1.216	0.822
	NO_3	4.429	0.226
	KNO_3	7.221	0.1385
	$Ca(NO_3)_2$	5.861	0.171
	$(NH_4)_2SO_4$	4.721	0.212
	NH_4NO_3	5.718	0.175
	$(NH_4)_2HPO_4$	4.718	0.212
P	P_2O_5	2.292	0.436
	PO_4	3.066	0.326
	KH_2PO_4	4.394	0.228
	$(NH_4)_2HPO_4$	4.255	0.235
	H_3PO_4	3.164	0.316
K	K_2O	1.205	0.83
	KNO_3	2.586	0.387
	KH_2PO_4	3.481	0.287
	Kcl	1.907	0.524
	K_2SO_4	2.229	0.449
Ca	CaO	1.399	0.715
	$Ca(NO_3)_2$	4.094	0.244
	$CaCl_2 \times 6H_2O$	5.467	0.183
	$CaSO_4 \times 2H_2O$	4.296	0.233
Mg	MgO	1.658	0.603
	$MgSO_4 \times 7H_2O$	1.014	0.0986
S	H_2SO_4	3.059	0.327
	$(NH_4)_2SO_4$	4.124	0.2425
	K_2SO_4	5.437	0.184
	$MgSO_4 \times 7H_2O$	7.689	0.13
	$CaSO_4 \times 2H_2O$	5.371	0.186

APPENDIX B

PIPE AND CONDUIT FLOW

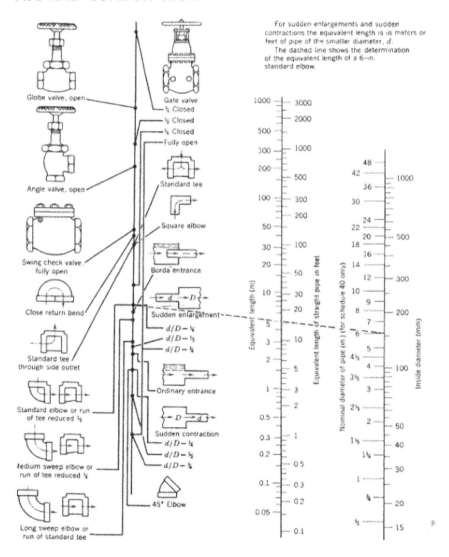

For sudden enlargements and sudden contractions the equivalent length is in meters or feet of pipe of the smaller diameter, d.

The dashed line shows the determination of the equivalent length of a 6-in. standard elbow.

APPENDIX C

Percentage of Daily Sunshine Hours: for North and South Hemispheres

Latitude	Jan	Feb	Mar	Apr	May	Jun	Jul	Aug	Sep	Oct	Nov	Dec
						North						
0	8.50	7.66	8.49	8.21	8.50	8.22	8.50	8.49	8.21	8.50	8.22	8.50
5	8.32	7.57	8.47	3.29	8.65	8.41	8.67	8.60	8.23	8.42	8.07	8.30
10	8.13	7.47	8.45	8.37	8.81	8.60	8.86	8.71	8.25	8.34	7.91	8.10
15	7.94	7.36	8.43	8.44	8.98	8.80	9.05	8.83	8.28	8.20	7.75	7.88
20	7.74	7.25	8.41	8.52	9.15	9.00	9.25	8.96	8.30	8.18	7.58	7.66
25	7.53	7.14	8.39	8.61	9.33	9.23	9.45	9.09	8.32	8.09	7.40	7.52
30	7.30	7.03	8.38	8.71	9.53	9.49	9.67	9.22	8.33	7.99	7.19	7.15
32	7.20	6.97	8.37	8.76	9.62	9.59	9.77	9.27	8.34	7.95	7.11	7.05
34	7.10	6.91	8.36	8.80	9.72	9.70	9.88	9.33	8.36	7.90	7.02	6.92
36	6.99	6.85	8.35	8.85	9.82	9.82	9.99	9.40	8.37	7.85	6.92	6.79
38	6.87	6.79	8.34	8.90	9.92	9.95	10.1	9.47	3.38	7.80	6.82	6.66
40	6.76	6.72	8.33	8.95	10.0	10.1	10.2	9.54	8.39	7.75	6.72	7.52
42	6.63	6.65	8.31	9.00	10.1	10.2	10.4	9.62	8.40	7.69	6.62	6.37
44	6.49	6.58	8.30	9.06	10.3	10.4	10.5	9.70	8.41	7.63	6.49	6.21
46	6.34	6.50	8.29	9.12	10.4	10.5	10.6	9.79	8.42	7.57	6.36	6.04
48	6.17	6.41	8.27	9.18	10.5	10.7	10.8	9.89	8.44	7.51	6.23	5.86
50	5.98	6.30	8.24	9.24	10.7	10.9	11.0	10.0	8.35	7.45	6.10	5.64
52	5.77	6.19	8.21	9.29	10.9	11.1	11.2	10.1	8.49	7.39	5.93	5.43

Percentage of Daily Sunshine Hours: for North and South Hemispheres (Continued)

Latitude	Jan	Feb	Mar	Apr	May	Jun	Jul	Aug	Sep	Oct	Nov	Dec
54	5.55	6.08	8.18	9.36	11.0	11.4	11.4	10.3	8.51	7.20	5.74	5.18
56	5.30	5.95	8.15	9.45	11.2	11.7	11.6	10.4	8.53	7.21	5.54	4.89
58	5.01	5.81	8.12	9.55	11.5	12.0	12.0	10.6	8.55	7.10	4.31	4.56
60	4.67	5.65	8.08	9.65	11.7	12.4	12.3	10.7	8.57	6.98	5.04	4.22
South												
0	8.50	7.66	8.49	8.21	8.50	8.22	8.50	8.49	8.21	8.50	8.22	8.50
5	8.68	7.76	8.51	8.15	8.34	8.05	8.33	8.38	8.19	8.56	8.37	8.68
10	8.86	7.87	8.53	8.09	8.18	7.86	8.14	8.27	8.17	8.62	8.53	8.88
15	9.05	7.98	8.55	8.02	8.02	7.65	7.95	8.15	8.15	8.68	8.70	9.10
20	9.24	8.09	8.57	7.94	7.85	7.43	7.76	8.03	8.13	8.76	8.87	9.33
25	9.46	8.21	8.60	7.74	7.66	7.20	7.54	7.90	8.11	8.86	9.04	9.58
30	9.70	8.33	8.62	7.73	7.45	6.96	7.31	7.76	8.07	8.97	9.24	9.85
32	9.81	8.39	8.63	7.69	7.36	6.85	7.21	7.70	8.06	9.01	9.33	9.96
34	9.92	8.45	8.64	7.64	7.27	6.74	7.10	7.63	8.05	9.06	9.42	10.1
36	10.0	8.51	8.65	7.59	7.18	6.62	6.99	7.56	8.04	9.11	9.35	10.2
38	10.2	8.57	8.66	7.54	7.08	6.50	6.87	7.49	8.03	9.16	9.61	10.3
40	10.3	8.63	8.67	7.49	6.97	6.37	6.76	7.41	8.02	9.21	9.71	10.5
42	10.4	8.70	8.68	7.44	6.85	6.23	6.64	7.33	8.01	9.26	9.8	10.6
44	10.5	8.78	8.69	7.38	6.73	6.08	6.51	7.25	7.99	9.31	9.94	10.8
46	10.7	8.86	8.90	7.32	6.61	5.92	6.37	7.16	7.96	9.37	10.1	11.0

APPENDIX D

Psychometric Constant (γ) for Different Altitudes (Z)

$\gamma = 10^{-3} \, [(C_p \cdot P) \div (\varepsilon \cdot \lambda)] = (0.00163) \times [P \div \lambda]$

γ, psychrometric constant [kPa C^{-1}] c_p, specific heat of moist air = 1.013 [kJ kg^{-1}°C^{-1}] P, atmospheric pressure [kPa].

ε, ratio molecular weight of water vapor/dry air = 0.622 λ, latent heat of vaporization [MJ kg^{-1}] = 2.45 MJ kg^{-1} at 20°C.

z (m)	γ kPa/°C	z (m)	γ kPa/°C	z (m)	γ kPa/°C	z (m)	γ kPa/°C
0	0.067	1000	0.060	2000	0.053	3000	0.047
100	0.067	1100	0.059	2100	0.052	3100	0.046
200	0.066	1200	0.058	2200	0.052	3200	0.046
300	0.065	1300	0.058	2300	0.051	3300	0.045
400	0.064	1400	0.057	2400	0.051	3400	0.045
500	0.064	1500	0.056	2500	0.050	3500	0.044
600	0.063	1600	0.056	2600	0.049	3600	0.043
700	0.062	1700	0.055	2700	0.049	3700	0.043
800	0.061	1800	0.054	2800	0.048	3800	0.042
900	0.061	1900	0.054	2900	0.047	3900	0.042
1000	0.060	2000	0.053	3000	0.047	4000	0.041

APPENDIX E

Saturation Vapor Pressure [e_s] for Different Temperatures (T)

Vapor Pressure Function = e_s = [0.6108]*exp{[17.27*T]/[T + 237.3]}

T °C	e_s kPa	T °C	e_s kPa	T °C	e_s kPa	T °C	e_s kPa
1.0	0.657	13.0	1.498	25.0	3.168	37.0	6.275
1.5	0.681	13.5	1.547	25.5	3.263	37.5	6.448
2.0	0.706	14.0	1.599	26.0	3.361	38.0	6.625
2.5	0.731	14.5	1.651	26.5	3.462	38.5	6.806
3.0	0.758	15.0	1.705	27.0	3.565	39.0	6.991
3.5	0.785	15.5	1.761	27.5	3.671	39.5	7.181
4.0	0.813	16.0	1.818	28.0	3.780	40.0	7.376
4.5	0.842	16.5	1.877	28.5	3.891	40.5	7.574
5.0	0.872	17.0	1.938	29.0	4.006	41.0	7.778
5.5	0.903	17.5	2.000	29.5	4.123	41.5	7.986
6.0	0.935	18.0	2.064	30.0	4.243	42.0	8.199
6.5	0.968	18.5	2.130	30.5	4.366	42.5	8.417
7.0	1.002	19.0	2.197	31.0	4.493	43.0	8.640
7.5	1.037	19.5	2.267	31.5	4.622	43.5	8.867
8.0	1.073	20.0	2.338	32.0	4.755	44.0	9.101
8.5	1.110	20.5	2.412	32.5	4.891	44.5	9.339
9.0	1.148	21.0	2.487	33.0	5.030	45.0	9.582
9.5	1.187	21.5	2.564	33.5	5.173	45.5	9.832
10.0	1.228	22.0	2.644	34.0	5.319	46.0	10.086
10.5	1.270	22.5	2.726	34.5	5.469	46.5	10.347
11.0	1.313	23.0	2.809	35.0	5.623	47.0	10.613
11.5	1.357	23.5	2.896	35.5	5.780	47.5	10.885
12.0	1.403	24.0	2.984	36.0	5.941	48.0	11.163
12.5	1.449	24.5	3.075	36.5	6.106	48.5	11.447

APPENDIX F

Slope of Vapor Pressure Curve (Δ) For Different Temperatures (T)

$\Delta = [4098.\ e°(T)] \div [T + 237.3]^2 = 2504\{\exp[(17.27T) \div (T + 237.2)]\} \div [T + 237.3]^2$

T °C	Δ kPa/°C	T °C	Δ kPa/°C	T °C	Δ kPa/°C	T °C	Δ kPa/°C
1.0	0.047	13.0	0.098	25.0	0.189	37.0	0.342
1.5	0.049	13.5	0.101	25.5	0.194	37.5	0.350
2.0	0.050	14.0	0.104	26.0	0.199	38.0	0.358
2.5	0.052	14.5	0.107	26.5	0.204	38.5	0.367
3.0	0.054	15.0	0.110	27.0	0.209	39.0	0.375
3.5	0.055	15.5	0.113	27.5	0.215	39.5	0.384
4.0	0.057	16.0	0.116	28.0	0.220	40.0	0.393
4.5	0.059	16.5	0.119	28.5	0.226	40.5	0.402
5.0	0.061	17.0	0.123	29.0	0.231	41.0	0.412
5.5	0.063	17.5	0.126	29.5	0.237	41.5	0.421
6.0	0.065	18.0	0.130	30.0	0.243	42.0	0.431
6.5	0.067	18.5	0.133	30.5	0.249	42.5	0.441
7.0	0.069	19.0	0.137	31.0	0.256	43.0	0.451
7.5	0.071	19.5	0.141	31.5	0.262	43.5	0.461
8.0	0.073	20.0	0.145	32.0	0.269	44.0	0.471
8.5	0.075	20.5	0.149	32.5	0.275	44.5	0.482
9.0	0.078	21.0	0.153	33.0	0.282	45.0	0.493
9.5	0.080	21.5	0.157	33.5	0.289	45.5	0.504
10.0	0.082	22.0	0.161	34.0	0.296	46.0	0.515
10.5	0.085	22.5	0.165	34.5	0.303	46.5	0.526
11.0	0.087	23.0	0.170	35.0	0.311	47.0	0.538
11.5	0.090	23.5	0.174	35.5	0.318	47.5	0.550
12.0	0.092	24.0	0.179	36.0	0.326	48.0	0.562
12.5	0.095	24.5	0.184	36.5	0.334	48.5	0.574

APPENDIX G

Number of the Day in the Year (Julian Day)

Day	Jan	Feb	Mar	Apr	May	Jun	Jul	Aug	Sep	Oct	Nov	Dec
1	1	32	60	91	121	152	182	213	244	274	305	335
2	2	33	61	92	122	153	183	214	245	275	306	336
3	3	34	62	93	123	154	184	215	246	276	307	337
4	4	35	63	94	124	155	185	216	247	277	308	338
5	5	36	64	95	125	156	186	217	248	278	309	339
6	6	37	65	96	126	157	187	218	249	279	310	340
7	7	38	66	97	127	158	188	219	250	280	311	341
8	8	39	67	98	128	159	189	220	251	281	312	342
9	9	40	68	99	129	160	190	221	252	282	313	343
10	10	41	69	100	130	161	191	222	253	283	314	344
11	11	42	70	101	131	162	192	223	254	284	315	345
12	12	43	71	102	132	163	193	224	255	285	316	346
13	13	44	72	103	133	164	194	225	256	286	317	347
14	14	45	73	104	134	165	195	226	257	287	318	348
15	15	46	74	105	135	166	196	227	258	288	319	349
16	16	47	75	106	136	167	197	228	259	289	320	350
17	17	48	76	107	137	168	198	229	260	290	321	351
18	18	49	77	108	138	169	199	230	261	291	322	352
19	19	50	78	109	139	170	200	231	262	292	323	353
20	20	51	79	110	140	171	201	232	263	293	324	354
21	21	52	80	111	141	172	202	233	264	294	325	355
22	22	53	81	112	142	173	203	234	265	295	326	356
23	23	54	82	113	143	174	204	235	266	296	327	357
24	24	55	83	114	144	175	205	236	267	297	328	358
25	25	56	84	115	145	176	206	237	268	298	329	359
26	26	57	85	116	146	177	207	238	269	299	330	360
27	27	58	86	117	147	178	208	239	270	300	331	361
28	28	59	87	118	148	179	209	240	271	301	332	362
29	29	(60)	88	119	149	180	210	241	272	302	333	363
30	30	—	89	120	150	181	211	242	273	303	334	364
31	31	—	90	—	151	—	212	243	—	304	—	365

APPENDIX H

Stefan-Boltzmann Law at Different Temperatures (T):

$[\sigma*(T_K)^4] = [4.903 \times 10^{-9}]$, MJ K^{-4} m^{-2} day^{-1}

where: $T_K = \{T[°C] + 273.16\}$

T	$\sigma*(T_K)^4$	T	$\sigma*(T_K)^4$	T	$\sigma*(T_K)^4$
		Units			
°C	MJ m^{-2} d^{-1}	°C	MJ m^{-2} d^{-1}	°C	MJ m^{-2} d^{-1}
1.0	27.70	17.0	34.75	33.0	43.08
1.5	27.90	17.5	34.99	33.5	43.36
2.0	28.11	18.0	35.24	34.0	43.64
2.5	28.31	18.5	35.48	34.5	43.93
3.0	28.52	19.0	35.72	35.0	44.21
3.5	28.72	19.5	35.97	35.5	44.50
4.0	28.93	20.0	36.21	36.0	44.79
4.5	29.14	20.5	36.46	36.5	45.08
5.0	29.35	21.0	36.71	37.0	45.37
5.5	29.56	21.5	36.96	37.5	45.67
6.0	29.78	22.0	37.21	38.0	45.96
6.5	29.99	22.5	37.47	38.5	46.26
7.0	30.21	23.0	37.72	39.0	46.56
7.5	30.42	23.5	37.98	39.5	46.85
8.0	30.64	24.0	38.23	40.0	47.15
8.5	30.86	24.5	38.49	40.5	47.46
9.0	31.08	25.0	38.75	41.0	47.76
9.5	31.30	25.5	39.01	41.5	48.06
10.0	31.52	26.0	39.27	42.0	48.37
10.5	31.74	26.5	39.53	42.5	48.68
11.0	31.97	27.0	39.80	43.0	48.99
11.5	32.19	27.5	40.06	43.5	49.30
12.0	32.42	28.0	40.33	44.0	49.61
12.5	32.65	28.5	40.60	44.5	49.92
13.0	32.88	29.0	40.87	45.0	50.24
13.5	33.11	29.5	41.14	45.5	50.56
14.0	33.34	30.0	41.41	46.0	50.87

Stefan-Boltzmann Law at Different Temperatures (T)   Continued

T	$\sigma*(T_K)^4$	T	$\sigma*(T_K)^4$	T	$\sigma*(T_K)^4$
			Units		
°C	MJ m^{-2} d^{-1}	°C	MJ m^{-2} d^{-1}	°C	MJ m^{-2} d^{-1}
14.5	33.57	30.5	41.69	46.5	51.19
15.0	33.81	31.0	41.96	47.0	51.51
15.5	34.04	31.5	42.24	47.5	51.84
16.0	34.28	32.0	42.52	48.0	52.16
16.5	34,52	32.5	42.80	48.5	52.49

APPENDIX I

THERMODYNAMIC PROPERTIES OF AIR AND WATER

1. Latent Heat of Vaporization (λ)

$$\lambda = [2.501 - (2.361 \times 10^{-3})\,T]$$

where: λ = latent heat of vaporization [MJ kg^{-1}]; and T = air temperature [°C].

The value of the latent heat varies only slightly over normal temperature ranges. A single value may be taken (for ambient temperature = 20°C): $\lambda = 2.45$ MJ kg^{-1}.

2. Atmospheric Pressure (P)

$$P = P_o\,[\{T_{Ko} - \alpha(Z - Z_o)\} \div \{T_{Ko}\}]^{(g/(\alpha.R))}$$

where: P, atmospheric pressure at elevation z [kPa]

P_o, atmospheric pressure at sea level = 101.3 [kPa]

z, elevation [m]

z_o, elevation at reference level [m]

g, gravitational acceleration = 9.807 [m s^{-2}]

R, specific gas constant == 287 [J kg^{-1} K^{-1}]

α, constant lapse rate for moist air = 0.0065 [K m^{-1}]

T_{Ko}, reference temperature [K] at elevation z_o = 273.16 + T

T, means air temperature for the time period of calculation [°C]

When assuming P_o = 101.3 [kPa] at z_o = 0, and T_{Ko} = 293 [K] for T = 20 [°C], above equation reduces to:

$$P = 101.3[(293-0.0065Z) (293)]^{5.26}$$

3. Atmospheric Density (ρ)

$$\rho = [1000P] \div [T_{Kv} R] = [3.486P] \div [T_{Kv}], \text{ and } T_{Kv} = T_K[1-0.378(e_a)/P]^{-1}$$

where: ρ, atmospheric density [kg m^{-3}]
R, specific gas constant = 287 [J kg$^{-1 \, K-1}$]
T_{Kv}, virtual temperature [K]
T_K, absolute temperature [K]: T_K = 273.16 + T [°C]
e_a, actual vapor pressure [kPa]
T, mean daily temperature for 24-hour calculation time steps.

For average conditions (e_a in the range 1–5 kPa and P between 80–100 kPa), T_{Kv} can be substituted by: $T_{Kv} \approx 1.01$ (T + 273)

4. Saturation Vapor Pressure function (es)

$$e_s = [0.6108]*\exp\{[17.27*T]/[T + 237.3]\}$$

where: e_s, saturation vapor pressure function [kPa]
T, air temperature [°C]

5. Slope Vapor Pressure Curve (Δ)

$$\Delta = [4098. \, e°(T)] \div [T + 237.3]^2$$
$$= 2504\{\exp[(17.27T) \div (T + 237.2)]\} \div [T + 237.3]^2$$

where: Δ, slope vapor pressure curve [kPa C^{-1}]
T, air temperature [°C]
e°(T), saturation vapor pressure at temperature T [kPa]

In 24-hour calculations, Δ is calculated using mean daily air temperature. In hourly calculations T refers to the hourly mean, T_{hr}.

6. Psychrometric Constant (γ)

$$\gamma = 10^{-3} [(C_p.P) \div (\varepsilon.\lambda)] = (0.00163) \times [P \div \lambda]$$

where: γ, psychrometric constant [kPa C^{-1}]

c_p, specific heat of moist air = 1.013 [kJ kg^{-1}°C^{-1}]

P, atmospheric pressure [kPa]: equations 2 or 4

ε, ratio molecular weight of water vapor/dry air = 0.622

λ, latent heat of vaporization [MJ kg^{-1}]

7. Dew Point Temperature (T_{dew})

When data is not available, T_{dew} can be computed from e_a by:

$$T_{dew} = [\{116.91 + 237.3Log_e(e_a)\} \div \{16.78-Log_e(e_a)\}]$$

where: T_{dew}, dew point temperature [°C]

e_a, actual vapor pressure [kPa]

For the case of measurements with the Assmann psychrometer, T_{dew} can be calculated from:

$$T_{dew} = (112 + 0.9T_{wet})[e_a \div (e° T_{wet})]^{0.125}-[112-0.1T_{wet}]$$

8. Short Wave Radiation on a Clear-Sky Day (R_{so})

The calculation of R_{so} is required for computing net long wave radiation and for checking calibration of pyranometers and integrity of R_{so} data. A good approximation for R_{so} for daily and hourly periods is:

$$R_{so} = (0.75 + 2 \times 10^{-5} z)R_a$$

where: z, station elevation [m]

R_a, extraterrestrial radiation [MJ m^{-2} d^{-1}]

Equation is valid for station elevations less than 6000 m having low air turbidity. The equation was developed by linearizing Beer's radiation extinction law as a function of station elevation and assuming that the average angle of the sun above the horizon is about 50°.

For areas of high turbidity caused by pollution or airborne dust or for regions where the sun angle is significantly less than 50° so that the path length of radiation through the atmosphere is increased, an adoption of Beer's law can be employed where P is used to represent atmospheric mass:

$$R_{so} = (R_a) \exp[(-0.0018P) \div (K_t \sin(\Phi))]$$

where: K_t, turbidity coefficient, $0 < K_t < 1.0$ where $K_t = 1.0$ for clean air and $K_t = 1.0$ for extremely turbid, dusty or polluted air.

P, atmospheric pressure [kPa]

Φ, angle of the sun above the horizon [rad]

R_a, extraterrestrial radiation [MJ m^{-2} d^{-1}]

For hourly or shorter periods, Φ is calculated as:

$$\sin \Phi = \sin \varphi \sin \delta + \cos \varphi \cos \delta \cos \omega$$

where: φ, latitude [rad]

δ, solar declination [rad] (Eq. (24) in Chapter 3)

ω, solar time angle at midpoint of hourly or shorter period [rad]

For 24-hour periods, the mean daily sun angle, weighted according to R_a, can be approximated as:

$$\sin(\Phi_{24}) = \sin[0.85 + 0.3 \, \varphi \sin\{(2\pi J/365)-1.39\}-0.42 \, \varphi^2]$$

where: Φ_{24}, average Φ during the daylight period, weighted according to R_a [rad]

φ, latitude [rad]

J, day in the year

The Φ_{24} variable is used to represent the average sun angle during daylight hours and has been weighted to represent integrated 24-hour transmission effects on 24-hour R_{so} by the atmosphere. Φ_{24} should be limited to >0. In some situations, the estimation for R_{so} can be improved by modifying to consider the effects of water vapor on short wave absorption, so that: $R_{so} = (K_B + K_D) \, R_a$ where:

$$K_B = 0.98\exp[\{(-0.00146P) \div (K_t \sin \Phi)\}-0.091\{w/\sin \Phi\}^{0.25}]$$

where: K_B, the clearness index for direct beam radiation

K_D, the corresponding index for diffuse beam radiation

$K_D = 0.35-0.33 \, K_B$ for $K_B > 0.15$

$K_D = 0.18 + 0.82 \, K_B$ for $K_B < 0.15$

R_a, extraterrestrial radiation [MJ m^{-2} d^{-1}]

K_t, turbidity coefficient, $0 < K_t < 1.0$ where $K_t = 1.0$ for clean air and $K_t = 1.0$ for extremely turbid, dusty or polluted air.

P, atmospheric pressure [kPa]

Φ, angle of the sun above the horizon [rad]

W, perceptible water in the atmosphere [mm] = 0.14 e_a P + 2.1

e_a, actual vapor pressure [kPa]

P, atmospheric pressure [kPa]

APPENDIX J

PSYCHROMETRIC CHART AT SEA LEVEL

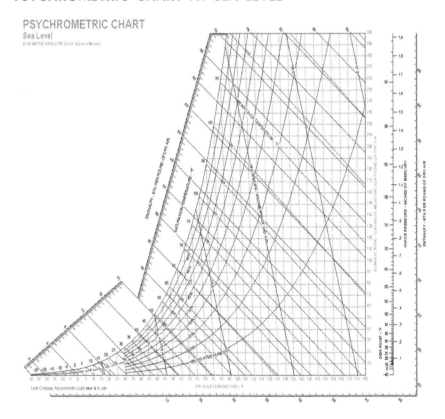

APPENDIX K

(<http://www.fao.org/docrep/T0551E/t0551e07.htm#5.5%20field%20 management%20practices%20in%20wastewater%20irrigation>)

1. Relationship between applied water salinity and soil water salinity at different leaching fractions (FAO, 1985)

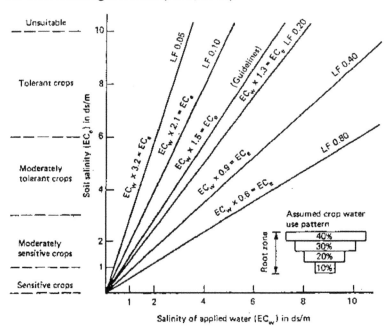

2. Schematic representations of salt accumulation, planting positions, ridge shapes and watering patterns.

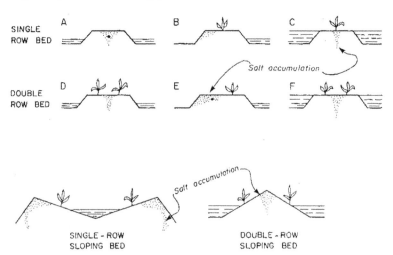

3. Main components of general planning guidelines for wastewater reuse (Cobham and Johnson, 1988)

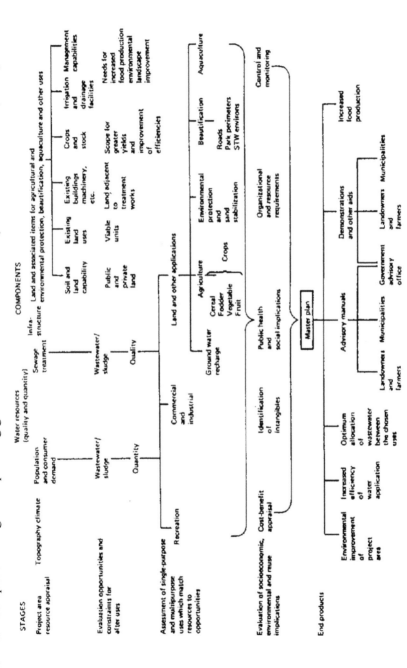

APPENDIX L

1. Uniformity classification.

Classification	Statistical Uniformity	Emission Uniformity
Excellent	For U = 100–95%	100–94%
Good	For U = 90–85%	87–81%
Fair	For U = 80–75%	75–68%
Poor	For U = 70–65%	62–56%
Not Acceptable	For U < 60%	<50%

2. Acceptable intervals of uniformity in a drip irrigation system.

Type of dripper	Slope	Uniformity interval, %
Point Source: located in planting distance > 3.9 m.	Level*	90–95
	Inclined**	85–90
Point Source: located in planting distance < 3.9 m.	Level*	85–90
	Inclined**	80–90
Drippers inserted in the lines for annual row crops.	Level*	80–90
	Inclined**	75–85

* Level = Slope less that 2%. ** Inclined = Slope greater than 2%.

3. Confidence limits for field uniformity (U).

Field uniformity	18 drippers		36 drippers		72 drippers	
	Confidence limit		Confidence limit		Confidence limit	
	N Sum*	%	N Sum	%	N Sum	%
100%	3	U ± 0.0	6	U ± 0.6%	12	U ± 0.0%
90%	3	U ± 2.9	6	U ± 2.0%	12	U ± 1.4%
80%	3	U ± 5.8	6	U ± 4.0%	12	U ± 2.8%
70%	3	U ± 9.4	6	U ± 6.5%	12	U ± 4.5%
60%	3	U ± 13.3	6	U ± 9.2%	12	U ± 6.5%

*N Sum = 1/6 part of the total measured drippers. This is a number of samples that will be added to calculate T_{max} and T_{min}.

(From: *Vincent F. Bralts*, 2015. Chapter 3: Evaluation of the uniformity coefficients. In: *Sustainable Micro Irrigation Management for Trees and Vines, Volume 3*, edited by Megh R. Goyal, Apple Academic Press Inc.).

4. Nomograph for statistical uniformity.

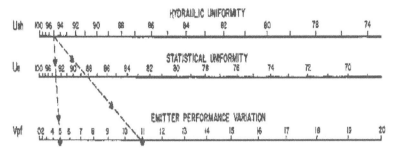

5. The field uniformity of an irrigation system based on the dripper times and the dripper flow rate.

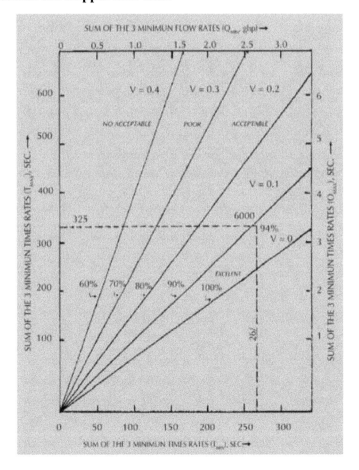

6. **The field uniformity of a drip irrigation system based on the time to collect a known quantity of water or based on pressure for hydraulic uniformity.**

7.

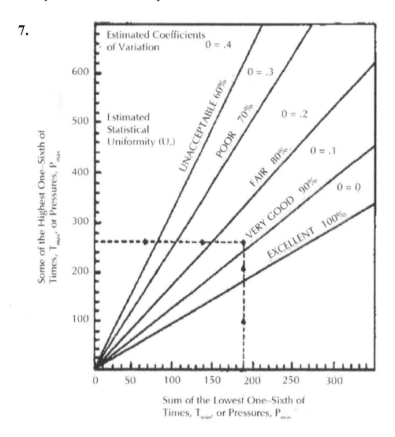

INDEX

Milton Keynes UK
Ingram Content Group UK Ltd.
UKHW022053141024
449569UK00031B/1621

9 781774 635421